T0156085

Lecture Notes in Physics

Volume 966

The Lecture Notes in Physics

The series Lecture Notes in Physics (LNP), founded in 1969, reports new developments in physics research and teaching-quickly and informally, but with a high quality and the explicit aim to summarize and communicate current knowledge in an accessible way. Books published in this series are conceived as bridging material between advanced graduate textbooks and the forefront of research and to serve three purposes:

- to be a compact and modern up-to-date source of reference on a well-defined topic
- to serve as an accessible introduction to the field to postgraduate students and nonspecialist researchers from related areas
- to be a source of advanced teaching material for specialized seminars, courses and schools

Both monographs and multi-author volumes will be considered for publication. Edited volumes should however consist of a very limited number of contributions only. Proceedings will not be considered for LNP.

Volumes published in LNP are disseminated both in print and in electronic formats, the electronic archive being available at springerlink.com. The series content is indexed, abstracted and referenced by many abstracting and information services, bibliographic networks, subscription agencies, library networks, and consortia.

Proposals should be sent to a member of the Editorial Board, or directly to the managing editor at Springer:

Dr Lisa Scalone
Springer Nature
Physics Editorial Department
Tiergartenstrasse 17
69121 Heidelberg, Germany
lisa.scalone@springernature.com

More information about this series at http://www.springer.com/series/5304

Edoardo Lauria • Antoine Van Proeyen

$\mathcal{N} = 2$ Supergravity in $D = 4, 5, 6$ Dimensions

Edoardo Lauria
CPHT
Ecole Polytechnique
Palaiseau, France

Antoine Van Proeyen
Institute for Theoretical Physics
KU Leuven
Leuven, Belgium

ISSN 0075-8450   ISSN 1616-6361   (electronic)
Lecture Notes in Physics
ISBN 978-3-030-33755-1   ISBN 978-3-030-33757-5   (eBook)
https://doi.org/10.1007/978-3-030-33757-5

This Springer imprint is published by the registered company Springer Nature Switzerland AG.
The registered company address is: Gewerbestrasse 11, 6330 Cham, Switzerland

Preface

Supergravity theories play an important role in present-day research on fundamental interactions. A profound knowledge can only be gained by understanding how supergravity theories are constructed. To understand the necessary techniques is not easy starting from the published literature. Despite the presence of a few books, it turns out that a detailed explanation is still lacking and this is particularly the case for supergravity theories with matter couplings in so-called $\mathcal{N} = 2$ supergravity. In this book, we want to provide a detailed explanation of these theories and tools that researchers have used to construct supergravity theories with 8 real supercharges, in 4, 5 and 6 dimensions.

The basic technique is the superconformal calculus. We explain these methods from basic principles. Chapter 1 will introduce the basic ingredients: the multiplets and the symmetries that will be instrumental in the rest of the book. In Chap. 2 it will be explained how these symmetry groups can be promoted to gauge groups. The matter multiplets will enter the game in Chap. 3: first defining these multiplets in the superconformal context, and then actions will be constructed. These are reduced to super-Poincaré theories in Chap. 4. The beautiful geometric structure that these theories enjoy is discussed and characterized in Chap. 5. These geometries go together under the name of special geometries, which include special Kähler geometry and quaternionic-Kähler manifolds.

For readers who want to progress soon to final results, Chap. 6 gives a shorter account.

The text grew from lecture notes during the semester 'Supergravity, superstrings and M-theory' at Institut Henri Poincaré, Paris, in November 2000. Parts of it have been available on internet sites and were appreciated by many researchers. We hope that this book will be useful as well for students as for researchers who want to know how these standard theories are constructed and what is their geometric structure.

Palaiseau, France
Leuven, Belgium
September 2019

Edoardo Lauria
Antoine Van Proeyen

v

Acknowledgements

The notes for this book received input from many people in our field. We first of all thank Robin De Roover and Gabriele Tartaglino-Mazzucchelli, who collaborated for some time to write this book.

The text is based on several fruitful collaborations on $N = 2$ supergravities, and we thank Eric Bergshoeff, Marco Billó, Anna Ceresole, Piet Claus, Frederik Coomans, Ben Craps, Eugène Cremmer, Sorin Cucu, Riccardo D'Auria, Gianguido Dall'Agata, Mees de Roo, Bernard de Wit, Tim de Wit, Jean-Pierre Derendinger, Martijn Derix, Sergio Ferrara, Pietro Frè, Dan Freedman, Jos Gheerardyn, Luciano Girardello, Rein Halbersma, Renata Kallosh, Costas Kounnas, Paul Lauwers, Marián Lledó, Óscar Maciá, Roeland Philippe, Tullio Regge, Diederik Roest, Frederik Roose, Jan Rosseel, Ergin Sezgin, Paolo Soriani, S.-Q. Su, Walter Troost, François Vanderseypen, Stefan Vandoren, Jan-Willem van Holten, Kor Van Hoof, Veeravalli Varadarajan and Bert Vercnocke who contributed significantly to the content in this book. We thank the many colleagues whose remarks have been helpful to improve the text: Marco Baggio, Nikolay Bobev, Alessio Celi, Jan De Rydt, Alessandra Gnecchi, Dietmar Klemm, Joris Van den Bergh and Marco Zagermann. Special thanks also to Stefan Theisen who encouraged us to write this review.

Most of this review has been prepared in Leuven in the Institute for Theoretical Physics. We thank the staff and the researchers for the ideal environment that was always present there. We have enjoyed the presence of inspiring and very friendly colleagues.

EL is supported by the Simons Foundation grant #488659 (Simons Collaboration on the non-perturbative bootstrap).

Last but not least, we thank our partners Giuliana and Laura for their support during this work.

Contents

Chapter 1
Basic Ingredients

Abstract We give an introduction to the book, discussing the role of the $\mathcal{N} = 2$ theories, its geometric structure, and the superconformal tensor calculus. We also refer to other treatments. We then set out the plan of the book.

In the second part of the chapter we introduce tools that are useful for the construction of superconformal gauge theory and multiplets. We first discuss the catalogue of supersymmetric theories with 8 supercharges (Sect. 1.2) and their multiplets (Sect. 1.2.1). After a short Sect. 1.2.2 with the strategy, we discuss the conformal (Sect. 1.2.3) and then superconformal (Sect. 1.2.4) groups. The transformations of the fields under the conformal symmetry are also given in Sect. 1.2.3, while for the fermionic symmetries, this is discussed in a short Sect. 1.2.5.

1.1 Introduction

Theories with 8 supercharges, i.e. $\mathcal{N} = 2$ in $D = 4, 5$ and 6, occupy a special place in the atlas of supersymmetric quantum field theories and supergravities. They lead to very interesting geometries called 'special manifolds', including special Kähler and quaternionic-Kähler geometries.

Why are the theories with 8 supersymmetries so interesting? The maximal supergravities[1] contain 32 supersymmetries. These are the $\mathcal{N} = 8$ theories in 4 dimensions, and exist in spaces of Lorentzian signature with at most 11 dimensions, i.e. (10, 1) spacetime dimensions. If one allows more time directions, 32 supersymmetries are possible in 12 dimensions with (10, 2) or (6, 6) signature. However, these theories allow no matter multiplets[2] For the geometry, determined by the kinetic terms of the scalars, this means that the manifold is fixed once the

---

[1] The restriction is due to interacting field theory descriptions, which e.g. in 4 dimensions does not allow fields with spin larger than 2.

[2] We distinguish the multiplet that contains the graviton and gravitini, and is determined by specifying the dimension and the number of supersymmetries, and other multiplets, which we call 'matter multiplets'.

© Springer Nature Switzerland AG 2020

E. Lauria, A. Van Proeyen, *N = 2 Supergravity in D = 4, 5, 6 Dimensions*, Lecture Notes in Physics 966, https://doi.org/10.1007/978-3-030-33757-5_1

dimension is given. For all theories with 32 supersymmetries this is a symmetric space.

Matter multiplets are possible if one limits the number of supersymmetries to 16 (thus $\mathcal{N} = 4$ in 4 dimensions). Theories with 16 real supersymmetries exist up to 10 dimensions with Lorentzian signature. In this case, the geometry is fixed to a particular coset geometry once the number of matter multiplets that are coupled to supergravity is given.

The situation becomes more interesting if the number of supersymmetries is 8. Now there are functions, which can be varied continuously, that determine the geometry. This makes the geometries much more interesting. Of course, if one further restricts to 4 supersymmetries, more geometries would be possible. In 4 dimensions, e.g., general Kähler manifolds appear. For 8 supersymmetries, these are restricted to 'special Kähler manifolds', determined by a holomorphic prepotential [1]. This restriction makes the class of manifolds very interesting and manageable. The holomorphicity is a useful ingredient, and was e.g. essential to allow the solution of the theory in the Seiberg–Witten model [2, 3]. The theories with 8 supersymmetries are thus the maximally supersymmetric ones that are not completely determined by the number of fields in the model, but allow arbitrary functions in their definition, i.e. continuous deformations of the metric of the manifolds.

A short account of the necessary techniques and results of this book has been given in Chaps. 20 and 21 of [4]. The scope of this work is to elaborate further on these results, as well as allowing a more complete and detailed explanation of these techniques. Specifically, we will consider $\mathcal{N} = 2$ supergravities in 4, 5 and 6 dimensions, and at the end pay special attention to the $D = 4$ framework. We use the terminology $\mathcal{N} = 2$ in 6 dimensions for what is also called $(1, 0)$ in 6 dimensions. Indeed, in 6 dimensions one can have chiral and antichiral real supersymmetry generators, and the nomenclature can thus be done by giving multiples of 8 real chiral, namely the minimal spinor of Spin(5, 1). Also for $\mathcal{N} = 2$ in 5 dimensions, the minimal spinor of Spin(4, 1) has 8 real components. We denote these theories still as $\mathcal{N} = 2$ because for practical work we always use doublets of 4-component spinors. Note also that for other signatures (2 or 3 time directions in $D = 5$) one can impose Majorana conditions such that only 4 of them survive ($\mathcal{N} = 1$). But for Minkowski signature,[3] one can only have (symplectic) Majorana conditions, which need doublets of 4-component spinors. The basic properties of the spinors are repeated in Appendix A.3. As well for $D = 4$, 5 and 6 we have an SU(2) automorphism group of the spinors, and for 4 dimensions the automorphism group has an extra U(1), important for the Kähler geometry that we will discuss.

---

[3]In this review, we only consider Minkowski signature of spacetime. In the literature, also other signatures are discussed, e.g. Euclidean signature in a series of papers [5–8], and special geometry has also been defined for other signatures [9–12].

The geometry described by the scalar fields in these theories is a main topic in this book and for these theories in general. This geometry is a direct product of two parts: the geometry of the scalars of vector multiplets and the one of the scalars of hypermultiplets. Even the supergravity couplings do not mix these independent parts [13]. That hypermultiplets in $\mathcal{N} = 2$ rigid supersymmetric theories lead to hyper-Kähler manifolds was already known since [14]. In supergravity theories these geometries are upgraded to quaternionic [15] geometries.[4] The fact that actions for vector multiplets can be constructed from an arbitrary holomorphic function in supersymmetry has been found first for rigid supersymmetry in [17]. For the coupling in supergravity this appeared first in [1]. In the following years it was called 'Kähler manifolds of restricted type', see e.g. [18]. Later, Strominger [19] gave it the name 'special Kähler geometry'. It was then also recognized that this is the geometry of the moduli of Calabi–Yau threefolds. In the first years only special coordinates were used. In [20] a start was made to formulate the geometry in arbitrary coordinates.

At the start of the second superstring revolution, people first considered models with $\mathcal{N} = 2$ and special geometry [2, 3] because of the restrictive nature (holomorphicity) of these theories. The concept of duality became very important at that time. These dualities in $\mathcal{N} = 2$, $D = 4$ have a natural formulation in the context of superconformal tensor calculus.

Superconformal tensor calculus is another main topic of this book. It has been the basis of the first constructions of general matter couplings. It was initiated in $\mathcal{N} = 1$ due to the work of S. Ferrara, M. Kaku, P.K. Townsend and P. van Nieuwenhuizen [21–23]. The extra symmetries of the superconformal group give an advantage over the direct super-Poincaré approach since many aspects of the theory get a clear structure. In fact, the natural vectors in which the dualities have to be formulated are the multiplets of the superconformal tensor calculus. In this approach, the superconformal symmetry is used as a tool to obtain the theories that have super-Poincaré symmetry. All the super-Poincaré theories are constructed as broken superconformal theories. Lately the conformal symmetry has gained importance. There is of course its interest in AdS/CFT correspondence, although in that context one mostly considers rigid conformal symmetry. But it may also be interesting to consider which parts of the supergravity theory are explicitly determined by the breaking of the superconformal invariance, and which parts are generically determined by the superconformal structure. Thus the superconformal approach may be more interesting than just as a tool to obtain super-Poincaré results.

The tensor calculus clarifies also the off-shell structure, and as such has been used in the developments of 'supersymmetric localization', where the auxiliary fields in the conformal calculus play an important role. We refer to [24] for an overview.

---

[4]Strictly speaking: to quaternionic-Kähler geometry, which means that there is a metric. For the difference between these manifolds and the used terminology we refer to the review [16].

The conformal tensor calculus is not the only one to obtain the theories that we consider. There are several superspace methods[5] developed for $\mathcal{N} = 2$. Standard superspace techniques for $D = 4$, $\mathcal{N} = 2$ conformal supergravity have been described long ago, see [25, 26] and references therein. For $D = 5$ and $D = 6$ they have been developed in [27, 28]. An approach that is closer to the superconformal methods is the 'conformal superspace' method. This has been developed in [29] first for $\mathcal{N} = 1$, and extended to $D = 4$, $\mathcal{N} = 2$ in [30, 31], for $D = 5$ in [32] and for $D = 6$ in [33, 34]. In this method the entire superconformal symmetry is gauged as in the conformal tensor calculus, and the connection with the approach in components was developed in detail in [31, 32, 34]. Especially for hypermultiplets, the harmonic and projective superspace approaches are alternatives allowing off-shell matter couplings at the expense of an infinite number of auxiliary fields, packed in dependences on extra superspace coordinates. The harmonic superspace for rigid $\mathcal{N} = 2$ supersymmetry has been developed in [35] and reviewed in [36], while projective superspace was developed in [37–39] and is reviewed in [40, 41]. For supergravity the harmonic and projective approaches have been developed in [42–45, 27, 46, 26, 47–49].

Furthermore there is the rheonomic approach that has been used for another formulation of the general $\mathcal{N} = 2$ theories in four dimensions [50, 51], and in five dimensions [52]. A relation with the conformal approach has been obtained in [53]. These methods (including the conformal tensor calculus) do not prove uniqueness of the $\mathcal{N} = 2$ theories. Recently some uniqueness proofs (at least for the pure supergravities) have been given in [54] based on BRST-BV methods [55–58]. They are consistent with the results found with other methods.

The plan of the book is as follows. In the rest of this chapter we will introduce some basic concepts: the supersymmetries and multiplets and the superconformal groups. In Chap. 2 we gauge these groups by a gauge multiplet, which is called the Weyl multiplet. This gauging of the spacetime symmetries is not straightforward and needs curvature constraints.

In Chap. 3 the matter multiplets are introduced: the vector multiplets in 6, 5 and 4 dimensions in that order, and the hypermultiplets, whose structure is not dependent on the spacetime dimension. We have chosen to first present the multiplets independent of the actions. We thus give the definition of the fields, possible reality conditions, and how they represent the superconformal algebra. Then we present the construction of the superconformal-invariant actions for vector and for hypermultiplets. Another alternative order is possible in Chap. 3: those mainly interested in vector multiplets can go from Sect. 3.2.1 directly to Sect. 3.3.1 for $D = 4$ or Sect. 3.3.2 for $D = 5$. On the other hand after discussing hypermultiplets in Sects. 3.2.3 and 3.2.4 one can also directly go to the discussion of its actions in Sect. 3.3.3.

---

[5]We thank G. Tartaglino-Mazzucchelli for his assistance in this overview.

After the partial gauge fixing (i.e. gauge fixing of the symmetries that are not in the super-Poincaré algebra) the actions of vector- and hypermultiplets are connected. The total action is considered in Chap. 4. We first discuss in Sect. 4.2 pure supergravity in 4 dimensions, obtained by adding to the minimal field representation (Weyl multiplet + a compensating vector multiplet) a second compensating multiplet. Here we consider three possibilities. In the other parts of the book, we always use the hypermultiplet as second compensating multiplet. We also consider the reduction of these pure supergravity theories to $\mathcal{N} = 1$ in Sect. 4.3. Then we go back to the complete matter-coupled supergravities. We discuss convenient coordinates for the scalars in these theories. The vector multiplets are discussed more thoroughly in preparation of the following chapter. The $D = 5$ and $D = 6$ theories are discussed shortly, referring to the techniques explained in detail for $D = 4$.

The structure of the special geometries that result are studied in Chap. 5. This structure is also characterized independently of its supergravity construction. The isometries (gauged or non-gauged) are discussed. In Sect. 5.5 the (charged) black holes are discussed as an application and the role of the symplectic transformations in this context becomes apparent. In Sect. 5.6 we will also give an introduction to the properties of quaternionic-Kähler spaces independent of supergravity. The matter couplings with $\mathcal{N} = 2$ in 5 and 6 dimensions are very much related to those in four dimensions as we discuss in Sect. 5.7. We discuss the differences and specific properties of the $\mathcal{N} = 2$ theories in these dimensions. Very special geometry will then show up in relations between the scalar manifolds defined by these theories. This is further clarified in relations between the homogeneous spaces.

Chapter 6 gives the final super-Poincaré theory. In Sect. 6.1 the actions, supersymmetry transformations and algebra are written explicitly for $D = 4$. Section 6.2 is a bit similar for $D = 5$, though we are less complete there, referring to [59] for the full results in the same notation. In the final remarks in Sect. 6.3 we stress the choices and restrictions that we made throughout the book. This includes then also references to papers that go beyond the present text, and the outlook to applications.

The notation in this book is the same as in [4]. But in Appendix A this is explained more explicitly for the theories that we consider. We repeat the use of indices, spinor notations, SU(2) conventions and the relations between spinors in $D = 6$, $D = 5$ and $D = 4$. A short summary of simple superalgebras is given in Appendix B. Finally for convenience of the reader who also wants to use the original papers, we give in Appendix C the translations of our notation to those of these papers.

Readers may also choose to start immediately with Chap. 6. There we repeat the main ingredients and that chapter contain references to earlier parts when necessary.

1.2 Supersymmetric Theories with 8 Real Supercharges

As mentioned in the introduction, we will be focusing on theories with 8 supercharges in $D = 4, 5, 6$, being the latter the maximal dimension for theories with
8 supercharges.[6] In fact, one could first construct $D = 6$ theories and then derive
several results for $D < 6$ from dimensional reduction. This programme has been
started for the superconformal theories in [60, 61]. There are many aspects that can
be treated at once for $D = 4, 5, 6$. The treatment of the Weyl multiplets is for a
large part the same in these dimensions. Hypermultiplets (multiplets with scalars
and spinors only) do not feel the difference of dimension. This difference is relevant
for vectors and tensors, which under dimensional reduction decompose in several
representations of the Lorentz algebra. However, only a subset of the couplings of
vectors multiplets in $D = 4$ and $D = 5$ can be obtained from dimensional reduction,
as we will explicitly demonstrate in Sect. 5.7. Spinors are also treated differently,
and this has an immediate consequence on the supersymmetry algebra.

In 4 dimensions, the supersymmetries are represented by Majorana spinors, but
in practice one can also use chiral spinors. We refer to [4, Chap. 3] for the notation
and definitions of the properties of the spinors. A supersymmetry operation is
represented as

$$\delta(\epsilon) = \bar{\epsilon}^i Q_i + \bar{\epsilon}_i Q^i \,,$$

$$\epsilon^i = \gamma_* \epsilon^i \,, \qquad Q_i = \gamma_* Q_i \,, \qquad \epsilon_i = -\gamma_* \epsilon_i \,, \qquad Q^i = -\gamma_* Q^i \,, \quad (1.1)$$

where $\gamma_* = i\gamma_0\gamma_1\gamma_2\gamma_3$ is Hermitian with $(\gamma_*)^2 = \mathbb{1}$ and Q is the supersymmetry
operator that acts on the fields, e.g. if $\delta(\epsilon)X = \frac{1}{2}\bar{\epsilon}^i\lambda_i$, then $Q_i X = \frac{1}{2}\lambda_i$ and
$Q^i X = 0$.

In 5 dimensions, one uses symplectic Majorana spinors. The reality rules of
Appendix A.3.2 imply that we have to insert a factor i in $\delta(\epsilon)$ (see also the different
meaning of the position of the indices i, as explained in Appendix A.3.2):

$$\delta(\epsilon) = i\bar{\epsilon}^i Q_i \,. \tag{1.2}$$

In 6 dimensions, symplectic Majorana–Weyl spinors can be used. In this case we
have

$$\delta(\epsilon) = \bar{\epsilon}^i Q_i \,,$$

$$\gamma_* \epsilon^i = \epsilon^i \,, \qquad \gamma_* Q_i = -Q_i \,, \tag{1.3}$$

where γ_* is given in (A.39).

---

[6]Of course one can introduce 8 supercharges in 1 or 2 dimensions, where the elementary spinors
have just one component. In 3 dimensions, gamma matrices are 2×2 matrices, and the theories
with 8 supercharges are $\mathcal{N} = 4$ theories where the spinors satisfy a reality condition (a Majorana
condition).

The properties of the spinors also imply how the translations appear in the anticommutator of two supersymmetries (the overall real factor is a matter of choice of normalization):

$$D = 4 : \left\{ Q_\alpha^i, Q_{\beta j} \right\} = -\tfrac{1}{2} (\gamma_a P_L)_{\alpha\beta} \, \delta_j^i \, P^a$$

$$D = 5 \text{ and } D = 6 : \left\{ Q_\alpha^i, Q_\beta^j \right\} = -\tfrac{1}{2} \varepsilon^{ij} (\gamma_a)_{\alpha\beta} \, P^a . \tag{1.4}$$

On the use of spinor indices, see [4, Sect. 3.2.2]. The normalization of the spinor generators is chosen to agree with much of the original literature in each case. See Appendix C.

Exercise 1.1 Verify that, given (1.4) one finds

$$[\delta(\epsilon_1), \delta(\epsilon_2)] = \xi^a (\epsilon_1, \epsilon_2) P_a , \tag{1.5}$$

with

$$D = 4 : \xi^a (\epsilon_1, \epsilon_2) = \tfrac{1}{2} \left(\bar\epsilon_2^i \gamma^a \epsilon_{1i} + \bar\epsilon_{2i} \gamma^a \epsilon_1^i \right)$$

$$D = 5 \text{ and } D = 6 : \xi^a (\epsilon_1, \epsilon_2) = \tfrac{1}{2} \bar\epsilon_2 \gamma^a \epsilon_1 . \tag{1.6}$$

□

1.2.1 Multiplets

On general grounds, in any supersymmetric theory, the bosonic and fermionic number of degrees of freedom must be equal.[7] More precisely, this rule applies to transformations between states where the Q-transformation of a fermionic state gives a bosonic state, and vice versa, with symbolically $\{Q, Q\} = P$, and P invertible. The invertible P can refer to translations in rigid supersymmetry or general coordinate transformations in supergravity. On the contrary, this rule does not apply to nonlinear realizations where some fields transform to constants.

Counting the number of degrees of freedom may be very subtle, and for this reason we need to be more precise about the correct procedure that we should adopt. Firstly, we will see that the anticommutator of supersymmetries often does not only contain the translations, but also gauge transformations. Therefore, the argument about equal number of bosonic and fermionic states is only correct up to gauge

---

[7]Different proofs of this statement can be given. A general argument has been given in [62]. This has been further discussed in [4, App. 6B]. For on-shell states in $D = 4$, see also [4, Sect. 6.4.1]. For solutions of field equations that preserve supersymmetry, a detailed explanation is in Appendix B of [63].

transformations. In other words, in general we should subtract the gauge degrees of freedom in order to perform the counting.[8]

Secondly, there are two possible countings of degrees of freedom: on-shell counting and off-shell counting. When the supersymmetry algebra is applied to the physical states determined by the equations of motion we denote this as '*on-shell counting*'. In that case, the consistency requires that there are as many bosonic as fermionic physical states. On the other hand, in order to couple many different actions to the multiplets that we are constructing, we usually want the supersymmetry algebra to be satisfied independently of the equations of motion. In other words, we consider an *off-shell closed* algebra and adopt *off-shell counting* of degrees of freedom. More details on the concepts of on-shell and off-shell degrees of freedom are explained in [4] in the beginning of Chap. 4 and in Appendix 6B.

Let us first consider on-shell states. The massless on-shell multiplets for extended supersymmetry, on which we will later impose the superconformal algebra, are classified in the classical reference [64]. The multiplets of $N = 2$ for the dimensions in which we are interested are given in Table 1.1. The on-shell components for massless fields form representations of $SO(D-2)$ with an internal $SU(2)$ group (which is the $R-$symmetry group of $N = 2$ theories and is labelled by i-type indices). The representation content is indicated then as its multiplicity in $(SO(D-2), SU(2))$.

The vectors (V_μ or W_μ) are a fundamental representation of $SO(D-2)$ and have therefore $D-2$ on-shell components (see the analysis in [4, Sect.4.1.2]). The gravitons e_μ^a are (after taking Lorentz symmetry into account) described as the symmetric $g_{\mu\nu}$. The on-shell components are symmetric traceless tensors so one counts $\frac{1}{2}(D-2+1)(D-2)-1 = D(D-3)/2$ (see the analysis in [4, Sect.8.2]).

In order to count the degrees of freedom for spinors one must take into account that their characteristics differ in different dimensions. It is useful to first discuss the off-shell degrees of freedom and then go to the on-shell degrees of freedom which just differ by a factor $\frac{1}{2}$. The number of complex components of a spinor in dimension D is given by $2^{[D/2]}$. In $D = 4$ we can furthermore impose consistent Majorana conditions, so that the minimal spinor has 4 real components, or, equivalently consider one helicity with 2 complex components (and the other helicity is then related by complex conjugation). For $N = 2$, spinors in 4 dimensions will be taken to be part of an $SU(2)$ doublet with index i, leading to 8 real components. For $D = 5$ it is not possible to define the Majorana condition in a consistent manner and one must use the symplectic Majorana condition. This does not lower the number of components but rewrites them as a symplectic pair of Majorana spinors related through charge conjugation. Thus one finds that a spinor in 5 dimensions has 4 complex components or 2×4 real components. For $D = 6$ one still needs to use symplectic Majorana spinors but, moreover, one can impose the Weyl condition, which projects half of the components of the spinors, thus leaving

---

[8]General coordinate transformations are also local gauge transformations that are more general than fixed translations. So the general coordinate-equivalent states should also be subtracted.

Table 1.1 Massless on-shell representations

$D=4$	e_μ^a	ψ_μ^i	V_μ		Gravity
SO(2)	(2,1)	(2,2)	(2,1)		
	W_μ	λ_i	X	X*	Vector multiplet
	(2,1)	(2,2)	(1,1)	(1,1)	
		ζ	q^i		Hypermultiplet
		(2,1)	(1,2)		
$D=5$	e_μ^a	ψ_μ^i	V_μ		Gravity
SO(3)	(5,1)	(4,2)	(3,1)		
	W_μ	λ^i	σ		Vector multiplet
	(3,1)	(2,2)	(1,1)		
		ζ	q^i		Hypermultiplet
		(2,1)	(1,2)		
$D=6$	e_μ^a	$\psi_{\mu L}^i$	$B_{\mu\nu}^-$		Gravity
SO(4)	(3,3;1)	(2,3;2)	(1,3;1)		
	$B_{\mu\nu}^+$	ψ_R^i	σ		Tensor multiplet
	(3,1;1)	(2,1;2)	(1,1;1)		
	W_μ	λ^i			Vector multiplet
	(2,2;1)	(1,2;2)			
		ζ	q^i		Hypermultiplet
		(2,1;1)	(1,1;2)		

The representation content is indicated as its multiplicity in $(\mathrm{SO}(D-2), \mathrm{SU}(2))$. The SU(2) representation can also be identified from the i-type indices. For 6 dimensions, $\mathrm{SO}(4) = \mathrm{SU}(2) \times \mathrm{SU}(2)$, and the corresponding decomposition is written. The non-symmetric representations are either chiral (chirality is indicated on the field) or they are antisymmetric tensors with self-dual ($+$) or anti-self-dual ($-$) field strengths. Of course, the two-form indicated here is not $\pm$ self-dual, only its field strength is

us with $\frac{1}{2}$ the number of complex components. So, one gets 8 complex components, which can be described by the symplectic condition as 2×8 real components, and due to the Weyl condition can be projected to 2×4 real components. Hence in all three cases this leads to 8 real off-shell degrees of freedom. Going from off-shell to on-shell degrees of freedom means multiplying the number of real components with $\frac{1}{2}$ leading to 4 real on-shell degrees of freedom for the three cases that we consider.

The gravitini are described by massless vector–spinors. Their off-shell degrees of freedom are reduced by the supersymmetry, so that for $D = 4, 5, 6$ we have $8(D-1)$ components. The on-shell degrees of freedom are γ-traceless in $D - 2$ dimensions (see [4, Sect.5.1]). Therefore one has $(D - 2 - 1)$ vector degrees of freedom where the $D - 2$ comes from the vector representation and the -1 comes from the γ-traceless condition. One must then multiply these vector degrees of freedom with the spinor degrees of freedom, which leads to $4(D - 3)$ degrees of freedom.

The vector multiplets in different dimensions are related by dimensional reduction. In the highest dimension ($D = 6$) one has a vector W_μ and a symplectic pair of spinors λ^i. The scalars in lower dimensions can be considered as some of the vector components of $D = 6$ after the reduction. For 5 dimensions one has an additional real scalar $\sigma = W_5$ and for 4 dimensions an additional complex scalar $X = \frac{1}{2}(W_4 - i\sigma)$. The notion that there are scalar fields present in the vector multiplet for $D = 4, 5$ will give rise to special geometries described by these scalar fields. We will describe this in detail in Chap. 5.

Looking at the hypermultiplets in Table 1.1 one notices that the scalars q^i transform as a doublet under SU(2). Following from the fact that the SU(2)-transformations are complex this means that the scalars cannot be real. Taking them complex is equivalent to taking 4 real scalars q^X, which will due to supersymmetry define 3 complex structures. The scalars naturally combine into quaternions as we will describe in Sect. 3.2.3. The corresponding geometry will thus be of the type of a quaternionic manifold, which we will exhibit in Sect. 5.6.

Remark that here the on-shell massless multiplets have been mentioned. The same multiplet may be represented by different off-shell multiplets. In 4 dimensions, physical scalars may be dualized to antisymmetric tensors. E.g. when one of the scalars of a vector multiplet is replaced by an antisymmetric tensor, we obtain the *vector–tensor multiplet* [65–67]

$$\text{vector–tensor multiplet: } V_\mu, \ \lambda^i, \ \phi, \ B_{\mu\nu}. \tag{1.7}$$

When one of the scalars of a hypermultiplet is replaced by an antisymmetric tensor, then this gives the so-called *linear multiplet* (the name is due to its relation with a superfield that has a linear constraint)

$$\text{linear multiplet: } \varphi^i, \ L^{(ij)}, \ E_a. \tag{1.8}$$

We write the antisymmetric tensor here as a vector E_a that satisfies a constraint $\partial^a E_a = 0$, such that $E_a = \varepsilon_{abcd}\partial^b E^{cd}$. In this way, it can be generalized to $D = 5$ and $D = 6$ in a similar way. In these cases the E_a is the field strength of a 3-form or 4-form, respectively.

In 5 dimensions a vector is dual to a 2-index antisymmetric tensor. Therefore the vector multiplets can be dualized to antisymmetric tensor multiplets:

$$D = 5 \text{ antisymmetric tensor multiplet: } H_{\mu\nu}, \ \lambda^i, \ \phi. \tag{1.9}$$

For non-abelian multiplets, the two formulations are not equivalent. This has been investigated in detail in a series of papers of Günaydin and Zagermann [68–70] and further worked out in [52, 59]. We will come back to this in Sect. 6.2. The above overview is not exhaustive.

1.2.2 The Strategy

Our aim is to study the transformation laws and actions for the $\mathcal{N} = 2$ multiplets, coupled to supergravity. As mentioned in the introduction, there are several ways to accomplish this. One possible way is to use in intermediate steps symmetries that will not be present in the final action. Auxiliary enhancement of symmetry has the particular advantage that it facilitates the construction of the theories, clarifying a lot of their structures.

The extra symmetry will be the superconformal symmetry and the method we use goes under the name of *superconformal tensor calculus*. In the classical work of Coleman and Mandula [71] and its supersymmetric extension by Haag–Łopuszański–Sohnius [72], it is proven that the superconformal groups is the largest spacetime symmetry allowed for a non-trivial quantum field theory.[9] Although our motivation and use of the group is completely different, this gives an indication that the use of this group may be the most advantageous strategy. Over the years we got more convinced that indeed the use of conformal symmetries is a very useful and clarifying method. Analyzing the steps that are taken in local superspaces, we see that after using part of the superspace constraints, the remaining part that leads to more insight is equivalent to the structure that we use in the 'superconformal tensor calculus'.

What we have in mind can be illustrated first for pure gravity. We show how Poincaré supergravity is obtained after gauge fixing a conformal invariant action. The details of this example will come back in Sect. 3.1. We now just give the general idea. The conformal invariant action for a scalar ϕ (in 4 dimensions) is[10]

$$\mathcal{L} = \sqrt{g} \left[\tfrac{1}{2} \left(\partial_\mu \phi \right) \left(\partial^\mu \phi \right) + \tfrac{1}{12} R \phi^2 \right],$$

$$\delta \phi = \lambda_D \phi, \qquad \delta g_{\mu\nu} = -2 \lambda_D g_{\mu\nu}, \tag{1.10}$$

where the second line gives the local dilatation symmetry that leaves this action invariant. Now, we can gauge fix this dilatation symmetry by choosing[11] the gauge

$$\phi = \frac{\sqrt{6}}{\kappa}. \tag{1.11}$$

---

[9]Coleman and Mandula prove that the largest spacetime group that is allowed without implying triviality of all scattering amplitudes is the conformal group. This theorem is valid under some assumptions, like the analyticity of the elastic two-body scattering amplitudes. Haag, Łopuszański and Sohnius base their analysis on the previous result and use group-theoretical arguments (essentially Jacobi identities). For an extension of the theorem to strongly coupled quantum field theories, see e.g. [73, 74].

[10]Note that the scalar here has negative kinetic energy, and the final gravity action has positive kinetic energy.

[11]A gauge fixing can be interpreted as choosing better coordinates such that only one field still transforms under the corresponding transformations. Then, the invariance is expressed as the absence of this field from the action. In this case we would use $g'_{\mu\nu} = \frac{\kappa^2}{6} g_{\mu\nu} \phi^2$ as D-invariant metric. One can check that this redefinition also leads to (1.12) in terms of the new field.

This leads to the pure Poincaré action

$$\mathcal{L} = \frac{1}{2\kappa^2}\sqrt{g}R.$$ (1.12)

Pure Poincaré is in this way obtained from a conformal action of a scalar after gauge fixing. This scalar, which we will denote further as 'compensating scalar', thus has no physical modes. Note also that the mass scale of the Poincaré theory is introduced through the gauge fixing (1.11).

What we have seen is (1) the use of conformal symmetry, (2) construction of a conformal invariant action, (3) gauge fixing of superfluous symmetries.

In the remaining part of this chapter, we will familiarize ourselves first with the conformal group, and first of all as a rigid symmetry. We will then take a look at possible superconformal groups, and repeat the basis rules for gauging symmetries. In Chap. 2 we will learn how to gauge the superconformal group. The step (2) of the above overview involves the superconformal construction of multiplets and their action. This will be the subject of Chap. 3. The step (3) will be taken in Chap. 4, which will allow us to obtain the physical theories, and be the starting point for analysing their physical and geometrical contents.

1.2.3 Rigid Conformal Symmetry

Conformal symmetry is defined as the symmetry that preserves angles. Under a conformal transformation, the metric may change by a Weyl factor. In terms of infinitesimal transformations $x'^\mu = x^\mu + \xi^\mu(x)$, for a given flat metric $\eta_{\mu\nu}$, the latter condition implies the following 'conformal Killing equation' for $\xi^\mu \equiv \xi^\mu(x)$

$$\partial_\mu \xi_\nu + \partial_\nu \xi_\mu - \frac{2}{D}\eta_{\mu\nu}\partial_\rho \xi^\rho = 0.$$ (1.13)

Exercise 1.2 Get more insight in the meaning of the statement that these are the transformations that preserve 'angles'. To consider angles, we should consider two variations of the same spacetime point. Consider the vectors from x to $x^\mu + (\Delta_1)^\mu$ and another one to $x^\mu + (\Delta_2)^\mu$, where the deformations are considered to be small. The angle between these two is

$$\cos^2\theta = \frac{(\Delta_1 \cdot \Delta_2)^2}{(\Delta_1 \cdot \Delta_1)(\Delta_2 \cdot \Delta_2)}.$$ (1.14)

Now we perform a spacetime transformation that takes a point x to $x'(x) = x + \xi(x)$. Then the first vector will be between $x'(x)$ and $x'(x + \Delta_1) = x'(x) + \Delta_1 \cdot \partial x'(x)$. The new vector is thus

$$\Delta_1' = \Delta_1 \cdot \partial x'(x) = \Delta_1 + \Delta_1 \cdot \partial \xi(x).$$ (1.15)

We thus find that

$$\Delta_1' \cdot \Delta_2' = \Delta_1 \cdot \Delta_2 + \Delta_1^\rho \left(\partial_\rho \xi^\mu(x)\right) \eta_{\mu\nu} \Delta_2^\nu + \Delta_1^\mu \eta_{\mu\nu} \Delta_2^\rho \left(\partial_\rho \xi^\mu(x)\right), \qquad (1.16)$$

where we added indices for clarity. If the last factor gives just a scaling, i.e. if

$$\Delta_1^\rho \left(\partial_\rho \xi^\mu(x)\right) \eta_{\mu\nu} \Delta_2^\nu + \Delta_1^\mu \eta_{\mu\nu} \Delta_2^\rho \left(\partial_\rho \xi^\mu(x)\right) = 2\Lambda_D(x) \Delta_1^\mu \eta_{\mu\nu} \Delta_2^\nu, \qquad (1.17)$$

then it is easy to see that (1.14) is invariant. Indeed, all factors scale with the same coefficient as the scale factor has to be evaluated at the same spacetime point.

The requirement (1.17) amounts to the scaling of the metric. Indeed, if a metric scales under spacetime transformations,

$$\delta(dx^\mu \eta_{\mu\nu} dx^\nu) \equiv 2dx^\rho \left(\partial_\rho \xi^\mu(x)\right) \eta_{\mu\nu} dx^\nu = 2\Lambda_D(x)(dx^\mu \eta_{\mu\nu} dx^\nu), \qquad (1.18)$$

then replace dx in the above by $(\Delta_1 + \Delta_2)$ and subtract the diagonal terms. This leads to (1.17) and hence to the invariance of the angle.

Angles are thus preserved by the transformations that scale the metric and these are the conformal transformations. □

In $D = 2$ with as non-zero metric[12] elements $\eta_{z\bar{z}} = 1$, the Killing equations are reduced to Cauchy–Riemann equations: $\partial_z \xi_{\bar{z}} = \partial_{\bar{z}} \xi_z = 0$. Therefore any holomorphic (and anti-holomorphic) function is locally a conformal transformation ($\xi_{\bar{z}}(z)$ and $\xi_z(\bar{z})$) and the conformal algebra becomes infinite dimensional (Virasoro algebra). In dimensions $D > 2$ the conformal algebra is finite-dimensional and the general solutions of (1.13) are

$$\xi^\mu(x) = a^\mu + \lambda^{\mu\nu} x_\nu + \lambda_D x^\mu + \left(x^2 \lambda_K^\mu - 2x^\mu x \cdot \lambda_K\right). \qquad (1.19)$$

They contain D translations P_μ corresponding to the parameters a^μ, one dilatation D related to the parameter λ_D, $D(D-1)/2$ (pseudo-)Lorentz rotations $M_{\mu\nu}$ corresponding to the parameters $\lambda^{\mu\nu}$ and a set of D 'special conformal transformations' K_μ related to the parameters λ_K^μ. One may notice that the full set of conformal transformations contains $(D+1)(D+2)/2$ parameters.

The most general conformal transformation, $\delta_C(\xi)$, can be expressed in terms of generators as follows:

$$\delta_C(\xi) = a^\mu P_\mu + \tfrac{1}{2}\lambda^{\mu\nu} M_{\mu\nu} + \lambda_D D + \lambda_K^\mu K_\mu. \qquad (1.20)$$

---

[12] In Minkowski space $z = \frac{1}{\sqrt{2}}\left(x^1 + x^0\right)$ and $\bar{z} = \frac{1}{\sqrt{2}}\left(x^1 - x^0\right)$ are not each other complex conjugates. In Euclidean signature they are complex conjugates.

With these transformations, one can obtain the conformal algebra. The non-zero commutators are

$$[M_{\mu\nu}, M^{\rho\sigma}] = -4\delta^{[\rho}_{[\mu} M_{\nu]}{}^{\sigma]},$$

$$[P_\mu, M_{\nu\rho}] = 2\eta_{\mu[\nu} P_{\rho]}, \quad [K_\mu, M_{\nu\rho}] = 2\eta_{\mu[\nu} K_{\rho]},$$

$$[P_\mu, K_\nu] = 2(\eta_{\mu\nu} D + M_{\mu\nu}),$$

$$[D, P_\mu] = P_\mu, \quad [D, K_\mu] = -K_\mu. \tag{1.21}$$

The conformal algebra defined on a flat space with signature (p, q), is isomorphic to the $SO(p + 1, q + 1)$ algebra. In what follows we will consider a D dimensional Minkowski space (whose signature is $(D - 1, 1)$ in our conventions), such that the conformal algebra is the SO$(D, 2)$ algebra.[13] Indeed one can define

$$M^{\hat{\mu}\hat{\nu}} = \begin{pmatrix} M^{\mu\nu} & \frac{1}{2}(P^\mu - K^\mu) & \frac{1}{2}(P^\mu + K^\mu) \\ -\frac{1}{2}(P^\nu - K^\nu) & 0 & -D \\ -\frac{1}{2}(P^\nu + K^\nu) & D & 0 \end{pmatrix}, \tag{1.22}$$

where indices are raised w.r.t. the rotation matrices $M^{\hat{\mu}}{}_{\hat{\nu}}$ with the metric

$$\hat{\eta} = \text{diag}(-1, 1, \ldots, 1, -1). \tag{1.23}$$

When applied on fields, conformal transformations have 'orbital parts' and may have 'intrinsic parts'. We absorb the orbital parts of all generators in the conformal Killing vector $\xi^\mu(x)$, given by (1.19). The general form of the conformal transformations of fields $\phi^i(x)$ is[14]

$$\delta_C(\xi)\phi^i(x) = \xi^\mu(x)\partial_\mu\phi^i(x) - \frac{1}{2}\Lambda_M{}^{\mu\nu}(x)\, m_{\mu\nu}{}^i{}_j\phi^j(x)$$

$$+\Lambda_D(x)\, k_D{}^i(\phi)(x) + \lambda^\mu_K k_\mu{}^i(\phi)(x), \tag{1.24}$$

where the x-dependent rotation $\Lambda_{M\,\mu\nu}(x)$ and x-dependent dilatation $\Lambda_D(x)$ are given by

$$\Lambda_{M\,\mu\nu}(x) = \partial_{[\nu}\xi_{\mu]} = \lambda_{\mu\nu} - 4x_{[\mu}\lambda_{K\,\nu]},$$

$$\Lambda_D(x) = \frac{1}{D}\partial_\rho\xi^\rho = \lambda_D - 2x \cdot \lambda_K, \tag{1.25}$$

---

[13]In the 2-dimensional case SO(2, 2) = SU(1, 1) × SU(1, 1) is realized by the finite subgroup of the infinite dimensional conformal group, and is well known in terms of $L_{-1} = \frac{1}{2}(P_0 - P_1)$, $L_0 = \frac{1}{2}(D + M_{10})$, $L_1 = \frac{1}{2}(K_0 + K_1)$, $\bar{L}_{-1} = \frac{1}{2}(P_0 + P_1)$, $\bar{L}_0 = \frac{1}{2}(D - M_{10})$, $\bar{L}_1 = \frac{1}{2}(K_0 - K_1)$. Higher order L_n, $|n| \geq 2$ have no analogs in $D > 2$.

[14]The expression ξ always determines the parameters $\{a^\mu, \lambda^{\mu\nu}, \lambda_D, \lambda^\mu_K\}$ as in (1.19).

and $m_{[\mu\nu]}{}^i{}_j \phi^j(x)$, $k_D{}^i(\phi)(x)$ and $k^i_\mu(\phi)(x)$ determine the intrinsic part of the conformal transformations on the fields.

To specify for each field ϕ^i its transformations under conformal group one has to specify the intrinsic part of the conformal transformation, namely

1. Transformations under the Lorentz group, encoded into the matrix $(m_{\mu\nu})^i{}_j$. The Lorentz transformation matrix $m_{\mu\nu}$ should satisfy

$$m_{\mu\nu}{}^i{}_k m_{\rho\sigma}{}^k{}_j - m_{\rho\sigma}{}^i{}_k m_{\mu\nu}{}^k{}_j = 2\eta_{\mu[\rho} m_{\sigma]\nu}{}^i{}_j - 2\eta_{\nu[\rho} m_{\sigma]\mu}{}^i{}_j. \qquad (1.26)$$

The explicit form for Lorentz transformation matrices is for vectors (the indices i and j are of the same kind as μ and ν)

$$m_{\mu\nu}{}^\rho{}_\sigma = 2\delta^\rho_{[\mu} \eta_{\nu]\sigma}\,, \qquad \text{e.g. } \delta V^\mu = -\lambda^{\mu\nu} V_\nu\,, \qquad (1.27)$$

while for spinors, (where i and j are (unwritten) spinor indices)

$$m_{\mu\nu} = \tfrac{1}{2}\gamma_{\mu\nu}\,, \qquad \text{e.g. } \delta\Psi = -\tfrac{1}{4}\lambda^{\mu\nu}\gamma_{\mu\nu}\Psi\,. \qquad (1.28)$$

2. Transformation under dilatation specified by $k_D{}^i(\phi)$. In most cases (and for all non-scalar fields), we just have

$$k_D{}^i(\phi) = w\phi^i\,, \qquad (1.29)$$

where w is a real number called the *Weyl weight* of the field ϕ^i (in principle, different for each field). However, for scalars in a non-trivial manifold with affine connection $\Gamma_{ij}{}^k$ (torsionless, i.e. symmetric in (ij)), these are closed homothetic Killing vectors, see [4, Sect. 15.7], i.e. solutions of

$$\nabla_i k_D{}^j \equiv \partial_i k_D{}^j + \Gamma_{ik}{}^j k_D{}^k = w\delta_i^j\,, \qquad (1.30)$$

where again w is the 'Weyl weight' and the derivatives ∂_i, ∇_i are with respect to the field ϕ^i. For $\Gamma_{ij}{}^k = 0$, this reduces to the simple case (1.29).

3. Special conformal transformations can have extra parts apart from those in (1.19) and (1.25), connected to translations, rotations and dilatations. These are denoted as $k^i_\mu(\phi)$. From the commutator $[D, K_\mu] = -K_\mu$ and (1.29) it is easy to see that[15]

$$k_\mu{}^j \partial_j k_D{}^i - k_D{}^j \partial_j k_\mu{}^i = k_\mu{}^i\,. \qquad (1.31)$$

---

[15]Remember that, for the intrinsic part, $\delta_K \delta_D \phi^i = \delta_K(k^i_D(\phi)) = \partial_j k^i_D(\phi)\delta_K\phi^j$.

For the simple form of the dilatations (1.29), this means that $k_\mu{}^i(\phi)$ must have Weyl weight one less than that of ϕ^i. Also, for consistency of the $[K, K]$ commutators, the $k_\mu^i \partial_i$ should be mutually commuting operators.

In this way, the algebra (1.21) is realized on the fields as

$$[\delta_C(\xi_1), \delta_C(\xi_2)]\phi = \delta_C \left(\xi^\mu = \xi_2^\nu \partial_\nu \xi_1^\mu - \xi_1^\nu \partial_\nu \xi_2^\mu\right)\phi, \tag{1.32}$$

where ξ^μ stands for the set $\{a^\mu, \lambda^{\mu\nu}, \lambda_D, \lambda_K^\mu\}$ building (1.19).

To understand fully the meaning of the order of the transformations, consider in detail the calculation of the commutator of transformations of fields. See e.g. for a field of zero Weyl weight ($k_D(\phi) = 0$), and notice how the transformations act only on fields, not on explicit spacetime points x^μ:

$$\begin{aligned}
\lambda_D a^\mu[D, P_\mu]\phi(x) &= \left(\delta_D(\lambda_D)\delta_P(a^\mu) - \delta_P(a^\mu)\delta_D(\lambda_D)\right)\phi(x) \\
&= \delta_D(\lambda_D)a^\mu \partial_\mu \phi(x) - \delta_P(a^\mu)\lambda_D x^\mu \partial_\mu \phi(x) \\
&= a^\mu \partial_\mu \left(\lambda_D x^\nu\right)\partial_\nu \phi(x) \\
&= a^\mu \lambda_D \partial_\mu \phi(x) = \lambda_D a^\mu P_\mu \phi(x).
\end{aligned} \tag{1.33}$$

It is important to notice that the derivative of a field of Weyl weight w has weight $w + 1$. E.g. for a scalar with dilatational transformation determined by the vector k_D (and without extra special conformal transformations) we obtain, using $\delta_C(\xi)(\partial_\mu \phi(x)) = \partial_\mu(\delta_C(\xi)\phi(x))$

$$\begin{aligned}
\delta_C(\xi)\partial_\mu \phi(x) = \xi^\nu(x)\partial_\nu \partial_\mu \phi(x) + \Lambda_D(x)\,\partial_\mu k_D(\phi) \\
-\Lambda_{M\mu}{}^\nu(x)\partial_\nu \phi(x) + \Lambda_D(x)\partial_\mu \phi(x) - 2\lambda_{K\mu}k_D(\phi).
\end{aligned} \tag{1.34}$$

The first term on the second line says that it behaves as a vector under Lorentz transformations. Furthermore, this equation implies that the dilatational transformation of $\partial_\mu \phi$ is determined by $\partial_\mu(k_D(\phi) + \phi)$. For the simple transformation (1.29), this means that the derivative also satisfies such a simple transformation with weight $w + 1$. Furthermore, the derivative $\partial_\mu \phi$ has an extra part in the special conformal transformation of the form $\lambda_K^\nu k_\nu(\partial_\mu \phi) = -2\lambda_{K\mu}k_D(\phi)$, as can be seen by looking at the special conformal transformation part in $\delta_C(\xi)\phi(x)$.

With these rules the conformal algebra is satisfied. The question remains when an action is conformally invariant. We consider local actions for fields ϕ^i with at most first derivatives of the fields:

$$S = \int d^D x \, \mathcal{L}(\phi^i(x), \partial_\mu \phi^i(x)). \tag{1.35}$$

The fields ϕ^i are scalars or other representations of the Lorentz group, and have dilation transformation (1.29) with weight w_i for each field (for an extension, see further Exercise 1.4). For invariance of the action under P_μ and $M_{\mu\nu}$ one has the usual requirements of a covariant action. For invariance under dilatation one needs that the total weight of each term should be equal to the dimension D. Here ∂_μ counts also for 1, as can be seen from (1.34). Then

$$\delta_D \mathcal{L} = \xi^\mu \partial_\mu \mathcal{L} + D \Lambda_D(x) \mathcal{L}$$
$$= \partial_\mu \left(\xi^\mu \mathcal{L} \right) - \left(\partial_\mu \xi^\mu \right) \mathcal{L} + D \Lambda_D(x) \mathcal{L} = \partial_\mu \left(\xi^\mu \mathcal{L} \right), \qquad (1.36)$$

so the remaining term is a total derivative which vanishes in the action.

Finally, we consider the condition for invariance of an action under special conformal transformation. The special conformal transformations sit in the last term of (1.24), and hidden in $\Lambda_D(x)$ and $\Lambda_{M\mu\nu}(x)$ in (1.25). The contributions when the latter are not differentiated in the transformation of $\partial_\mu \phi^i$, are canceled by the steps mentioned above where the Lorentz transformations and dilatational transformations are considered (e.g. in (1.36)). When they are differentiated they give rise to extra terms

$$\delta_{K,\text{extra}} \partial_\mu \phi^i = w_i (\partial_\mu \Lambda_D) \phi^i - \tfrac{1}{2} (\partial_\mu \Lambda_M^{\rho\sigma}) m_{\rho\sigma}{}^i{}_j \phi^j . \qquad (1.37)$$

E.g. for scalars, the last term in (1.34) is the contribution due to the first term here. The contributions due to the last term of (1.24) act, using the chain rule, as a field equation of ϕ^i times the transformation $k_\mu{}^i(\phi)(x)$. We thus remain with

$$\delta_K S = 2\lambda_K^\mu \int d^D x \, \frac{\mathcal{L} \overleftarrow{\partial}}{\partial(\partial_\nu \phi^i)} \left(-\eta_{\mu\nu} w_i \phi^i - m_{\mu\nu}{}^i{}_j \phi^j \right) + \lambda_K^\mu \frac{S \overleftarrow{\delta}}{\delta \phi^i(x)} k_\mu{}^i(\phi)(x) , \qquad (1.38)$$

where $\overleftarrow{\partial}$ indicates a right derivative. In many cases the first terms already cancel and no $k_\mu(\phi)$ are necessary. In fact, the latter are often excluded because of the requirement that they should have Weyl weight $w_i - 1$, and in many cases there are no such fields available.

Exercise 1.3 There are typical cases in which the first two terms of (1.38) cancel. Check the following ones

1. scalars with Weyl weight 0.
2. spinors appearing as $\partial\!\!\!/\lambda$ if their Weyl weight is $(D-1)/2$. This is also the appropriate weight for actions as $\bar\lambda \partial\!\!\!/\lambda$.
3. Vectors or antisymmetric tensors whose derivatives appear only as field strengths $\partial_{[\mu_1} B_{\mu_2 ... \mu_p]}$ if their Weyl weight is $p - 1$. This value of the Weyl weight is what we need also in order that their gauge invariances and their zero modes commute with the dilatations. Then scale invariance of the usual square of the field strengths will fix $p = \frac{d}{2}$.

4. Scalars X^i with Weyl weight $\frac{d}{2} - 1$ and

$$\mathcal{L} = \left(\partial_\mu X^i\right) A_{ij} \left(\partial^\mu X^j\right), \tag{1.39}$$

where A_{ij} are constants.

□

Exercise 1.4 When the scalars transform under dilatations and special conformal transformations according to

$$\delta\phi^i = \xi^\mu(x)\partial_\mu\phi^i + \Lambda_D(x)k_D^i(\phi), \tag{1.40}$$

with $k_D^i(\phi)$ arbitrary, check that the conformal algebra is satisfied. Consider now the action for scalars

$$S_{sc} = -\frac{1}{2}\int d^D x\, \partial_\mu\phi^i\, g_{ij}(\phi)\, \partial^\mu\phi^j. \tag{1.41}$$

It is invariant under translations and Lorentz rotations. Check that the dilatational and special conformal transformations leave us with

$$\delta S_{sc} = -\int d^D x\, \Big\{\Lambda_D(x)\partial_\mu\phi^i\, \partial^\mu\phi^j\Big[g_{k(i}\left(\partial_{j)}k_D^k + \Gamma_{j)\ell}^k k_D^\ell\right) - \tfrac{1}{2}(D-2)g_{ij}\Big]$$

$$-2\lambda_K^\mu\partial_\mu\phi^i\, g_{ij}k_D^j\Big\}, \tag{1.42}$$

if one identifies the affine connection with the Levi-Civita connection of the metric, similar to (A.4). The invariance under rigid dilatations is thus obtained if

$$\nabla_{(i}k_{j)D} = \tfrac{1}{2}(D-2)g_{ij}, \tag{1.43}$$

with the usual definition of a covariant derivative ∇_i. Vectors satisfying this equation are called 'homothetic Killing vectors'.[16] However, to obtain special conformal invariance, the last term of (1.42), originating from a contribution $\partial_\mu\Lambda_D(x)$, should be a total derivative. Thus one requires that

$$k_{iD} = \partial_i k, \tag{1.44}$$

for some k. Then $\nabla_i k_{jD}$ is already symmetric, and thus with the requirement (1.43) one obtains that (1.30) should be satisfied with $w = (D-2)/2$. The vectors

---

[16]The terminology reflects that the right-hand side is a *constant* times g_{ij}. For a function times g_{ij} it is just a 'conformal Killing vector'.

satisfying (1.30) are 'exact homothetic Killing vectors'. One can find that the scalar k in (1.44) is

$$(D - 2)k = k_D{}^i g_{ij} k_D{}^j \,. \tag{1.45}$$

A systematic investigation of conformal actions for scalars in gravity can be found in [75]. □

Exercise 1.5 Check that for a Lagrangian of the form

$$\mathcal{L} = \left(\partial_\mu \phi^1\right) \left(\partial^\mu \phi^2\right) , \tag{1.46}$$

the conformal Killing equation has as solution

$$k_D{}^1 = w_1 \phi^1 \,, \qquad k_D{}^2 = w_2 \phi^2 \,, \qquad w_1 + w_2 = D - 2 \,. \tag{1.47}$$

However, the Eq. (1.30) gives also that $w_1 = w_2 = \frac{1}{2}(D - 2)$. Check that this is necessary for special conformal transformations.

Modifying the Lagrangian to

$$\mathcal{L} = \left(1 + \frac{\phi^1}{\phi^2}\right) \left(\partial_\mu \phi^1\right) \left(\partial^\mu \phi^2\right) , \tag{1.48}$$

one finds as only non-zero Levi-Civita connections

$$\Gamma_{11}^1 = \frac{1}{\phi^1 + \phi^2} \,, \qquad \Gamma_{22}^2 = -\frac{\phi^1/\phi^2}{\phi^1 + \phi^2} \,. \tag{1.49}$$

The solutions of the conformal Killing equations (1.43) already fix (1.47) with $w_1 = w_2 = \frac{1}{2}(D - 2)$. However, now the Eq. (1.30) gives no solution. Hence, this model can only have rigid dilatations, but no rigid special conformal transformations. □

1.2.4 Superconformal Groups

A classical work on the classification of superconformal groups is the paper of Nahm [76]. In the groups that he classified, the bosonic subgroup is a direct product of the conformal group and the R-symmetry group, the automorphism group of the supersymmetries. The latter are bosonic symmetries that are not in the conformal algebra, hence are spacetime scalars, as motivated by the works of [71] and [72].[17]

---

[17] However, with branes the assumptions of these papers may be too constrained. Other examples have been considered first in 10 and 11 dimensions in [77].

Table 1.2 Superconformal algebras, with the two parts of the bosonic subalgebra: one that contains the conformal algebra and the other one is the R-symmetry

D	Supergroup	Bosonic group	
3	$OSp(\mathcal{N}|4)$	$Sp(4) = SO(3, 2)$	$SO(\mathcal{N})$
4	$SU(2, 2|\mathcal{N})$	$SU(2, 2) = SO(4, 2)$	$SU(\mathcal{N}) \times U(1)$
5	$OSp(8^*|\mathcal{N})$	$SO^*(8) \supset SO(5, 2)$	$USp(\mathcal{N})$
	$F(4)$	$SO(5, 2)$	$SU(2)$
6	$OSp(8^*|\mathcal{N})$	$SO^*(8) = SO(6, 2)$	$USp(\mathcal{N})$
7	$OSp(16^*|\mathcal{N})$	$SO^*(16) \supset SO(7, 2)$	$USp(\mathcal{N})$
8	$SU(8, 8|\mathcal{N})$	$SU(8, 8) \supset SO(8, 2)$	$SU(\mathcal{N}) \times U(1)$
9	$OSp(\mathcal{N}|32)$	$Sp(32) \supset SO(9, 2)$	$SO(\mathcal{N})$
10	$OSp(\mathcal{N}|32)$	$Sp(32) \supset SO(10, 2)$	$SO(\mathcal{N})$
11	$OSp(\mathcal{N}|64)$	$Sp(64) \supset SO(11, 2)$	$SO(\mathcal{N})$

In the cases $D = 4$ and $D = 8$, the U(1) factor in the R-symmetry group can be omitted for $\mathcal{N} \neq 4$ and $\mathcal{N} \neq 16$, respectively

Another classification has appeared in [78] from which we can extract[18] Table 1.2 for dimensions from 3 to 11. The bosonic subgroup contains always two factors: the conformal group (one should consider its covering that allows the spinor representation) and the R-symmetry group. When the first factor is just the conformal group, then the algebra appears in Nahm's classification. Note that 5 dimensions is a special case. There is a generic superconformal algebra for any extension. But for the case $\mathcal{N} = 2$ there exists a smaller superconformal algebra that is in Nahm's list. Note that for $D = 6$ or $D = 10$, where one can have chiral spinors, only the case that all supersymmetries have the same chirality has been included. Non-chiral supersymmetry can be obtained from the reduction in one more dimension, so e.g. from the $D = 7$ algebra $OSp(16^*|\mathcal{N})$ we obtain $(\mathcal{N}, \mathcal{N})$ supersymmetry in $D = 6$ [79]. So far, superconformal tensor calculus has only been based on algebras of Nahm's type.[19]

In the case of our interest $\mathcal{N} = 2$ in $D = 4$, $D = 5$ and $D = 6$ the superconformal algebras are, respectively, $SU(2, 2|2)$, $F^2(4)$ and $OSp(8^*|2)$ and with R-symmetry groups $SU(2) \times U(1)$, $SU(2)$, $SU(2)$. In the practical treatment, we will not see any fundamental difference in the structure of these supergroups. The fermionic part of the superconformal group is generated by the supersymmetries Q^i (superpartners of the translations P_a) and the special supersymmetries S^i (superpartners of the special conformal transformation K_a).

---

[18]See Appendix B for the notations of groups and supergroups.

[19]Note that the superalgebras that are relevant for quantum field theories, according to Coleman and Mandula [71] and Haag–Łopuszański–Sohnius [72], only exist for $D \leq 6$. The maximal number of supercharges of the corresponding superconformal QFTs is bounded by the requirement that they admit a suitable stress-tensor multiplet, see e.g. [80] for a recent discussion.

The dilatations provide a 3-grading of the conformal algebra and a 5-grading of the superconformal algebra. In the conformal algebra, the translations P_a have weight 1, in the sense that $[D, P_a] = P_a$. The special conformal transformations K_a have weight -1, while Lorentz generators M_{ab} and dilatations have all weight 0. Of course, the R-symmetry algebra commutes with the conformal algebra and thus has weight 0. The supersymmetries have weight $+\frac{1}{2}$ (see already that this is consistent with (1.4)), and the special supersymmetries have weight $-\frac{1}{2}$. Thus, there is a clear structure in the superconformal algebra, ordering them according to the Weyl (dilatational) weight:

$$1 : P_a$$

$$\tfrac{1}{2} : Q$$

$$0 : D,\ M_{ab},\ U_i{}^j,\ T$$

$$-\tfrac{1}{2} : S$$

$$-1 : K_a . \tag{1.50}$$

This grading already determines the structure of many commutators.

Let us now present some details of the superconformal algebras.[20] First, we have the conformal algebra, as given in (1.21). The other part of the bosonic algebra is the R-symmetry. We denote $U_i{}^j$ as the anti-hermitian generators of SU(2) and T as the real generator of U(1), the latter only for $D = 4$. The SU(2) algebra is

$$\left[U_i{}^j, U_k{}^\ell \right] = \delta_i{}^\ell U_k{}^j - \delta_k{}^j U_i{}^\ell . \tag{1.51}$$

The antihermiticity and tracelessness properties of the SU(2) generators are

$$U_i{}^j = -\left(U_j{}^i \right)^* , \qquad U_i{}^i = 0 ,$$

$$U_i{}^j = \mathbf{U} \cdot \boldsymbol{\tau} , \tag{1.52}$$

with $\tau_i{}^j = i\sigma_i{}^j$ in terms of the three Pauli matrices $\sigma_i{}^j$. The 3-vectors $\mathbf{U}$ have real components U_1, U_2 and U_3 are real operators, for which the commutators (1.51) imply

$$[U_1, U_2] = U_3 , \qquad [U_2, U_3] = U_1 , \qquad [U_3, U_1] = U_2 . \tag{1.53}$$

---

[20]To prepare these formulae, we made use of [81] for $D = 4$, apart from a sign change in the choice of charge conjugation, such that the anticommutators of fermionic generators have all opposite sign. For $D = 5$, we made use of [82], and for $D = 6$ of [83], but replacing there K by $-K$ and U_{ij} by $-\frac{1}{2}U_{ij}$.

More on the use of the SU(2) indices is collected in Appendix A.2 and specifically for the triplet notation in Appendix A.2.2.

The equations of the first line of (1.52) are valid as well in $D = 4$, where raising and lowering of i-indices is done by complex conjugation, and in $D = 5, 6$ where they are raised and lowered by the ε^{ij} as in (A.15). In 4 dimensions, the first equation is written as

$$D = 4 \; : \; U_i{}^j = -U^j{}_i \, . \tag{1.54}$$

In 5 and 6 dimensions the one-index raised or lowered matrices are (e.g. $\tau_{1ij} = \tau_{1i}{}^k \varepsilon_{kj} = i\sigma_1 i\sigma_2 = -i\sigma_3$)

$$D = 5, 6 \; : \; U_{ij} = -iU_1\sigma_3 - U_2\mathbb{1} + i\sigma_1 U_3 \, , \qquad U^{ij} = iU_1\sigma_3 - U_2\mathbb{1} - i\sigma_1 U_3 \, . \tag{1.55}$$

The first equation of (1.52) implies with the charge conjugation rule (see Appendices A.3.2 and A.3.3) $M^C = \sigma_2 M^* \sigma_2$ that U is a C-invariant matrix. The tracelessness translates in symmetry of $U_{(ij)}$ and $U^{(ij)}$. Thus, in conclusion, U is C-invariant and symmetric (and thus also $U_i{}^j = U^j{}_i$ here, different from (1.54)).

Then we include the fermionic generators. In 6 dimensions Q^i and S^i are symplectic Majorana–Weyl spinors with opposite chirality

$$Q^i = P_R Q^i = -\gamma_* Q^i, \qquad S^i = P_L S^i = \gamma_* S^i \, . \tag{1.56}$$

In 4 dimensions, the Q_i and S_i have opposite chirality: i.e. Q_i and S^i are left-handed (compare with (1.1))

$$S^i = P_L S^i = \gamma_* S^i \, , \qquad S_i = P_R S_i = -\gamma_* S_i \, . \tag{1.57}$$

The commutators between the bosonic and fermionic generators are (we leave implicit the trivial ones)

$$[M_{ab}, Q_\alpha^i] = -\tfrac{1}{2}(\gamma_{ab} Q^i)_\alpha \, , \qquad\qquad [M_{ab}, S_\alpha^i] = -\tfrac{1}{2}(\gamma_{ab} S^i)_\alpha \, ,$$

$$[D, Q_\alpha{}^i] = \tfrac{1}{2} Q_\alpha{}^i \, , \qquad\qquad\qquad [D, S_\alpha{}^i] = -\tfrac{1}{2} S_\alpha{}^i \, ,$$

$$[U_i{}^j, Q_\alpha{}^k] = \delta_i{}^k Q_\alpha{}^j - \tfrac{1}{2}\delta_i{}^j Q_\alpha{}^k \, , \qquad [U_i{}^j, S_\alpha{}^k] = \delta_i{}^k S_\alpha{}^j - \tfrac{1}{2}\delta_i{}^j S_\alpha{}^k \, ,$$

$$[U_i{}^j, Q_{\alpha k}] = -\delta_k{}^j Q_{\alpha i} + \tfrac{1}{2}\delta_i{}^j Q_{\alpha k} \, , \qquad [U_i{}^j, S_{\alpha k}] = -\delta_k{}^j S_{\alpha i} + \tfrac{1}{2}\delta_i{}^j S_{\alpha k} \, ,$$

$$[T, Q_\alpha{}^i] = \tfrac{1}{2} i Q_\alpha{}^i \, , \qquad\qquad\qquad [T, S_\alpha{}^i] = \tfrac{1}{2} i S_\alpha{}^i \, ,$$

$$\qquad\qquad\qquad\qquad\qquad\qquad\qquad\qquad T \text{ exists only for } D = 4 \, ,$$

$$[K_a, Q_\alpha^i] = s_D \, (\gamma_a S^i)_\alpha \, , \qquad\qquad [P_a, S_\alpha^i] = \bar{s}_D (\gamma_a Q^i)_\alpha \, ,$$

$$\text{with} \qquad\qquad\qquad \begin{cases} D = 4 : & s_4 = 1, \\ D = 5 : & s_5 = i, \\ D = 6 : & s_6 = -1. \end{cases}$$

$$\tag{1.58}$$

and $\bar{s}_D$ is the complex conjugate of s_D (only different for $D = 5$). Note that the commutators $[K, Q]$ and $[P, S]$ have a different form in different dimensions due to the difference in reality and chirality conditions. The factors i are necessary in 5 dimensions if we use the same symplectic Majorana condition for all the spinors. This can be seen easily from the C-conjugation rules in Appendix A.3.2.

The anticommutation relations between the fermionic generators are

$$
D = 4 : \{Q_\alpha{}^i, Q_j{}^\beta\} = -\tfrac{1}{2}\delta_j{}^i (\gamma^a)_\alpha{}^\beta P_a , \qquad \{Q_\alpha{}^i, Q^{j\beta}\} = 0 ,
$$

$$
\{S_\alpha{}^i, S_j{}^\beta\} = -\tfrac{1}{2}\delta_j{}^i (\gamma^a)_\alpha{}^\beta K_a , \qquad \{S_\alpha{}^i, S^{j\beta}\} = 0 ,
$$

$$
\{Q_\alpha{}^i, S^{j\beta}\} = 0 ,
$$

$$
\{Q_\alpha{}^i, S_j{}^\beta\} = -\tfrac{1}{2}\delta_j{}^i \delta_\alpha{}^\beta D - \tfrac{1}{4}\delta_j{}^i (\gamma^{ab})_\alpha{}^\beta M_{ab} - \tfrac{1}{2}\mathrm{i}\delta_j{}^i \delta_\alpha{}^\beta T + \delta_\alpha{}^\beta U_j{}^i ,
$$

$$
D = 5, 6 : \{Q_{i\alpha}, Q^{j\beta}\} = -\tfrac{1}{2}\delta_i{}^j (\gamma^a)_\alpha{}^\beta P_a , \qquad \{S_{i\alpha}, S^{j\beta}\} = -\tfrac{1}{2}\delta_i{}^j (\gamma^a)_\alpha{}^\beta K_a ,
$$

$$
D = 5 : \{Q_{i\alpha}, S^{j\beta}\} = -\tfrac{1}{2}\mathrm{i}\left(\delta_i{}^j \delta_\alpha{}^\beta D + \tfrac{1}{2}\delta_i{}^j (\gamma^{ab})_\alpha{}^\beta M_{ab} + 3\delta_\alpha{}^\beta U_i{}^j\right) ,
$$

$$
D = 6\{Q_{i\alpha}, S^{j\beta}\} = \tfrac{1}{2}\left(\delta_i{}^j \delta_\alpha{}^\beta D + \tfrac{1}{2}\delta_i{}^j (\gamma^{ab})_\alpha{}^\beta M_{ab} + 4\delta_\alpha{}^\beta U_i{}^j\right) . \tag{1.59}
$$

Note that in 6 dimensions, the spinors have a chiral projection, and the gamma matrices in the right-hand side should also be understood as their chiral projection, thus e.g. in the last line $\delta_\alpha{}^\beta$ stands for $\tfrac{1}{2}(\mathbb{1} - \gamma_*)_\alpha{}^\beta$.

Exercise 1.6 Check that in 5 and 6 dimensions the $[U, Q]$ commutator can be written as $[U_{ij}, Q^k] = \delta_{(i}{}^k Q_{j)}$. $\qquad\qquad \square$

1.2.5 Rigid Superconformal Symmetry

To upgrade a realization of supersymmetry to superconformal transformations, we thus have to define also the S-supersymmetry transformations on the fields on top of the Q-supersymmetry. At the bosonic side, we saw in (1.24) that conformal transformations of fields have 'orbital' parts and 'intrinsic' parts. The same will be true for the supersymmetries. Though the word 'orbital' is not really clear for supersymmetries, we will use the same terminology since the procedure is similar. The orbital parts at the bosonic side were chosen such that the first term in (1.24) with the definition (1.19) already satisfies the conformal algebra. We can do the same for supersymmetry: we replace the constant supersymmetry parameters ϵ of a rigid supersymmetric theory by

$$
\epsilon(x) = \epsilon + s_D \gamma_\mu x^\mu \eta , \tag{1.60}
$$

where the ϵ and η at the right-hand side are the constant parameters for rigid Q- and S-supersymmetry, respectively. The constants s_D are those in (1.58), such that the algebra is satisfied.

Finally, in parallel with the special conformal transformations, we should expect intrinsic contributions from S-supersymmetry transformations. As follows from (1.50), these can only transform a field A to a quantity B that has dilatation weight $1/2$ less than A.

References

1. B. de Wit, A. Van Proeyen, Potentials and symmetries of general gauged $N = 2$ supergravity—Yang–Mills models. Nucl. Phys. **B245**, 89–117 (1984). https://doi.org/10.1016/0550-3213(84)90425-5
2. N. Seiberg, E. Witten, Electric-magnetic duality, monopole condensation, and confinement in $N = 2$ supersymmetric Yang–Mills theory. Nucl. Phys. **B426**, 19–52 (1994). https://doi.org/10.1016/0550-3213(94)90124-4, 10.1016/0550-3213(94)00449-8, arXiv:hep-th/9407087 [hep-th]. [Erratum: Nucl. Phys.B430,485(1994)]
3. N. Seiberg, E. Witten, Monopoles, duality and chiral symmetry breaking in $N = 2$ supersymmetric QCD. Nucl. Phys. **B431**, 484–550 (1994). https://doi.org/10.1016/0550-3213(94)90214-3, arXiv:hep-th/9408099 [hep-th]
4. D.Z. Freedman, A. Van Proeyen, *Supergravity* (Cambridge University, Cambridge, 2012). http://www.cambridge.org/mw/academic/subjects/physics/theoretical-physics-and-mathematical-physics/supergravity?format=AR
5. V. Cortés, C. Mayer, T. Mohaupt, F. Saueressig, Special geometry of Euclidean supersymmetry. I: vector multiplets. J. High Energy Phys. **03**, 028 (2004). https://doi.org/10.1088/1126-6708/2004/03/028, arXiv:hep-th/0312001 [hep-th]
6. V. Cortés, C. Mayer, T. Mohaupt, F. Saueressig, Special geometry of Euclidean supersymmetry. II: Hypermultiplets and the c-map. J. High Energy Phys. **06**, 025 (2005). https://doi.org/10.1088/1126-6708/2005/06/025, arXiv:hep-th/0503094 [hep-th]
7. V. Cortés, T. Mohaupt, Special geometry of Euclidean supersymmetry III: the local r-map, instantons and black holes. J. High Energy Phys. **07**, 066 (2009). https://doi.org/10.1088/1126-6708/2009/07/066, arXiv:0905.2844 [hep-th]
8. V. Cortés, P. Dempster, T. Mohaupt, O. Vaughan, Special geometry of Euclidean supersymmetry IV: the local c-map. J. High Energy Phys. **10**, 066 (2015). https://doi.org/10.1007/JHEP10(2015)066, arXiv:1507.04620 [hep-th]
9. M.A. Lledó, Ó. Maciá, A. Van Proeyen, V.S. Varadarajan, Special geometry for arbitrary signatures, in *Handbook on Pseudo-Riemannian Geometry and Supersymmetry*, ed. by V. Cortés. IRMA Lectures in Mathematics and Theoretical Physics, vol. 16, chap. 5 (European Mathematical Society, Zürich, 2010). hep-th/0612210
10. W.A. Sabra, Special geometry and space-time signature. Phys. Lett. **B773**, 191–195 (2017). https://doi.org/10.1016/j.physletb.2017.08.021, arXiv:1706.05162 [hep-th]
11. L. Gall, T. Mohaupt, Five-dimensional vector multiplets in arbitrary signature. J. High Energy Phys. **09**, 053 (2018). https://doi.org/10.1007/JHEP09(2018)053, arXiv:1805.06312 [hep-th]
12. V. Cortés, L. Gall, T. Mohaupt, Four-dimensional vector multiplets in arbitrary signature. arXiv:1907.12067 [hep-th]
13. B. de Wit, P.G. Lauwers, R. Philippe, S.Q. Su, A. Van Proeyen, Gauge and matter fields coupled to $N = 2$ supergravity. Phys. Lett. **B134**, 37–43 (1984). https://doi.org/10.1016/0370-2693(84)90979-1

14. L. Alvarez-Gaumé, D.Z. Freedman, Geometrical structure and ultraviolet finiteness in the supersymmetric σ-model. Commun. Math. Phys. **80**, 443 (1981). https://doi.org/10.1007/BF01208280

15. J. Bagger, E. Witten, Matter couplings in $N = 2$ supergravity. Nucl. Phys. **B222**, 1–10 (1983). https://doi.org/10.1016/0550-3213(83)90605-3

16. E. Bergshoeff, S. Vandoren, A. Van Proeyen, The identification of conformal hypercomplex and quaternionic manifolds. Int. J. Geom. Meth. Mod. Phys. **3**, 913–932 (2006). https://doi.org/10.1142/S0219887806001521, arXiv:math/0512084 [math.DG]

17. G. Sierra, P.K. Townsend, An introduction to $N = 2$ rigid supersymmetry, in *Supersymmetry and supergravity 1983*, ed. by B. Milewski (World Scientific, Singapore, 1983)

18. E. Cremmer, A. Van Proeyen, Classification of Kähler manifolds in $N = 2$ vector multiplet–supergravity couplings. Class. Quant. Grav. **2**, 445 (1985). https://doi.org/10.1088/0264-9381/2/4/010

19. A. Strominger, Special geometry. Commun. Math. Phys. **133**, 163–180 (1990). https://doi.org/10.1007/BF02096559

20. L. Castellani, R. D'Auria, S. Ferrara, Special geometry without special coordinates. Class. Quant. Grav. **7**, 1767–1790 (1990). https://doi.org/10.1088/0264-9381/7/10/009

21. S. Ferrara, M. Kaku, P.K. Townsend, P. van Nieuwenhuizen, Unified field theories with U(N) internal symmetries: gauging the superconformal group. Nucl. Phys. **B129**, 125–134 (1977). https://doi.org/10.1016/0550-3213(77)90023-2

22. M. Kaku, P.K. Townsend, P. van Nieuwenhuizen, Properties of conformal supergravity. Phys. Rev. **D17**, 3179–3187 (1978). https://doi.org/10.1103/PhysRevD.17.3179

23. M. Kaku, P.K. Townsend, Poincaré supergravity as broken superconformal gravity. Phys. Lett. **76B**, 54–58 (1978). https://doi.org/10.1016/0370-2693(78)90098-9

24. V. Pestun, M. Zabzine, Localization techniques in quantum field theories. J. Phys. **A50**(44), 440301 (2017). https://doi.org/10.1088/1751-8121/aa63c1, arXiv:1608.02952 [hep-th]

25. P.S. Howe, Supergravity in superspace. Nucl. Phys. **B199**, 309–364 (1982). https://doi.org/10.1016/0550-3213(82)90349-2

26. S.M. Kuzenko, U. Lindström, M. Roček, G. Tartaglino-Mazzucchelli, On conformal supergravity and projective superspace. J. High Energy Phys. **08**, 023 (2009) . https://doi.org/10.1088/1126-6708/2009/08/023, arXiv:0905.0063 [hep-th]

27. S.M. Kuzenko, G. Tartaglino-Mazzucchelli, Super-Weyl invariance in 5D supergravity. J. High Energy Phys. **04**, 032 (2008). https://doi.org/10.1088/1126-6708/2008/04/032, arXiv:0802.3953 [hep-th]

28. W.D. Linch, III, G. Tartaglino-Mazzucchelli, Six-dimensional supergravity and projective superfields. J. High Energy Phys. **08**, 075 (2012). https://doi.org/10.1007/JHEP08(2012)075, arXiv:1204.4195 [hep-th]

29. D. Butter, $N = 1$ conformal superspace in four dimensions. Ann. Phys. **325**, 1026–1080 (2010). https://doi.org/10.1016/j.aop.2009.09.010, arXiv:0906.4399 [hep-th]

30. D. Butter, $N = 2$ conformal superspace in four dimensions. J. High Energy Phys. **10**, 030 (2011). https://doi.org/10.1007/JHEP10(2011)030, arXiv:1103.5914 [hep-th]

31. D. Butter, J. Novak, Component reduction in $N = 2$ supergravity: the vector, tensor, and vector-tensor multiplets. J. High Energy Phys. **05**, 115 (2012). https://doi.org/10.1007/JHEP05(2012)115, arXiv:1201.5431 [hep-th]

32. D. Butter, S.M. Kuzenko, J. Novak, G. Tartaglino-Mazzucchelli, Conformal supergravity in five dimensions: new approach and applications. J. High Energy Phys. **02**, 111 (2015). https://doi.org/10.1007/JHEP02(2015)111, arXiv:1410.8682 [hep-th]

33. D. Butter, S.M. Kuzenko, J. Novak, S. Theisen, Invariants for minimal conformal supergravity in six dimensions. J. High Energy Phys. **12**, 072 (2016). https://doi.org/10.1007/JHEP12(2016)072, arXiv:1606.02921 [hep-th]

34. D. Butter, J. Novak, G. Tartaglino-Mazzucchelli, The component structure of conformal supergravity invariants in six dimensions. J. High Energy Phys. **05**, 133 (2017). https://doi.org/10.1007/JHEP05(2017)133, arXiv:1701.08163 [hep-th]

35. A. Galperin, E. Ivanov, S. Kalitsyn, V. Ogievetsky, E. Sokatchev, Unconstrained $N = 2$ matter, Yang–Mills and supergravity theories in harmonic superspace. Class. Quant. Grav. **1**, 469–498 (1984). https://doi.org/10.1088/0264-9381/1/5/004. [Erratum: Class. Quant. Grav.2,127(1985)]

36. A.S. Galperin, E.A. Ivanov, V.I. Ogievetsky, E.S. Sokatchev, Harmonic superspace, in *Cambridge Monographs on Mathematical Physics* (Cambridge University, Cambridge, 2007). https://doi.org/10.1017/CBO9780511535109, http://www.cambridge.org/mw/academic/subjects/physics/theoretical-physics-and-mathematical-physics/harmonic-superspace?format=PB

37. A. Karlhede, U. Lindström, M. Rocek, Selfinteracting tensor multiplets in $N = 2$ superspace. Phys. Lett. **147B**, 297–300 (1984). https://doi.org/10.1016/0370-2693(84)90120-5

38. U. Lindström, M. Roček, New hyperkähler metrics and new supermultiplets. Commun. Math. Phys. **115**, 21 (1988). https://doi.org/10.1007/BF01238851

39. U. Lindström, M. Roček, $N = 2$ super Yang–Mills theory in projective superspace. Commun. Math. Phys. **128**, 191 (1990). https://doi.org/10.1007/BF02097052

40. U. Lindström, M. Roček, Properties of hyperkähler manifolds and their twistor spaces. Commun. Math. Phys. **293**, 257–278 (2010). https://doi.org/10.1007/s00220-009-0923-0, arXiv:0807.1366 [hep-th]

41. S.M. Kuzenko, Lectures on nonlinear sigma-models in projective superspace. J. Phys. **A43**, 443001 (2010). https://doi.org/10.1088/1751-8113/43/44/443001, arXiv:1004.0880 [hep-th]

42. A.S. Galperin, N.A. Ky, E. Sokatchev, $N = 2$ supergravity in superspace: solution to the constraints. Class. Quant. Grav. **4**, 1235 (1987). https://doi.org/10.1088/0264-9381/4/5/022

43. A.S. Galperin, E.A. Ivanov, V.I. Ogievetsky, E. Sokatchev, $N = 2$ supergravity in superspace: different versions and matter couplings. Class. Quant. Grav. **4**, 1255 (1987). https://doi.org/10.1088/0264-9381/4/5/023

44. S.M. Kuzenko, G. Tartaglino-Mazzucchelli, Five-dimensional superfield supergravity. Phys. Lett. **B661**, 42–51 (2008). https://doi.org/10.1016/j.physletb.2008.01.055, arXiv:0710.3440 [hep-th]

45. S.M. Kuzenko, G. Tartaglino-Mazzucchelli, 5D Supergravity and projective superspace. J. High Energy Phys. **02**, 004 (2008). https://doi.org/10.1088/1126-6708/2008/02/004, arXiv:0712.3102 [hep-th]

46. S.M. Kuzenko, U. Lindström, M. Rocek, G. Tartaglino-Mazzucchelli, 4D $N = 2$ supergravity and projective superspace. J. High Energy Phys. **09**, 051 (2008). https://doi.org/10.1088/1126-6708/2008/09/051, arXiv:0805.4683 [hep-th]

47. D. Butter, New approach to curved projective superspace. Phys. Rev. **D92**(8), 085004 (2015). https://doi.org/10.1103/PhysRevD.92.085004, arXiv:1406.6235 [hep-th]

48. D. Butter, Projective multiplets and hyperkähler cones in conformal supergravity. J. High Energy Phys. **06**, 161 (2015). https://doi.org/10.1007/JHEP06(2015)161, arXiv:1410.3604 [hep-th]

49. D. Butter, On conformal supergravity and harmonic superspace. J. High Energy Phys. **03**, 107 (2016). https://doi.org/10.1007/JHEP03(2016)107, arXiv:1508.07718 [hep-th]

50. P. Fré, P. Soriani, *The $N = 2$ Wonderland: From Calabi–Yau Manifolds to Topological Field Theories* (World Scientific, Singapore, 1995)

51. L. Andrianopoli, M. Bertolini, A. Ceresole, R. D'Auria, S. Ferrara, P. Frè T. Magri, $N = 2$ supergravity and $N = 2$ super Yang–Mills theory on general scalar manifolds: symplectic covariance, gaugings and the momentum map. J. Geom. Phys. **23**, 111–189 (1997). https://doi.org/10.1016/S0393-0440(97)00002-8, arXiv:hep-th/9605032 [hep-th]

52. A. Ceresole, G. Dall'Agata, General matter coupled $N = 2$, $D = 5$ gauged supergravity. Nucl. Phys. **B585**, 143–170 (2000). https://doi.org/10.1016/S0550-3213(00)00339-4, arXiv:hep-th/0004111 [hep-th]

53. N. Cribiori, G. Dall'Agata, On the off-shell formulation of $N = 2$ supergravity with tensor multiplets. J. High Energy Phys. **08**, 132 (2018). https://doi.org/10.1007/JHEP08(2018)132, arXiv:1803.08059 [hep-th]

54. N. Boulanger, B. Julia, L. Traina, Uniqueness of $N = 2$ and 3 pure supergravities in 4D. J. High Energy Phys. **04**, 097 (2018). https://doi.org/10.1007/JHEP04(2018)097, arXiv:1802.02966 [hep-th]

55. I.A. Batalin, G.A. Vilkovisky, Quantization of gauge theories with linearly dependent generators, Phys. Rev. **D28**, 2567–2582 (1983). https://doi.org/10.1103/PhysRevD.28.2567,10.1103/PhysRevD.30.508. [Erratum: Phys. Rev.D30,508(1984)]

56. M. Henneaux, Lectures on the antifield—BRST formalism for gauge theories. Nucl. Phys. Proc. Suppl. **18A**, 47–106 (1990). https://doi.org/10.1016/0920-5632(90)90647-D

57. G. Barnich, M. Henneaux, Consistent couplings between fields with a gauge freedom and deformations of the master equation. Phys. Lett. **B311**, 123–129 (1993). https://doi.org/10.1016/0370-2693(93)90544-R, arXiv:hep-th/9304057 [hep-th]

58. J. Gomis, J. París, S. Samuel, Antibracket, antifields and gauge theory quantization. Phys. Rept. **259**, 1–145 (1995). https://doi.org/10.1016/0370-1573(94)00112-G, arXiv:hep-th/9412228 [hep-th]

59. E. Bergshoeff, S. Cucu, T. de Wit, J. Gheerardyn, S. Vandoren, A. Van Proeyen, $N = 2$ supergravity in five dimensions revisited. Class. Quant. Grav. **21**, 3015–3041 (2004). https://doi.org/10.1088/0264-9381/23/23/C01,10.1088/0264-9381/21/12/013, arXiv:hep-th/0403045[hep-th], erratum **23** (2006) 7149

60. T. Kugo, K. Ohashi, Supergravity tensor calculus in $5D$ from $6D$. Prog. Theor. Phys. **104**, 835–865 (2000). https://doi.org/10.1143/PTP.104.835, arXiv:hep-ph/0006231 [hep-ph]

61. T. Kugo, K. Ohashi, Off-shell $d = 5$ supergravity coupled to matter–Yang–Mills system. Prog. Theor. Phys. **105**, 323–353 (2001). https://doi.org/10.1143/PTP.105.323, arXiv:hep-ph/0010288 [hep-ph]

62. M.F. Sohnius, Introducing supersymmetry. Phys. Rept. **128**, 39–204 (1985). https://doi.org/10.1016/0370-1573(85)90023-7

63. P. Binétruy, G. Dvali, R. Kallosh, A. Van Proeyen, Fayet–Iliopoulos terms in supergravity and cosmology. Class. Quant. Grav. **21**, 3137–3170 (2004). https://doi.org/10.1088/0264-9381/21/13/005, arXiv:hep-th/0402046 [hep-th]

64. J. Strathdee, Extended Poincaré supersymmetry. Int. J. Mod. Phys. **A2**, 273 (1987). https://doi.org/10.1142/S0217751X87000120, [104(1986)]

65. M. Sohnius, K.S. Stelle, P.C. West, Off mass shell formulation of extended supersymmetric gauge theories. Phys. Lett. **92B**, 123–127 (1980). https://doi.org/10.1016/0370-2693(80)90319-6

66. B. de Wit, V. Kaplunovsky, J. Louis, D. Lüst, Perturbative couplings of vector multiplets in $N = 2$ heterotic string vacua. Nucl. Phys. **B451**, 53–95 (1995). https://doi.org/10.1016/0550-3213(95)00291-Y, arXiv:hep-th/9504006 [hep-th]

67. P. Claus, B. de Wit, B. Kleijn, R. Siebelink, P. Termonia, $N = 2$ supergravity Lagrangians with vector–tensor multiplets. Nucl. Phys. **B512**, 148–178 (1998). https://doi.org/10.1016/S0550-3213(97)00781-5, arXiv:hep-th/9710212 [hep-th]

68. M. Günaydin, M. Zagermann, The gauging of five-dimensional, $N = 2$ Maxwell–Einstein supergravity theories coupled to tensor multiplets. Nucl. Phys. **B572**, 131–150 (2000). https://doi.org/10.1016/S0550-3213(99)00801-9, arXiv:hep-th/9912027 [hep-th]

69. M. Günaydin, M. Zagermann, The vacua of $5d$, $N = 2$ gauged Yang–Mills/Einstein/tensor supergravity: Abelian case. Phys. Rev. **D62**, 044028 (2000). https://doi.org/10.1103/PhysRevD.62.044028, arXiv:hep-th/0002228 [hep-th]

70. M. Günaydin, M. Zagermann, Gauging the full R-symmetry group in five-dimensional, $N = 2$ Yang–Mills/Einstein/tensor supergravity. Phys. Rev. **D63**, 064023 (2001). https://doi.org/10.1103/PhysRevD.63.064023, arXiv:hep-th/0004117 [hep-th]

71. S. Coleman, J. Mandula, All possible symmetries of the S matrix. Phys. Rev. **159**, 1251–1256 (1967). https://doi.org/10.1103/PhysRev.159.1251

72. R. Haag, J.T. Łopuszański, M. Sohnius, All possible generators of supersymmetries of the S-matrix. Nucl. Phys. **B88**, 257 (1975). https://doi.org/10.1016/0550-3213(75)90279-5, [257(1974)]

73. J. Maldacena, A. Zhiboedov, Constraining conformal field theories with a higher spin symmetry. J. Phys. **A46**, 214011 (2013). https://doi.org/10.1088/1751-8113/46/21/214011, arXiv:1112.1016 [hep-th]
74. V. Alba, K. Diab, Constraining conformal field theories with a higher spin symmetry in $d > 3$ dimensions. J. High Energy Phys. **03**, 044 (2016). https://doi.org/10.1007/JHEP03(2016)044, arXiv:1510.02535 [hep-th]
75. E. Sezgin, Y. Tanii, Superconformal sigma models in higher than two dimensions. Nucl. Phys. **B443**, 70–84 (1995). https://doi.org/10.1016/0550-3213(95)00081-3, arXiv:hep-th/9412163 [hep-th]
76. W. Nahm, Supersymmetries and their representations. Nucl. Phys. **B135**, 149 (1978). https://doi.org/10.1016/0550-3213(78)90218-3, [7(1977)]
77. J.W. van Holten, A. Van Proeyen, $N = 1$ supersymmetry algebras in $d = 2, 3, 4$ mod. 8. J. Phys. **A15**, 3763 (1982). https://doi.org/10.1088/0305-4470/15/12/028
78. R. D'Auria, S. Ferrara, M.A. Lledó, V.S. Varadarajan, Spinor algebras. J. Geom. Phys. **40**, 101–128 (2001). https://doi.org/10.1016/S0393-0440(01)00023-7, arXiv:hep-th/0010124 [hep-th]
79. M.A. Lledó, V.S. Varadarajan, Spinor algebras and extended superconformal algebras, in *Proceedings of 2nd International Symposium on Quantum Theory and Symmetries (QTS-2): Cracow, Poland, July 18-21, 2001*, pp. 463–472 (2002). https://doi.org/10.1142/9789812777850_0057, arXiv:hep-th/0111105 [hep-th]
80. C. Cordova, T.T. Dumitrescu, K. Intriligator, Multiplets of superconformal symmetry in diverse dimensions. J. High Energy Phys. **03**, 163 (2019). https://doi.org/10.1007/JHEP03(2019)163, arXiv:1612.00809 [hep-th]
81. P. Claus, *Conformal Supersymmetry in Supergravity and on Branes*, Ph.D. thesis, Leuven, 2000
82. E. Bergshoeff, S. Cucu, M. Derix, T. de Wit, R. Halbersma, A. Van Proeyen, Weyl multiplets of $N = 2$ conformal supergravity in five dimensions. J. High Energy Phys. **06**, 051 (2001). https://doi.org/10.1088/1126-6708/2001/06/051, arXiv:hep-th/0104113 [hep-th]
83. E. Bergshoeff, E. Sezgin, A. Van Proeyen, Superconformal tensor calculus and matter couplings in six dimensions. Nucl. Phys. **B264**, 653 (1986). https://doi.org/10.1016/0550-3213(86)90503-1, [Erratum: Nucl. Phys.B598,667(2001)]

Chapter 2
Gauging Spacetime Symmetries: The Weyl Multiplet

Abstract In this chapter, we will discuss the gauge multiplets of the superconformal algebra, called Weyl multiplets. We start by repeating the basic transformation rules for gauge fields and curvatures, and discuss then the modifications necessary for spacetime symmetries. We will see that one needs constraints on curvatures, and will learn how to deal with them. At the end of this chapter we obtain a Weyl multiplets for $\mathcal{N} = 2$ theories in $D = 4$, $D = 5$ and $D = 6$.

2.1 Rules of (Super)Gauge Theories, Gauge Fields and Curvatures

Consider a general (super)algebra with commutators

$$[\delta(\epsilon_1), \delta(\epsilon_2)] = \delta \left(\epsilon_3^C = c_2^B \epsilon_1^A f_{AB}{}^C \right). \tag{2.1}$$

In general $f_{AB}{}^C$ may be structure functions, i.e. depend on the fields. Moreover, the equality above may be satisfied only modulo equations of motion, as we will see below. The $f_{AB}{}^C$ are related to the abstract algebra introduced in Sect. 1.2.4:

$$D = 4, 6 : [T_A, T_B] = T_A T_B - (-)^{AB} T_B T_A = f_{AB}{}^C T_C ,$$

$$D = 5 : [T_A, T_B] = T_A T_B - (-)^{AB} T_B T_A = (-)^{AB} f_{AB}{}^C T_C , \tag{2.2}$$

where $(-)^A$ is a minus sign if T_A is fermionic. The extra sign factor for $D = 5$ is due to the factors i in (1.2). It is assumed that in all other cases the transformation is generated by $\epsilon^A T_A$.

The algebra above can be realized locally by introducing gauge fields and curvatures. This means that for every generator there is a gauge field, $B_\mu{}^A$, which

© Springer Nature Switzerland AG 2020

E. Lauria, A. Van Proeyen, $\mathcal{N} = 2$ *Supergravity in D = 4, 5, 6 Dimensions,*
Lecture Notes in Physics 966, https://doi.org/10.1007/978-3-030-33757-5_2

Table 2.1 Superconformal gauge symmetries, their gauge fields and parameters

T_A	P_a	M_{ab}	D	K_a	$U_i{}^j$	T	Q	S
$B_\mu{}^A$	e_μ^a	$\omega_\mu{}^{ab}$	b_μ	$f_\mu{}^a$	$V_{\mu j}{}^i$	A_μ	ψ_μ	ϕ_μ
ϵ^A	ξ^a	λ^{ab}	λ_D	λ_K^a	$\lambda_j{}^i$	λ_T	ϵ	η

transforms as follows[1]:

$$\delta(\epsilon)B_\mu{}^A = \partial_\mu \epsilon^A + \epsilon^C B_\mu{}^B f_{BC}{}^A . \tag{2.3}$$

Covariant derivatives have a term involving the gauge field for every gauge transformation

$$\mathcal{D}_\mu = \partial_\mu - \delta(B_\mu), \tag{2.4}$$

and their commutators are new transformations with as parameters the curvatures:

$$[\mathcal{D}_\mu, \mathcal{D}_\nu] = -\delta(R_{\mu\nu}),$$
$$R_{\mu\nu}{}^A = 2\partial_{[\mu}B_{\nu]}{}^A + B_\nu{}^C B_\mu{}^B f_{BC}{}^A, \tag{2.5}$$

which transform 'covariantly':

$$\delta(\epsilon)R_{\mu\nu}{}^A = \epsilon^C R_{\mu\nu}{}^B f_{BC}{}^A, \tag{2.6}$$

and, with the definitions (2.4), satisfy Bianchi identities

$$\mathcal{D}_{[\mu}R_{\nu\rho]}{}^A = 0. \tag{2.7}$$

Having in mind the motivations presented in Sect. 1.2.2, we now start gauging the superconformal group. In Table 2.1 we give names to the gauge fields and parameters for the generators that appeared in Sect. 1.2.4. The relation between parameters (or gauge fields) and generators varies in different dimensions due to the different spinor properties and notations for raising and lowering i indices. This

---

[1]Note that the order of the fields and parameters is relevant here. For fermionic fields, the indices contain spinor indices and one may use the conventions of [1, Sect. 3.2.2]. Although the objects may be fermionic or bosonic, you do not see many sign factors. The trick to avoid most sign factors is to keep objects with contracted indices together. For example, you see here the B index of the gauge field next to the B index in the structure constants, and then the C contracted indices do not have other uncontracted indices between them.

is encoded in the transformation rule $\delta = \epsilon^A T_A$:

$$\delta = \epsilon^A T_A = \tfrac{1}{2}\lambda^{ab} M_{[ab]} + \lambda_{\mathrm{D}} D + \lambda_{\mathrm{K}}^a K_a +$$

$$\begin{cases} +\lambda_i{}^j U_j{}^i + \lambda_T\, T + \bar{\epsilon}^i Q_i + \bar{\epsilon}_i Q^i + \bar{\eta}^i S_i + \bar{\eta}_i S^i & \text{for } D = 4 \\ +\lambda^{ij} U_{ij} + \mathrm{i}\bar{\epsilon}^i Q_i + \mathrm{i}\bar{\eta}^i S_i & \text{for } D = 5 \quad (2.8) \\ +\lambda^{ij} U_{ij} + \bar{\epsilon}^i Q_i + \bar{\eta}^i S_i & \text{for } D = 6 \,. \end{cases}$$

The same correspondence holds for the gauge fields in expressions $B_\mu{}^A T_A$. With this correspondence, the commutators can be written in terms of parameters in commutators of transformations, e.g. for $D = 4$, we have

$$\left[\delta_S(\eta), \delta_Q(\epsilon)\right] = \delta_{\mathrm{D}}(\lambda_{\mathrm{D}}(\epsilon, \eta)) + \delta_{\mathrm{M}}(\lambda^{ab}(\epsilon, \eta)) + \delta_{\mathrm{U}}(\lambda_i{}^j(\epsilon, \eta))\,,$$

$$\lambda_{\mathrm{D}}(\epsilon, \eta) = \tfrac{1}{2}(\epsilon^i \eta_i + \text{h.c.})\,,$$

$$\lambda^{ab}(\epsilon, \eta) = \tfrac{1}{4}(\epsilon^i \gamma^{ab} \eta_i + \text{h.c.})\,,$$

$$\lambda_T(\epsilon, \eta) = \tfrac{1}{2}\mathrm{i}(-\bar{\epsilon}^i \eta_i + \bar{\epsilon}_i \eta^i)\,,$$

$$\lambda_i{}^j(\epsilon, \eta) = \bar{\epsilon}^j \eta_i - \bar{\epsilon}_i \eta^j - \tfrac{1}{4}\delta_i^j(\bar{\epsilon}^k \eta_k - \bar{\epsilon}_k \eta^k)\,, \qquad (2.9)$$

and

$$\left[\delta_T(\lambda_T), \delta_Q(\epsilon)\right] = \delta_Q(\epsilon'(\epsilon, \lambda_T))\,, \qquad \epsilon_i'(\epsilon, \lambda_T) = \tfrac{1}{2}\mathrm{i}\lambda_T \epsilon_i\,,$$

$$\left[\delta_{\mathrm{SU}(2)}(\lambda), \delta_Q(\epsilon)\right] = \delta_Q(\epsilon'(\epsilon, \lambda))\,, \qquad \epsilon_i'(\epsilon, \lambda) = \lambda_i{}^j \epsilon_j\,. \qquad (2.10)$$

Exercise 2.1 Check that the first equation of (A.2) corresponds to the definition (2.5) for $R_{\mu\nu}{}^{ab}$ if we just consider the Lorentz group, i.e. the first equation of (1.21), and define $\omega_\mu{}^{ab}$ as the gauge field of M_{ab}. □

2.2 Gauge Theory of Spacetime Symmetries

The sole procedure explained in Sect. 2.1 is not sufficient to define a suitable theory for local spacetime symmetries. In the following we will explain the necessary steps to improve the recipe given in Sect. 2.1.

2.2.1 General Considerations

The main problem is that in general relativity one should have general coordinate transformations (gct) as a local symmetry rather than the local translations defined by rules of the previous section. The action of gct on spacetime scalar fields and gauge vectors is, by definition

$$\delta_{\text{gct}}(\xi)\phi(x) = \xi^\mu(x)\partial_\mu\phi(x),\tag{2.11}$$

$$\delta_{\text{gct}}(\xi)B_\mu{}^A \equiv \xi^\nu\partial_\nu B_\mu{}^A + (\partial_\mu\xi^\nu)B_\nu{}^A = \delta_B(\xi^\nu B_\nu{}^B)B_\mu{}^A - \xi^\nu R_{\mu\nu}{}^A,\tag{2.12}$$

where in the last line we have rewritten the transformation using (2.3) and (2.5). On the other hand, the definition (2.3) with the algebra (1.21) would lead e.g. to $\delta_P b_\mu = 2\xi_a f_\mu{}^a$, which is not yet a general coordinate transformation. We will take a few steps to relate gct to local translations as defined by the algebra (1.21) and (1.58).

Step 1   First of all we distinguish the translations from all the other transformations. All these others will be denoted as *standard gauge transformations*. We therefore split the range of indices A in the following way:

$$T_A = \left(\underbrace{P_a}_{\text{Translations}}, \underbrace{T_I}_{\text{Standard Gauge}}\right).\tag{2.13}$$

The gauge field of P_a, denoted as $e_\mu{}^a$, is required to be invertible as a matrix, and it is interpreted as the frame field.

Step 2   When we consider the parameter a^μ in (1.19) as a local function, this absorbs all the other terms in (1.19), and we further denote it as $\xi^\mu(x)$. In other words, a change of basis in the set of the gauge transformations is performed such that all the orbital parts of Lorentz rotations, dilatations and special conformal transformations are reabsorbed into the general coordinate transformation. In the same way the special conformal transformations in (1.25) are absorbed in the local parameters $\lambda^{ab}(x)$ and $\lambda_D(x)$ and the S-supersymmetry part in (1.60) is absorbed in $\epsilon(x)$.

Step 3   A further basis change is performed using *covariant general coordinate transformations* (cgct) [2, 3]:

$$\delta_{\text{cgct}}(\xi) = \delta_{\text{gct}}(\xi) - \delta_I\left(\xi^\mu B_\mu{}^I\right).\tag{2.14}$$

Note that (2.14) is a combination of general coordinate transformations and all the non-translation transformations (standard gauge transformations) whose parameter ϵ^I has been replaced by $\xi^\mu B_\mu{}^I$. As we require that the final action is invariant both under generic δ_{gct} and δ_I, it should be invariant also under cgct. To summarize, we replace

$$\delta(\epsilon) = \epsilon^A T_A = \xi^a P_a + \epsilon^I T_I,\tag{2.15}$$

by a different transformation that leaves the action invariant:

$$\bar{\delta} = \delta_{\text{cgct}}(\xi) + \epsilon^I T_I = \delta_{\text{gct}}(\xi) + \delta_I \left(\epsilon^I - \xi^\mu B_\mu{}^I \right). \tag{2.16}$$

Below, we will discuss the action of cgct in more detail for the different types of fields.

Step 4 In order to identify the cgct transformations with gauged translations we further need constraints on some curvatures. As we will see in Sect. 2.2.2, the first one will be the curvature of translations that will be put to zero. This will imply that the gauge field of Lorentz rotations $\omega_\mu{}^{ab}$ will become a function of e_μ^a as is common in general relativity, where it is then often called the 'spin connection'. The expression in the supersymmetric theory will also involve other fields such as the gravitino. This is related to torsion, see e.g. [1, Sect. 7.9]. We will see in Sect. 2.4 that more constraints will be imposed such that also other gauge fields of Table 2.1 are composite rather than independent fields.

Step 5 When general coordinate transformations are properly implemented, the numbers of bosonic and fermionic degrees of freedom in a supersymmetric theory should match, as we discussed in the beginning of Sect. 1.2.1. In the $\mathcal{N} = 2$ theories, this will involve the introduction of additional (auxiliary) degree of freedom, which we will discuss in Sect. 2.6.

In summary the essential modifications of the procedure outlined in Sect. 2.1 are

1. Translations are replaced by general coordinate transformations, which are further combined with other symmetries to covariant general covariant transformations (cgct) (2.14).
2. Some gauge fields will turn out to be 'composite', i.e. functions of the other fields in the multiplet.
3. The multiplet that gauges the superconformal group will contain auxiliary fields.
4. The structure constants will be replaced by structure functions.

In the following we present the general form of cgct acting on frame fields, other gauge fields[2] and matter fields, which will be useful for our considerations. At this point, it is important to keep in mind that the fundamental distinction between

---

[2]The part on gauge fields could be generalized to p-form fields. We do not include a general setup for these, but the essential characteristics of cgct apply in the same way.

'gauge fields' and 'matter fields' is that the former have a coordinate index μ,[3] while matter fields have no coordinate indices. In other terms,

> **Matter vs Gauge Fields**
> The transformation of matter fields does not involve a derivative of a gauge parameter, while the transformation of gauge fields does have the $\partial_\mu \epsilon^A$ term.

2.2.2 Transformations of the Frame Fields

Let us derive the explicit form of the transformations of the frame field. For a cgct the last term of (2.14) cancels the first term of the second expression in (2.12), apart from the translation part (where B takes only the values corresponding to translations, i.e. b). As a result we obtain

$$\delta_{\text{cgct}}(\xi)e_\mu{}^a = \partial_\mu \xi^a + \xi^b B_\mu{}^B f_{Bb}{}^a - \xi^\nu R_{\mu\nu}\left(P^a\right) . \tag{2.17}$$

The first two terms in the expression above are just (2.3), with ϵ replaced by ξ^a. In particular *after imposing the constraint* $R_{\mu\nu}(P^a) = 0$, *the cgct of the frame field is equal to its P_a transformation as it would directly follow from (2.3).*

Let us now consider this first constraint using the explicit expressions of the commutators that are of the form $[P, \cdot] = P$ in the algebra (1.21) and (1.58).

$$R_{\mu\nu}(P^a) = 2\left(\partial_{[\mu} + b_{[\mu}\right)e_{\nu]}^a + 2\omega_{[\mu}{}^{ab}e_{\nu]b} + \xi^a(\psi_\mu, \psi_\nu) = 0 , \tag{2.18}$$

where ξ^a is the function introduced in (1.6). This constraint implies that $\omega_\mu{}^{ab}$ is the connection such that for pure gravity (hence $\psi_\mu = 0$) the spacetime manifold with metric $g_{\mu\nu} = e_\mu^a e_{\nu a}$ is torsionless, while the gravitino terms define a torsion. We will come back to this constraint and its consequences below, and explain its solution for $\omega_\mu{}^{ab}$ in general. The expression (2.17) then collapses to

$$\delta_{\text{cgct}}(\xi)e_\mu{}^a = \left(\partial_\mu + b_\mu\right)\xi^a + \omega_\mu{}^{ab}\xi_b . \tag{2.19}$$

---

[3]Of course one could change it to a frame index a by multiplication with $e_a{}^\mu$, but we consider it in the form B_μ^A as the basic field and B_a^A as the composite of $e_a{}^\mu B_\mu^A$. This field B_μ^A may still be composite by itself, but that is not important at this point.

The standard gauge transformations of the frame field follow straightforwardly from the rule (2.3). That is, using the notation $\xi^a(\cdot, \cdot)$ from (1.6),

$$\delta_I(\epsilon^I)e_\mu{}^a = \epsilon^I B_\mu^B f_{BI}{}^a = -\lambda_D e_\mu{}^a - \lambda^{ab} e_{\mu b} + \xi^a(\psi_\mu, \epsilon),$$

$$D = 4 : \xi^a(\psi_\mu, \epsilon) = \tfrac{1}{2}\bar{\epsilon}^i \gamma^a \psi_{\mu i} + \tfrac{1}{2}\bar{\epsilon}_i \gamma^a \psi_\mu^i,$$

$$D = 5, 6 : \xi^a(\psi_\mu, \epsilon) = \tfrac{1}{2}\bar{\epsilon}^i \gamma^a \psi_{\mu i}. \tag{2.20}$$

2.2.3 Transformations of the Other Gauge Fields

Transformations (2.3) of gauge fields other than $e_\mu{}^a$ are often deformed by the presence of matter fields in the multiplet. That is, the expression (2.3) is often not complete. This will be the case after other fields have been added as mentioned above under 'Step 5', and will be discussed in full in Sect. 2.6. Here we will already discuss how this modifies some general rules that were presented in Sect. 2.1. A simple example of this phenomenon appears in the supersymmetry transformation of the gauge field W_μ in the abelian vector multiplet, which contains a 'gaugino', λ^i:

$$\delta_Q(\epsilon)W_\mu = -\tfrac{1}{2}\varepsilon_{ij}\bar{\epsilon}^i \gamma_\mu \lambda^j + \text{h.c.}, \qquad (D = 4). \tag{2.21}$$

Clearly the r.h.s. of the equation above cannot be seen as a part of (2.3). Therefore, in order to account for these additional fields, we need to modify some of the general rules presented in Sect. 2.1. We allow in general a modification of (2.3) by considering the following general form of standard gauge transformations:

$$\delta_J\left(\epsilon^J\right) B_\mu^I = \partial_\mu \epsilon^I + \epsilon^J B_\mu^A f_{AJ}{}^I + \epsilon^J M_{\mu J}{}^I. \tag{2.22}$$

The expression $M_{aJ}{}^I$ is a function of 'matter fields', which should be a covariant quantity (i.e. not transforming with a derivative of a parameter, see Sect. 2.3). As an example, when the field B_μ^I is the U(1) gauge field W_μ, the term written in (2.21) is of this form where the supersymmetry index is $J = (\alpha i)$ and I refers to this gauge field, say $I = \boldsymbol{\cdot}$:

$$M_{\mu\alpha i}{}^{\boldsymbol{\cdot}} = -\tfrac{1}{2}\varepsilon_{ij}\left(\gamma_\mu \lambda^j\right)_\alpha. \tag{2.23}$$

As mentioned above, such terms will also appear for the gauge fields in Table 2.1 after auxiliary fields have been added.

In (2.22) there is still a sum over A, which we want to split in the standard gauge transformations, and the contribution when B_μ^A is the frame field $e_\mu{}^a$. Therefore we rewrite (2.22) as

$$\delta_J\left(\epsilon^J\right) B_\mu^I = \partial_\mu \epsilon^I + \epsilon^J B_\mu^K f_{KJ}{}^I + \epsilon^J M_{\mu J}{}^I,$$

$$\mathcal{M}_{aJ}{}^I = f_{aJ}{}^I + M_{aJ}{}^I . \tag{2.24}$$

While the terms containing $M_{aJ}{}^I$ are determined case by case, depending on the multiplet, those proportional to $f_{aJ}{}^I$ are fixed by the superconformal algebra (in particular by the commutators $[P_a, T_J] = f_{aJ}{}^I T_I$). Explicitly,

$$\delta_I\left(\epsilon^I\right) \psi_\mu^i = \cdots - s_D \gamma_\mu \eta^i ,$$

$$\delta_I\left(\epsilon^I\right) b_\mu = \cdots + 2\lambda_{K\mu},$$

$$\delta_I\left(\epsilon^I\right) \omega_\mu{}^{ab} = \cdots - 4\lambda_K^{[a} e_\mu{}^{b]} . \tag{2.25}$$

The constants s_D are those that appear in the algebra (1.58). Including the matter terms as in (2.24), the cgct (2.14) on the gauge fields B_μ^I are

$$\delta_{\text{cgct}}(\xi) B_\mu^I = -\xi^\nu R_{\mu\nu}{}^I + \xi^a B_\mu^J f_{Ja}{}^I - \xi^\nu B_\nu^J M_{\mu J}{}^I$$

$$= -\xi^\nu \widehat{R}_{\mu\nu}{}^I - \xi^a B_\mu^J \mathcal{M}_{aJ}{}^I . \tag{2.26}$$

The second term in the first line only occurs for the transformation of the gauge fields of supersymmetry, dilatations and Lorentz rotations. This is the original P_a transformation of the gauge field. In the second line appears a new covariant curvature, which takes the transformations of the gauge fields to matter fields into account. Indeed, the last term of (2.22) implies that $R_{\mu\nu}{}^I$ transforms in the derivative of a parameter, i.e. there is a term $2\partial_{[\mu} \epsilon^J M_{\nu]J}{}^I$. In the last line we introduced the modified curvature

$$\widehat{R}_{\mu\nu}{}^I = R_{\mu\nu}{}^I - 2B_{[\mu}{}^J M_{\nu]J}{}^I = r_{\mu\nu}{}^I - 2B_{[\mu}{}^J \mathcal{M}_{\nu]J}{}^I , \tag{2.27}$$

which does not transform to a derivative of a parameter and we defined

$$r_{\mu\nu}{}^I = 2\partial_{[\mu} B_{\nu]}{}^I + B_\nu^K B_\mu^J f_{JK}{}^I , \tag{2.28}$$

by stripping all contributions from translations out of the curvatures.

Exercise 2.2 We saw already in (2.24)–(2.25) the explicit form of the terms that make the difference between M and $\mathcal{M}$. Using (2.27), this should allow you to determine that the only ones where these play a role are

$$R_{\mu\nu}\left(M^{ab}\right) = r_{\mu\nu}\left(M^{ab}\right) + 8 f_{[\mu}{}^{[a} e_{\nu]}{}^{b]},$$

$$R_{\mu\nu}(D) = r_{\mu\nu}(D) - 4 f_{[\mu}{}^{a} e_{\nu]a},$$

$$R_{\mu\nu}(Q) = r_{\mu\nu}(Q) - 2 s_D \gamma_{[\mu} \phi_{\nu]}. \tag{2.29}$$

$\square$

The cgct of a standard gauge field, given by (2.26), takes into account the matter fields as well as modifications of the gauge curvatures by the latter. On the frame field, upon imposing the constraint (2.18), a cgct gives a local translation as dictated by (2.3). In the case of standard gauge fields, the relation between local translations and these covariant general coordinate transformations is less obvious. First of all, we cannot simply put all curvatures equal to zero to use (2.12). This would impose derivative constraints on all fields, and thus restrict their dynamics. It could be done for $R(P)$ because in that case, this constraint will just determine the spin connection in terms of the frame field, without imposing further dynamical constraints.

We will only choose constraints that in the same way can be solved for gauge fields. In this respect, (2.29) will be useful, which already shows that with such constraints $f_\mu{}^a$ and ϕ_μ may be determined in terms of other fields. How this is done exactly and what are the consequences will be the subject of Sect. 2.4. Suppose that this is done, let us count the number of off-shell degrees of freedom associated with the independent fields of Table 2.1. This is done in Table 2.2, where for comparison, we also included $N = 1$ in $D = 4$. As explained in the beginning of Sect. 1.2.1, the number of off-shell degrees of freedom is given by total number of components minus the gauge degrees of freedom. From the components we have a factor D for the index μ and a, a factor 3 for the SU(2) gauge fields, and a factor 8 for the spinors of the $N = 2$ theories (4 for $N = 1$). Upon subtracting the gauge degrees of freedom corresponding to the superconformal group, we see that for $N = 1$ there remain an equal number of bosonic and fermionic degrees of freedom. It turns out that indeed the gauged translations can be identified with general coordinate transformations. A detailed analysis can be found in [4]. On the other hand, the numbers do not match for the $N = 2$ theories. A standard procedure to cure this mismatch introduces other (auxiliary) fields, which will be new 'matter fields'. This solution will be presented in Sect. 2.6.

Table 2.2 Off-shell degrees of freedom in gauge fields

	bosonic										fermionic		
	fields				symm.						#	field	symm.	#
	e_μ^a	b_μ	$V_{\mu j}{}^i$	A_μ	P_a	M_{ab}	D	K_a	$U_i{}^j$	T	dof	ψ_μ	Q S	dof
$N = 1\, D = 4$	16	4		4	-4	-6	-1	-4		-1	8	16	-4 -4	8
$N = 2\, D = 4$	16	4	12	4	-4	-6	-1	-4	-3	-1	17	32	-8 -8	16
$N = 2\, D = 5$	25	5	15		-5	-10	-1	-5	-3		21	40	-8 -8	24
$N = 2\, D = 6$	36	6	18		-6	-15	-1	-6	-3		29	48	-8 -8	32

2.2.4   Transformations of Matter Fields

Matter fields transform by definition without derivatives on the parameters. Using this, we can rewrite the definition (2.14) of the cgct, extracting the ξ^μ, as

$$\delta_{\text{cgct}}(\xi)\phi = \xi^\mu \partial_\mu \phi - \delta_I \left(\xi^\mu B_\mu^I\right)\phi$$

$$= \xi^\mu \mathcal{D}_\mu \phi, \qquad \mathcal{D}_\mu \phi \equiv \partial_\mu \phi - \delta_I \left(B_\mu^I\right)\phi. \tag{2.30}$$

Note that the new derivative $\mathcal{D}_\mu$ such defined is not a covariant quantity. Indeed, under general transformations its transformation includes $\partial_\mu \xi^\nu \mathcal{D}_\nu \phi$, hence a derivative on a parameter. This can be avoided by defining $\mathcal{D}_a \phi$, as we will discuss below. We will then also prove that this is a covariant quantity, and then the action of translations on matter fields is $P_a \phi = \mathcal{D}_a \phi$.

2.3   Covariant Quantities and Covariant Derivatives

We reserve the present section for a more specific discussion on covariant quantities in gauge theories. Several important steps for this were obtained in [5, Sect. 3.2]. We start from the following definition:

> **Covariant Quantity**
> A covariant quantity is a field whose transformation under *any* local symmetry has no derivative on a transformation parameter.

The Lie derivative (general coordinate transformation) of a world scalar (namely an object with only frame indices) does not involve a derivative on the parameter, while all other fields that have coordinate indices (components of forms, world vectors, ...) transform under general coordinate transformations with derivatives of ξ^μ. Therefore, any covariant quantity must be a world scalar. Using this prescription, we have immediately two ways to build covariant quantities:

1. A covariant derivative on a covariant quantity with its index turned to a frame index:

$$\mathcal{D}_a \phi = e_a{}^\mu \mathcal{D}_\mu \phi. \tag{2.31}$$

2. Covariant curvatures with their indices turned to frame indices:

$$\widehat{R}_{ab}{}^I = e_a{}^\mu e_b{}^\nu \widehat{R}_{\mu\nu}{}^I. \tag{2.32}$$

The elementary matter fields transform under the symmetries in other covariant quantities. As will become clearer soon, this is not a general property of covariant quantities. For quantities that do not transform in covariant quantities, the expression of the covariant derivative is not as in (2.30). Let us first look to covariant quantities that do transform in covariant quantities. For those we have the lemma

Lemma on Covariant Derivatives *If a covariant quantity ϕ transforms into covariant quantities under standard gauge transformations, its covariant derivative $\mathcal{D}_a\phi$ given by (2.31), with $\mathcal{D}_\mu\phi$ given by (2.30), is a covariant quantity. Moreover, if the algebra closes on the field ϕ then the standard gauge transformations of $\mathcal{D}_a\phi$ involve only covariant quantities.*

An immediate consequence of the lemma is that under these conditions the cgct $\mathcal{D}_a\phi = P_a\phi$ gives a covariant quantity.

In this lemma a difference is made whether the algebra is 'open' or 'closed' on the original field. We refer for more information to [1, Sect. 11.1.3]), but repeat here the essential statement. In general, the commutator of two transformations on a field ϕ is of the form

$$\delta_I\left(\epsilon^I\right)\phi = \epsilon^I\chi_I \,,$$

$$\left[\delta_I\left(\epsilon_1^I\right), \delta_J\left(\epsilon_2^J\right)\right]\phi = \epsilon_2^J\epsilon_1^I\left(f_{IJ}{}^K\chi_K + f_{IJ}{}^a\mathcal{D}_a\phi + \eta_{IJ}\right). \quad (2.33)$$

The last tensor, η_{IJ} is the non-closure function for the field ϕ. It is not always possible to close the algebra on all fields without using equations of motion. If these transformations occur for a theory based on an invariant action, the tensor η_{IJ} is proportional to equations of motion, and the algebra is still 'on-shell' closed, i.e. using trivial on-shell symmetries.[4] In conclusion if, for the field ϕ, $\eta_{IJ} \neq 0$, then $\mathcal{D}_a\phi$ is still a covariant quantity, but its transformation $\delta_I(\epsilon^I)\mathcal{D}_a\phi$ is not. Vice versa, if $\eta_{IJ} = 0$, then the algebra closes off-shell, and the lemma says that $\delta_I(\epsilon^I)\mathcal{D}_a\phi$ is again a covariant quantity.

A similar lemma holds for curvatures, which are in fact the generalization of the covariant derivatives for gauge fields:

Lemma on Covariant Curvatures *For a gauge field with transformation law as in (2.22) where $M_{aJ}{}^I$ is a covariant quantity, the covariant curvature (2.32), with (2.27) is a covariant quantity.*

The rest of this section is organized as follows. We first give a proof of the lemma of covariant derivatives in Sect. 2.3.1. In Sects. 2.3.2 and 2.3.3 we discuss the example of the $D = 6$ abelian vector multiplet, in which we see in practice how the cgct indeed needs the seemingly non-covariant last term in (2.26) and how the

---

[4]Note that in principle, η_{IJ} may have contributions for all symmetries denoted by the indices I, J, but in practice (due to the engineering dimensions of the transformations) non-closure functions are related only to $[\delta_Q, \delta_Q]$.

lemma on the covariant curvature is realized: $\widehat{R}_{\mu\nu}$ does not transform to a covariant quantity, but $\widehat{R}_{ab}$ does. As it will be clear soon, computing these variations is in general long and painful. Useful simplifications happen if we strategically make use the lemmas above, provided the algebra is closed, and this will be in fact the content of Sect. 2.3.4, the 'easy way'. In Sect. 2.3.5 we will also consider an example with an open algebra.

2.3.1 Proof of Lemma on Covariant Derivatives

Let us start from computing $\delta_J(\epsilon^J)\mathcal{D}_\mu\phi$. Using the definitions (2.30), (2.24) and (2.33)

$$
\begin{aligned}
\delta_J\left(\epsilon^J\right)\mathcal{D}_\mu\phi &= \epsilon^I\partial_\mu\chi_I - \epsilon^J B_\mu^K f_{KJ}{}^I\chi_I - \epsilon^J M_{\mu J}{}^I\chi_I - B_\mu^I\delta_J\left(\epsilon^J\right)\chi_I \\
&= \epsilon^I\left(\mathcal{D}_\mu + \delta_J\left(B_\mu^J\right)\right)\chi_I - \epsilon^J B_\mu^K f_{KJ}{}^I\chi_I \\
&\quad -\epsilon^J M_{\mu J}{}^I\chi_I - \delta_J\left(\epsilon^J\right)\delta_I\left(B_\mu^I\right)\phi .
\end{aligned}
$$
(2.34)

For the last term, we used the notation in the way it is used in calculating a commutator. That means, the first $\delta(\epsilon)$ does not act on the B_μ^I within the $\delta_I(B_\mu^I)\phi$, as it is done when one calculates a commutator. Then from this piece plus $\epsilon^I\delta_J(B_\mu^J)\chi_I$ in the second line one reconstructs a commutator with ϵ_1^I replaced by B_μ^I and ϵ_2^J by ϵ^J. Using then (2.33), we obtain

$$
\begin{aligned}
\delta_J(\epsilon^J)\mathcal{D}_\mu\phi &= \epsilon^I\mathcal{D}_\mu\chi_I - \epsilon^J M_{\mu J}{}^I\chi_I + \epsilon^J B_\mu^I\left(f_{IJ}{}^a\mathcal{D}_a\phi + \eta_{IJ}\right) \\
&= \epsilon^I\mathcal{D}_\mu\chi_I - \epsilon^J M_{\mu J}{}^I\chi_I + \epsilon^J B_\mu^I\eta_{IJ} \\
&\quad + \left(\delta_I(\epsilon^I)e_\mu{}^a - \epsilon^I e_\mu^b f_{bI}{}^a\right)\mathcal{D}_a\phi .
\end{aligned}
$$
(2.35)

At the end, we made use of (2.20). The first term in the bracket has an explicit ψ_μ, but is canceled when reverting to the transformation of $\mathcal{D}_a\phi$:

$$
\delta_J\left(\epsilon^J\right)\mathcal{D}_a\phi = \epsilon^I\mathcal{D}_a\chi_I - \epsilon^J M_{aJ}{}^I\chi_I + \epsilon^J B_a^I\eta_{IJ} - \epsilon^I f_{aI}{}^b\mathcal{D}_b\phi ,
$$
(2.36)

namely $\mathcal{D}_a\phi$ is a covariant quantity, since $\delta_J(\epsilon^J)\mathcal{D}_a\phi$ does not contain derivatives on the parameters. If $\eta_{IJ} = 0$, the explicit gauge fields disappear from (2.36) and the transformation of the derivative is again a covariant quantity. This proves the lemma on covariant derivatives. We will come back to the meaning of the last term of (2.36) in Sect. 2.3.4.

2.3.2 Example: $D = 6$ Abelian Vector Multiplet

We will use an example to illustrate the closure of the algebra on gauge fields. In Sect. 2.3.3 this example is used to show how the transformation of curvatures gives a covariant result. Finally, in Sect. 2.3.4 it is used to illustrate how to facilitate calculations using the lemmas.

The $D = 6$ abelian vector multiplet consists of a vector W_μ and a spinor λ^i. The transformations[5] under standard gauge transformations are

$$\delta_I\left(\epsilon^I\right) W_\mu = \partial_\mu \theta + \tfrac{1}{2}\bar{\epsilon}\gamma_\mu\lambda \,,$$

$$\delta_I\left(\epsilon^I\right) \lambda^i = \left(\tfrac{3}{2}\lambda_\mathrm{D} - \tfrac{1}{4}\gamma^{ab}\lambda_{ab}\right)\lambda^i + \lambda^{ij}\lambda_j - \tfrac{1}{4}\gamma^{ab}\widehat{F}_{ab}\epsilon^i \,, \qquad (2.37)$$

where θ is the parameter for the U(1) transformation that W_μ gauges. This U(1) commutes with all other symmetries. W_μ is thus one of the B_μ^I in the general treatment, and comparing with (2.22), we can identify

$$M_{\mu(\alpha i)}{}^{\mathrm{U}(1)} = \tfrac{1}{2}\left(\gamma_\mu\lambda_i\right)_\alpha \,, \qquad (2.38)$$

where (αi) stands for the combined index (spinor + extension) indicating a supersymmetry. The i-index (and also the spinor index α) is implicit in the first line of (2.37), as explained in (A.18). The modified curvature, which is called $\widehat{F}_{\mu\nu}$, is

$$\widehat{F}_{\mu\nu} = F_{\mu\nu} - \bar{\psi}_{[\mu}\gamma_{\nu]}\lambda, \qquad F_{\mu\nu} = 2\partial_{[\mu}W_{\nu]}. \qquad (2.39)$$

Instead of using (2.27), just recognize that derivatives in $F_{\mu\nu}$ have to be completed to covariant derivatives.

Let us now comment on $\delta_I(\epsilon^I)\lambda^i$. The λ_D-term states that the fermion λ^i has 'Weyl weight' $\tfrac{3}{2}$. We will explain in Sect. 3.2 how this weight can be easily obtained. The Lorentz transformation is valid for all the spinors. Its form can in fact already be seen from the first commutator in (1.58). Similarly, the SU(2) transformation is general for any doublet. Finally, under supersymmetry transformations, the matter field λ^i transforms to $\widehat{F}_{ab}$, which is a covariant quantity (as will be shown explicitly in Sect. 2.3.3). This should be the case, since λ^i is a matter field and should thus transform to a covariant quantity.

---

[5]We apologize for the many occurrences of λ here. λ^i is the spinor, while λ^{ab} and λ^{ij} are the parameters appearing in Table 2.1.

Transformations under cgct follow from (2.26)

$$\delta_{\text{cgct}}(\xi)W_\mu = -\xi^\nu \widehat{F}_{\mu\nu} - \tfrac{1}{2}\xi^a \bar{\psi}_\mu \gamma_a \lambda$$

$$\delta_{\text{cgct}}(\xi)\lambda^i = \xi^a \mathcal{D}_a \lambda^i, \qquad \mathcal{D}_\mu \lambda^i = D_\mu \lambda^i + \tfrac{1}{4}\bar{\psi}_\mu^i \gamma^{ab} \widehat{F}_{ab},$$

$$D_\mu \lambda^i = \left(\partial_\mu - \tfrac{3}{2}b_\mu + \tfrac{1}{4}\gamma^{ab}\omega_{\mu ab}\right)\lambda^i - V_\mu{}^{ij}\lambda_j. \tag{2.40}$$

Let us first comment on the last expression. The D_μ introduced here only includes the D, M_{ab} and R-symmetry, i.e. the linearly realized symmetries.[6]

Note that the last term in the first line is necessary for a correct result of the anticommutator of two supersymmetries. Indeed, consider the exercise of calculating the commutator of two supersymmetries on W_μ. In order to close the algebra, this commutator should give a cgct.

$$\delta_Q(\epsilon_1)\delta_Q(\epsilon_2)W_\mu = \tfrac{1}{2}\bar{\epsilon}_2 \gamma_\mu \delta_Q(\epsilon_1)\lambda + \tfrac{1}{2}\bar{\epsilon}_2 \gamma_a \lambda \, \delta_Q(\epsilon_1)e_\mu{}^a. \tag{2.41}$$

The first term gives clearly the covariant curvature, and the second one leads to the second term in the last expression of (2.26). The $\{Q, Q\}$ algebra is thus closed on W_μ.

Exercise 2.3 Check further that one obtains indeed the right coefficient for the transformations, using (1.6), symmetries using [1, (3.51)], Fierz formulae (A.48), gamma manipulations using (A.42), and (A.17).   □

2.3.3 Illustration of Full Calculation of the Transformation of a Curvature

We will now show that $\widehat{F}_{ab}$ is a covariant quantity. We first calculate the super-symmetry transformation of $\widehat{F}_{\mu\nu}$, defined in (2.39). Therefore we need $\delta_Q \psi_\mu$. The matter terms are collected into the symbols Υ_μ. As we have seen, the latter are in principle determined by the gauge algebra. For the sake of simplicity we will leave the matter terms arbitrary in the following derivation. The starting point is then

$$\delta_Q \psi_\mu = D_\mu \epsilon + \Upsilon_\mu, \qquad D_\mu \epsilon^i \equiv \left(\partial_\mu + \tfrac{1}{2}b_\mu + \tfrac{1}{4}\gamma^{ab}\omega_{\mu ab}\right)\epsilon^i - V_\mu{}^{ij}\epsilon_j, \tag{2.42}$$

thus

$$\delta_Q(\epsilon)\widehat{F}_{\mu\nu} = \partial_{[\mu}\left(\bar{\epsilon}\gamma_{\nu]}\lambda\right) - \overline{\left(D_{[\mu}\epsilon\right)}\gamma_{\nu]}\lambda - \bar{\Upsilon}_{[\mu}\gamma_{\nu]}\lambda - \bar{\psi}_{[\mu}\delta_Q(\epsilon)\left(\gamma_{\nu]}\lambda\right). \tag{2.43}$$

---

[6]That turns out to be convenient in computations, as will be illustrated below.

The first ∂_μ can be replaced by a covariant derivative D_μ with only Yang–Mills gauge connections, as it acts on a SU(2) scalar, Lorentz scalar, and D-invariant quantity. To say so, we consider for now the parameter ϵ as an SU(2) doublet, Lorentz spinor, and of dilatational weight $-\frac{1}{2}$ (check that λ got weight $\frac{3}{2}$ and the implicit $e_\nu{}^a$ has weight -1, see (2.20)). That is implicit in the definition of $D_\mu \epsilon^i$ in (2.42). Then we can 'distribute' this covariant derivative, and terms with D_μ on ϵ cancel. This illustrates the convenience to work with the D_μ derivatives. We are left with

$$\delta_Q(\epsilon)\widehat{F}_{\mu\nu} = \bar{\epsilon}\gamma_{[\nu}D_{\mu]}\lambda + \bar{\epsilon}\gamma_a\lambda\, D_{[\mu}e_{\nu]}{}^a - \bar{\Upsilon}_{[\mu}\gamma_{\nu]}\lambda - 2\delta_Q(\epsilon)\delta_Q(\psi_{[\mu})W_{\nu]}. \quad (2.44)$$

The writing of the last term is similar to what was done in Sect. 2.3.1, and the following manipulations are similar to those that we did there. Replacing the first derivative by a full covariant derivative, this can be written as $D_\mu\lambda = \mathcal{D}_\mu\lambda + \delta_Q(\psi_\mu)\lambda$. The latter term nearly leads to $\delta_Q(\psi_\mu)\delta_Q(\epsilon)W_\nu$, apart from that we have to be careful that also the frame field transforms in the latter expression. We obtain:

$$\delta_Q(\epsilon)\widehat{F}_{\mu\nu} = \bar{\epsilon}\gamma_{[\nu}\mathcal{D}_{\mu]}\lambda + \bar{\epsilon}\gamma_a\lambda\, D_{[\mu}e_{\nu]}{}^a - \bar{\Upsilon}_{[\mu}\gamma_{\nu]}\lambda$$
$$+ 2\left[\delta_Q(\psi_{[\mu}), \delta_Q(\epsilon)\right]W_{\nu]} - \bar{\epsilon}\gamma_a\lambda\delta_Q(\psi_{[\mu})e_{\nu]}{}^a. \quad (2.45)$$

We already calculated the commutator on W_μ, checking that it gives the covariant general coordinate transformation. The parameter that we have to use here is, see (1.6):

$$\xi^a(\psi_\mu, \epsilon) = \tfrac{1}{2}\bar{\epsilon}\gamma^a\psi_\mu = \delta_Q(\epsilon)e_\mu{}^a. \quad (2.46)$$

That can be inserted in (2.40) and one can use the constraint (2.18) to obtain

$$0 = 2D_{[\mu}e_{\nu]}{}^a + \xi^a(\psi_\mu, \psi_\nu). \quad (2.47)$$

We leave to you that after further Fierz manipulations as in Exercise 2.3, one arrives at

$$\delta_Q(\epsilon)\widehat{F}_{\mu\nu} = \bar{\epsilon}\gamma_{[\nu}\mathcal{D}_{\mu]}\lambda - \bar{\Upsilon}_{[\mu}\gamma_{\nu]}\lambda - 2\widehat{F}_{a[\nu}\delta_Q(\epsilon)e_{\mu]}{}^a. \quad (2.48)$$

The last term contains an explicit ψ_μ. This is canceled when calculating the transformation

$$\delta_Q(\epsilon)\widehat{F}_{ab} = \bar{\epsilon}\gamma_{[b}\mathcal{D}_{a]}\lambda - \bar{\Upsilon}_{[a}\gamma_{b]}\lambda, \quad (2.49)$$

which does not contain any explicit gauge fields. The gauge fields are hidden in the covariant derivative.

2.3.4 The Easy Way

This was enough to show that the transformation works out as expected. But the calculation was complicated, and we were only looking to a simple example! However, now comes the good news: once you know some tricks, you never have to do all the computations. These tricks involve the knowledge of some basic facts:

1. $\mathcal{D}_a$ on a covariant quantity and $\widehat{R}_{ab}$ are covariant quantities;
2. The transformation of a covariant quantity does not involve a derivative of a parameter (definition).
3. *If the algebra closes on the fields*, then the transformation of a covariant quantity is again a covariant quantity, i.e. gauge fields only appear either included in covariant derivatives or in curvatures.

Let us consider again the calculation of $\delta_Q \widehat{F}_{ab}$ in the $D = 6$ abelian vector multiplet, where from (2.41) we know that the algebra closes on the gauge fields. The third principle implies that $\delta_Q \widehat{F}_{ab}$ should be a covariant quantity. Let us start from the definition (2.39)

$$\widehat{F}_{\mu\nu} = F_{\mu\nu} - \bar{\psi}_{[\mu}\gamma_{\nu]}\lambda, \qquad F_{\mu\nu} = 2\partial_{[\mu}W_{\nu]}. \tag{2.50}$$

First of all, consider the transformation $\delta_Q F_{\mu\nu} \simeq 2\partial_{[\mu}\delta_Q W_{\nu]}$, where $\delta_Q W_\mu = -\bar{\epsilon}\gamma_\mu\lambda$. When taking the ∂_μ derivative, we delete the term where the derivative acts on the parameters α, $\bar{\epsilon}$, because these have to disappear in the transformation of a covariant quantity. Hence,

$$\delta_Q F_{\mu\nu} \Rightarrow \bar{\epsilon}\gamma_{[\nu}\partial_{\mu]}\lambda. \tag{2.51}$$

Moreover we can forget about derivatives on a frame field. Indeed, a derivative on any gauge field can only appear as its curvature, but we know that the curvature for the frame field has been constrained to zero. Hence,

$$\delta_Q F_{ab} \Rightarrow \bar{\epsilon}\gamma_{[b}\partial_{a]}\lambda \Rightarrow \bar{\epsilon}\gamma_{[b}\mathcal{D}_{a]}\lambda. \tag{2.52}$$

The last implication is just the 'covariantization' of the term involving the partial derivative ∂_a, which is not covariant. The covariant quantity must involve $\mathcal{D}_a$, instead of ∂_a.

Let us now consider the second term, $\bar{\psi}_{[\mu}\gamma_{\nu]}\lambda$. One can realize that the interesting term arises from

$$\left(\delta_Q \bar{\psi}_{[\mu}\right)\gamma_{\nu]}\lambda. \tag{2.53}$$

Indeed, if we would act with δ_Q on the other factors, then a ψ_μ remains and we know that these must be included in those covariant derivatives that we have already considered in the first term. For the same reason, regarding the variation of ψ_μ, as for any B_μ^I gauge field, one can neglect the first and second terms in the first line of (2.24):

$$\delta_J \left(\epsilon^J\right) B_\mu^I \Rightarrow \epsilon^J M_{\mu J}^I. \tag{2.54}$$

In our case this M is the Υ-term in (2.42). These steps are quite general, leading to the conclusion that the only relevant part of the transformation of gauge fields is M.

In light of this smart strategy, we now review the result of the *transformation of a covariant derivative* on a covariant quantity, $\delta_J(\epsilon^J)\mathcal{D}_a\phi$, as obtained in (2.36). In order to apply this method we need to assume that the algebra closes on the field ϕ, since otherwise $\delta_J(\epsilon^J)\mathcal{D}_a\phi$ is not a covariant quantity and this 'easy' method cannot be applied straightforwardly.

The first term in (2.36) is the covariantization of the $\partial_\mu\phi$ term in (2.34).

The second term, the M term, is the only one that remains explicit from $\delta_J(\epsilon^J)B_\mu^I$, the rest being implicitly included in covariant derivatives. Thus, from a practical point of view, these non-gauge terms are the most interesting terms in the transformation of gauge fields. As shown in (2.24), M terms are of two types. The first ones are those from the gauge algebra where the gauge field was the frame field. These are the transformations explicitly given in (2.25). The others are the matter terms, which we still have to find for each case. For the gauge field in the vector multiplet, that was the relevant term, which gave its supersymmetry transformation to the gaugino.

The third term appears only in case of non-closure. We will consider below the fermionic field in the vector multiplet for which there is no closure.[7]

Now consider the fourth term, which finds its origin in transformations of the frame field to the frame field, (2.20). This term refers to the first two terms in the explicit expression of (2.20):

$$\delta_J \left(\epsilon^J\right) \mathcal{D}_a\phi = \cdots - \left(\lambda_D \delta_a{}^b + \lambda_a{}^b\right) \mathcal{D}_b\phi, \tag{2.55}$$

which amount to the following. The first one implies that the Weyl weight of $\mathcal{D}_a\phi$ is one higher than that of ϕ. The second one implies that $\mathcal{D}_a\phi$ is a Lorentz vector.

Exercise 2.4 Check that, whatever would be the S-transformation of λ (here in 6 dimensions it is zero, however the corresponding fermion in 4 and 5 dimensions has S-supersymmetry transformations), the S-variation of $\widehat{F}_{ab}$ is due to (2.25):

$$\delta_S(\eta)\widehat{F}_{ab} = \bar{\eta}\gamma_{ab}\lambda. \tag{2.56}$$

□

---

[7] Closure could be obtained if we would have introduced auxiliary fields, but for the didactical value of the example, it is good to consider the situation without the auxiliary fields.

Exercise 2.5 One can even give a general formula for the transformation of curvatures, correcting (2.6) for the effects that gauge fields transform with matter-like terms. To apply the methods explained earlier, the last decomposition in (2.27) is most useful. Indeed, explicit gauge fields appear only quadratically in $r_{ab}{}^I$. You can then determine that

$$\delta_J\left(\epsilon^J\right)\widehat{R}_{ab}{}^I = \epsilon^J\widehat{R}_{ab}{}^K f_{KJ}{}^I + 2\epsilon^J \mathcal{D}_{[a}M_{b]J}{}^I - 2\epsilon^K M_{[aK}{}^J M_{b]J}{}^I. \qquad (2.57)$$

Similarly, the Bianchi identity becomes

$$\mathcal{D}_{[a}\widehat{R}_{bc]}{}^I + \widehat{R}_{[ab}{}^J M_{c]J}{}^I = 0. \qquad (2.58)$$

□

2.3.5 Non-closure Terms in $D = 6$ Abelian Vector Multiplet

Finally, let us consider the fermionic field (gaugino) λ^i in the $D = 6$ vector multiplet, on which the supersymmetry algebra does not close. As we have already calculated the supersymmetry transformation of $\widehat{F}_{ab}$ in (2.49), it is easy to calculate the commutator of two supersymmetries on the gaugino. We do not take the Υ-term into account, as we do not know its form yet, and it is independent of the rest. The commutator is

$$\left[\delta_Q(\epsilon_1),\delta_Q(\epsilon_2)\right]\lambda^i = -\tfrac{1}{2}\mathcal{D}_c\lambda^i\,\bar{\epsilon}_1\gamma^c\epsilon_2$$

$$+ \tfrac{3}{16}\gamma_c\mathcal{D}\!\!\!/\lambda^i\,\bar{\epsilon}_1\gamma^c\epsilon_2 + \tfrac{1}{96}\gamma_{cde}\mathcal{D}\!\!\!/\lambda_j\,\bar{\epsilon}_1^{(i}\gamma^{cde}\epsilon_2^{j)}. \qquad (2.59)$$

The first term is the covariant general coordinate transformation. The others are the non-closure terms, all proportional to $\mathcal{D}\!\!\!/\lambda$. From these terms one finds that the supersymmetry variation of $\mathcal{D}_a\lambda$ is (neglecting again possible extra matter terms in transformations of gauge fields)

$$\delta_Q(\epsilon)\mathcal{D}_a\lambda^i = -\tfrac{1}{4}\gamma^{bc}\epsilon^i D_a\widehat{F}_{bc} + \tfrac{3}{16}\gamma_c\mathcal{D}\!\!\!/\lambda^i\,\bar{\psi}_a\gamma^c\epsilon + \tfrac{1}{96}\gamma_{cde}\mathcal{D}\!\!\!/\lambda_j\,\bar{\psi}_a^{(i}\gamma^{cde}\epsilon^{j)}. \qquad (2.60)$$

To conclude, we see explicitly that the transformation of $\mathcal{D}_a\lambda$ is not a covariant quantity. It is now not possible to define $\mathcal{D}_b\mathcal{D}_a\lambda$ such that it does not transform to a derivative of a gauge field. However, note that one can define it such that at least the antisymmetric part in $[ab]$ does not transform in a derivative. That is analogous to a curvature. Also on a gauge field we cannot define a covariant generalization of $\partial_{(\mu}W_{\nu)}$. The covariant $\mathcal{D}_{[b}\mathcal{D}_{a]}$ should just have extra factors $\tfrac{1}{2}$ for every term in which there are two gauge fields. Also that is similar to a curvature:

$$
\mathcal{D}_{[b}\mathcal{D}_{a]}\lambda^i = \partial_{[b}\mathcal{D}_{a]}\lambda^i + \tfrac{1}{4}\gamma^{cd}\psi^i_{[b}\mathcal{D}_{a]}\widehat{F}_{cd}
$$
$$
- \tfrac{3}{32}\gamma_c\mathcal{D}\lambda^i\,\bar{\psi}_a\gamma^c\psi_b - \tfrac{1}{192}\gamma_{cde}\mathcal{D}\lambda_j\,\bar{\psi}_a^{(i}\gamma^{cde}\psi_b^{j)}\,. \tag{2.61}
$$

But these are objects that one seldom needs.

2.4 Curvature Constraints

As we advocated in Sect. 2.2, we need to impose certain constraints in order that the translations on the gauge fields take the form of (covariant) general coordinate transformations. The kind of constraints we need are the so-called *conventional constraints*, which are algebraic (and not differential) conditions on some fields. At the classical level, conventional constraints do not impose restrictions on the dynamic itself. Importantly, this discussion is not specific for conformal supergravity.[8]

2.4.1 Constraint on $R(P)$

The vanishing of the P^a curvature has been already considered in (2.18) and it is immediately clear why this leads to a conventional constraint. Concerning representation content, (2.18) is a vector times an antisymmetric tensor, and the same holds for the spin connection $\omega_\mu{}^{ab}$. For this reason one can solve (2.18) algebraically for $\omega_\mu{}^{ab}$, which from now on will become a dependent field[9]:

$$
\omega_\mu{}^{ab} = \omega_\mu{}^{ab}(e, b) - \xi^{[a}\left(\psi_\mu, \psi^{b]}\right) - \tfrac{1}{2}\xi_\mu\left(\psi^a, \psi^b\right)\,,
$$
$$
\omega_\mu{}^{ab}(e, b) = \omega_\mu{}^{ab}(e) + 2e_\mu{}^{[a}b^{b]}\,, \tag{2.62}
$$

where $\omega_\mu{}^{ab}(e)$ is the usual expression of the spin connection, see (A.4), obtained by the antisymmetric part of (A.3), which is a simplification of the above constraint without b_μ and gravitini. The term with b^b can be understood as necessary to reproduce the special conformal invariance, see (2.25). The gravitino contribution is already known from pure Poincaré $\mathcal{N} = 1$ supergravity [6–8].

It is important to realize that the constraint (2.18) is not invariant under all the symmetries if one assumes for $\omega_\mu{}^{ab}$ the transformation rule that follows

---

[8]Indeed, one needs such constraints also in general relativity (as explained at the beginning of this chapter) where the connection $\omega_\mu{}^{ab}$ must be a function of e^a_μ, and not an independent field.

[9]This is what we were looking for (see the first item of the list of three shortcomings in Sect. 2.2).

from (2.3), such that the curvatures would just transform as in (2.6). Explicitly, the transformation $\delta_I(\epsilon^I)R_{ab}(P^c)$ can be calculated from the definition (2.18) using the 'easy method'. The terms quadratic in the gravitino do not play a role in the variation, as they will always leave a naked gravitino behind. Thus we should only consider the variations of the frame field in the derivative. According to (2.20), this involves only ∂e, i.e. $R_{ab}(P)$, and $\partial \psi$, i.e. $\widehat{R}_{ab}(Q)$. The former are zero, so there remains only a variation in $\widehat{R}_{ab}(Q)$:

$$\delta_I\left(\epsilon^I\right) R_{ab}(P^c) = \xi^c\left(\widehat{R}_{ab}(Q), \epsilon\right) . \tag{2.63}$$

One may therefore expect that there would be a new constraint $\widehat{R}_{ab}(Q) = 0$. However, since in the expression for $\widehat{R}_{ab}(Q)$ there is no field that is a spinor–antisymmetric tensor, this choice would be an unconventional constraint: it cannot be solved. Instead, one should define the transformation laws of the dependent field from its definition as in (2.62). As $\omega_\mu{}^{ab}$ is now defined by the vanishing of $R_{ab}(P)$, its transformations can be defined by its expression in terms of the other fields. Another way to say so, is that $\delta\omega_\mu{}^{ab}$ has extra terms to compensate for the non-invariance of $R_{ab}(P)$, calculated above.

To see how this method works concretely, let us be slightly more general and consider a constraint $C(\omega(\phi), \phi)$, where $\omega(\phi)$ is a composite of the independent field ϕ and we assume that C contains at least a curvature. The transformation of C consists of the pure gauge terms, from the gauge theory algebra and following the general rules, plus extra terms $\delta_M\omega$ like in (2.24):

$$\delta C = \delta_{\text{gauge}}C + \delta_M C = \delta_{\text{gauge}}C + \frac{\partial C}{\partial\omega}\delta_M\omega . \tag{2.64}$$

Because $\frac{\partial C}{\partial\omega}$ is invertible we can calculate the extra terms in the transformation of ω that are necessary in order that the constraint is invariant, $\delta C = 0$. The result for the constraint (2.18) has the same structure as the solution (2.62):

$$\delta_M\omega_\mu{}^{ab} = \xi^{[a}\left(\widehat{R}_\mu{}^{b]}(Q), \epsilon\right) + \tfrac{1}{2}\xi_\mu\left(\widehat{R}^{ab}(Q), \epsilon\right) , \tag{2.65}$$

and can be thus interpret as the additional M term in (2.24) for the ω gauge field.

A final consequence of the imposing (2.18) can be seen from the Bianchi identity (2.7) applied to translations: $\mathcal{D}_{[\mu}R_{\nu\rho]}(P^a) = 0$. Of course the covariant derivative used in this expression is not limited to the standard gauge transformations, and as such one will find covariantization terms corresponding to local translations, i.e. $e_\mu^b f_{bJ}{}^a R_{\nu\rho}{}^J$, where $f_{bJ}{}^a$ are structure constants in the commutator $[P_b, T_J]$ proportional to P^a. This commutator is only non-zero for T_J being dilatations and Lorentz rotations. According to the lemmas in Sect. 2.3, these covariantization terms (with all μ indices changed to tangent spacetime indices) are the only ones that we have to consider when we upgrade the curvatures to covariant quantities. Hence the Bianchi identity for translations immediately gives

$$- \delta_{[a}{}^{d} \widehat{R}_{bc]}(D) + \widehat{R}_{[ab} \left(M_{c]}{}^{d} \right) = 0 . \tag{2.66}$$

So far, this is independent of supersymmetry or even of the conformal group. In fact, the last identity, without $R(D)$ would then be the well-known equation that the M-curvature antisymmetrized in 3 indices is zero. Multiplying (2.66) with $\delta_d{}^a$ gives that the contracted M-curvature, i.e. the Ricci tensor, is symmetric if there is no $R_{ab}(D)$:

$$(2 - D) \widehat{R}_{ab}(D) = 2 \widehat{R}_{d[a} \left(M_{b]}{}^{d} \right) . \tag{2.67}$$

Exercise 2.6 The M-curvature can be calculated directly by taking the derivative of (2.62). As the final result has to be covariant, we can drop all other terms, and restrict to the $\partial \omega$ term, where the derivative acts either on $\omega(e)$ or on b_μ. This leads to a direct relation expressing $\widehat{R}_{\mu\nu}(M^{ab})$ in the covariantization of $\partial \omega(e)$, and $\widehat{R}_{\mu\nu}$. Check that in this terminology, (2.66) implies that the former part satisfies the cyclic identity $R_{[\mu\nu\rho]}{}^\sigma = 0$ and leads to a symmetric Ricci tensor.   □

2.4.2 Other Conventional Curvature Constraints

Now that we have seen the procedure how $\omega_\mu{}^{ab}$ became composite, we may consider whether this can be done for other gauge fields in the same way. If we look for the smallest possible multiplet, then we should use this procedure for all the gauge fields that we can algebraically solve. The clue is given in (2.29). These equations show which gauge fields appear 'linearly' in the curvatures, namely multiplied by a frame field. In particular, we may choose constraints that determine $f_\mu{}^a$ and ϕ_μ. This looks rather convenient. Indeed, without such a constraint $f_\mu{}^a$ would already appear in bosonic conformal gravity, while we do not know this field in the physical conformal gravity. At first it looks that we can define it either from a constraint on the dilatational or on the Lorentz curvature. But (2.67) implies that $\widehat{R}(D)$ is a function of $\widehat{R}(M)$, so that we can restrict ourselves to the Lorentz curvature. To be able to eliminate $f_\mu{}^a$ completely, the constraint should be a general $d \times d$ matrix:

$$\widehat{R}_{ac} \left(M^{bc} \right) + C_a{}^b = 0 , \tag{2.68}$$

where $C_a{}^c$ may be any covariant function that is independent of $f_\mu{}^a$. Similarly, there can be a third constraint that determines ϕ_μ from a constraint of the form

$$\gamma^b \widehat{R}(Q)_{ba} + \rho_a = 0 , \tag{2.69}$$

where ρ_a is a covariant spinor-vector. One might impose the constraints with $C_a{}^b$ and ρ_a vanishing, but we will see in Sect. 2.6 that modifying the constraints in this way depending on other (auxiliary) fields can be convenient. In any case, the addition or not of such extra parts amounts at the end only to a field redefinition of the dependent gauge fields.

We go now again through the consequences of these constraints. Let us repeat the names of the various covariantizations of curvatures (and introduce meanwhile another one: $\widehat{R}'$):

$$r_{\mu\nu}{}^I = 2\partial_{[\mu}B_{\nu]}{}^I + B_\nu{}^K B_\mu{}^J f_{JK}{}^I \, ,$$

$$R_{\mu\nu}^I = r_{\mu\nu}{}^I - 2B_{[\mu}{}^J f_{\nu]J}{}^I \, ,$$

$$\widehat{R}_{\mu\nu}{}^I = R_{\mu\nu}{}^I - 2B_{[\mu}{}^J M_{\nu]J}{}^I = r_{\mu\nu}{}^I - 2B_{[\mu}{}^J M_{\nu]J}{}^I \, ,$$

$$\widehat{R}'_{\mu\nu}{}^I = r_{\mu\nu}{}^I - 2B_{[\mu}{}^J M_{\nu]J}{}^I = \widehat{R}_{\mu\nu}{}^I + 2B_{[\mu}{}^J f_{\nu]J}{}^I \, . \tag{2.70}$$

Furthermore, as a completion of the Ricci tensor, see (A.2), we define

$$\mathcal{R}_{\mu\nu} = \widehat{R}'_{\rho\mu}\left(M^{ab}\right)e_a{}^\rho e_{vb} \, , \qquad \mathcal{R} = \mathcal{R}_\mu{}^\mu \, . \tag{2.71}$$

The solution of (2.68) and (2.69) is

$$2(D-2)f_\mu{}^a = -\mathcal{R}_\mu{}^a - C_\mu{}^a + \frac{1}{2(D-1)}e_\mu{}^a\left(\mathcal{R} + C_b{}^b\right) \, ,$$

$$-s_D(D-2)\phi_a = \gamma^b\widehat{R}'_{ab}(Q) - \rho_a - \frac{1}{2(D-1)}\gamma_a\left(\gamma^{bc}\widehat{R}'_{bc}(Q) - \gamma^b\rho_b\right) \tag{2.72}$$

This is thus the analogue of (2.62). Similar to what we found there, the constraints will have the consequence that the transformation law of the fields $f_\mu{}^a$ and ϕ_μ get extra matter-like contributions. We assume that the extra $C_a{}^b$ and ρ terms in the constraints have the same Lorentz structure, SU(2) structure, dilatational and U(1) weight as the curvature term. As we explained at the beginning of Sect. 2.4.1 (see e.g. around (2.64)), the constraints do not break supersymmetry (in general, this can be as well Q as S supersymmetry) as long as one modifies the transformations of the composite fields. The exact rules depend on the transformations of the matter terms in the constraint. The simplicity of the result, e.g. invariance of the constraint under S-transformations may even be an argument for the choice of the matter terms in the constraints.

The K transformations of these constraints vanish due to the constraint on $R_{\mu\nu}(P)$, at least when the extra matter fields do not transform under K. Indeed, this will not be the case due to a Weyl weight related argument that we will give in Sect. 2.6.1.

Furthermore, there are relations between the curvatures as a consequence of Bianchi identities, similar to (2.66). Those give relations for the S and K curvatures. The explicit form depends on the matter sector.

2.5   Example: Non-SUSY Sigma Model

Before continuing with the full theory, let us illustrate the construction of a conformal invariant action by looking at a scalar sigma model without supersymmetry. In this context, the scalars ϕ^i are coordinates of a Riemannian manifold, with metric g_{ij}. The latter defines the Levi-Civita connection Γ^k_{ij}. We assume that there is a closed homothetic Killing vector as in (1.30), which allows to define the conformal symmetry of the scalars. The gauge fields are $e_\mu{}^a$, $\omega_\mu{}^{ab}$, b_μ and $f_\mu{}^a$ (with conventions given in Table 2.1) and their transformation under local dilatations and special conformal transformations follow directly from (2.3):

$$\delta\phi^i = k_D{}^i \lambda_D \,,$$

$$\delta e_\mu{}^a = -\lambda_D e_\mu{}^a \,, \qquad \delta e^\mu{}_a = \lambda_D e^\mu{}_a \,,$$

$$\delta\omega_\mu{}^{ab} = -4\lambda_K^{[a} e_\mu{}^{b]} \,,$$

$$\delta b_\mu = \partial_\mu \lambda_D + 2\lambda_{K\mu} \,,$$

$$\delta f_\mu{}^a = \lambda_D f_\mu{}^a + \partial_\mu \lambda_K{}^a \,, \tag{2.73}$$

where as usual, $e^\mu{}_a$ is the inverse of $e_\mu{}^a$. The scalar fields are thus invariant under special conformal transformations.[10] The covariant derivative for the scalars follows from (2.30):

$$\mathcal{D}_\mu \phi^i = \partial_\mu \phi^i - b_\mu k_D{}^i \,, \qquad \mathcal{D}_a \phi^i = e_a^\mu \mathcal{D}_\mu \phi^i \,, \tag{2.74}$$

leading to

$$\delta\mathcal{D}_a\phi^i = \lambda_D \left[\mathcal{D}_a\phi^i + \left(\partial_j k_D{}^i\right)\mathcal{D}_a\phi^j\right] - 2\lambda_{Ka}k_D{}^i$$

$$= \lambda_D \left[(w+1)\mathcal{D}_a\phi^i - k_D{}^k \Gamma^i_{jk}\mathcal{D}_a\phi^j\right] - 2\lambda_{Ka}k_D{}^i \,. \tag{2.75}$$

The second (contracted) covariant derivative is therefore

$$\mathcal{D}_a\mathcal{D}^a\phi^i = \partial_a\mathcal{D}^a\phi^i + 2f_a^a k_D{}^i - b^a\left[(w+1)\mathcal{D}_a\phi^i - k^k\Gamma^i_{jk}\mathcal{D}_a\phi^j\right]$$

$$+ e^{\mu a}\omega_{\mu a}{}^b\mathcal{D}_b\phi^i \,. \tag{2.76}$$

The covariant box is

$$\Box^c\phi^i = \mathcal{D}^a\mathcal{D}_a\phi^i + \Gamma^i_{jk}\mathcal{D}_a\phi^j\mathcal{D}^a\phi^k \,. \tag{2.77}$$

---

[10]$k_\mu^i(\phi) = 0$ in the terminology of (1.24), and this can be expressed by saying that ϕ^i are *primary fields*.

Since

$$\delta_K \Box^c \phi^i = 2\lambda_K^a (D - 2 - 2w) \mathcal{D}_a \phi^i ,\tag{2.78}$$

the covariant box is conformally invariant only for $w = (D - 2)/2$.

We will show that the following action is conformal invariant:

$$e^{-1}\mathcal{L} = -\frac{1}{2} g_{ij} \mathcal{D}_a \phi^i \mathcal{D}^a \phi^j + \frac{1}{w} f_a^a g_{ij} k_D^i k_D^j .\tag{2.79}$$

Both terms on the right-hand side scale under dilatations separately with a factor $2(w + 1)$, while the left-hand side scales with weight D. So it is consistent under the same condition

$$w = \tfrac{1}{2}(D - 2) ,\tag{2.80}$$

which keeps the conformal box also K-invariant. Now we will show the K-invariance, which is why we need the second term in (2.79). The K transformations of the action give

$$e^{-1}\delta_K \mathcal{L} = 2\lambda_K^a k_D^i g_{ij} \mathcal{D}_a \phi^j + \frac{1}{w} \left(\partial_a \lambda_K^a\right) g_{ij} k_D^i k_D^j = \frac{1}{w} \partial_a \left(\Lambda_K^a g_{ij} k_D^i k_D^j\right) ,\tag{2.81}$$

where one uses that

$$\partial_a \left(g_{ij} k_D^i k_D^j\right) = 2w k_D^i g_{ij} \partial_a \phi^j .\tag{2.82}$$

This finishes the proof of the invariance of the action (2.79).

The equations of motion from (2.79) with (2.80) read

$$\frac{\delta}{\delta \phi^i} \int d^D \mathcal{L} = g_{ij} \Box^c \phi^j .\tag{2.83}$$

We now consider the constraint (2.68) with $C_a^b = 0$. Indeed, here we do not have any reason nor fields available to consider a different C_a^b. Then (2.72) leads to

$$f_a^a = -\frac{1}{4(D - 1)} R .\tag{2.84}$$

We now just write R, rather than $\mathcal{R}$ to indicate that there are no further corrections to the usual bosonic scalar curvature of the metric. Therefore, the Lagrangian is

$$e^{-1}\mathcal{L} = -\frac{1}{2} g_{ij} \mathcal{D}_a \phi^i \mathcal{D}^a \phi^j - \frac{1}{2(D - 1)(D - 2)} R g_{ij} k_D^i k_D^j .\tag{2.85}$$

In order to have positive energy for gravity, the metric g_{ij} should have a negative signature in the direction of the homothetic Killing vector $k_D{}^i$. In flat space, $k_D{}^i = w\phi^i$ and we obtain

$$e^{-1}\mathcal{L} = -\frac{1}{2}g_{ij}\mathcal{D}_a\phi^i\mathcal{D}^a\phi^j - \frac{D-2}{8(D-1)}Rg_{ij}\phi^i\phi^j. \qquad (2.86)$$

For $D = 4$, in the case of a single scalar with negative kinetic energy, this reduces to (1.10).

2.6 The Standard Weyl Multiplets

The aim of this section is to construct and describe the multiplet of pure superconformal gravity, called *Weyl multiplet*. As long as we are not interested in higher-derivatives supergravity theories, we do not need to consider an action for the Weyl multiplet. Instead, in the spirit of the superconformal calculus, results given in the present section will be very useful to build up actions for Poincaré supergravity.

2.6.1 Matter Fields Completing the Weyl Multiplet

So far, the problem mentioned in the fifth step of Sect. 2.2.1 has not been considered. We saw in Table 2.2 that the number of bosonic and fermionic components of the independent gauge fields do not match. As a consequence, the supersymmetry algebra cannot give rise to invertible coordinate transformations. We could expect this result, since given the general forms for the covariant general coordinate transformations and for the other gauge transformations, nothing guarantees that supersymmetries anticommute to the covariant general coordinate transformations. The matter terms M in the transformations of the gauge fields (2.22) should be chosen appropriately to obtain the right anticommutator, and they should be functions of new matter fields.

The solution is not unique and the arguments to obtain it are not so obvious. The first one of this nature was obtained in 4 dimensions from splitting a linearized Poincaré multiplet [9–11] that was found earlier. One way of constructing the complete set of fields is to make use of supercurrents. One starts with a multiplet that has rigid superconformal symmetry and considers the fields that couple to the Noether currents. This method has been used in various cases, see [12–18]. We will not go into this subject any further here.

We present here the solutions with the fields mentioned in Table 2.3, which are called the 'standard Weyl multiplets'. Alternative versions for Weyl multiplets have been constructed [5, 18, 19], which differ from the above ones in the choice of auxiliary fields (fields below the double line in Table 2.3). These multiplets contain a dilaton auxiliary field and are therefore called the 'dilaton Weyl multiplets'. The existence of a $D = 4$ version, recently found in [19] was already noticed in [20, Sect. 4.4].

Table 2.3 Number of off-shell components in the fields of the standard Weyl multiplet

	$D=4$	$D=5$	$D=6$	w	c	γ_*	gauge transf.	subtracted
$e_\mu{}^a$	5	9	14	-1	0		P^a	M_{ab}, D
b_μ	compensating K^a			0	0		D	K^a
$\omega_\mu{}^{ab}$	composite			0	0		M^{ab}	
$f_\mu{}^a$	composite			1	0		K^a	
$V_{\mu i}{}^j$	9	12	15	0	0		SU(2)	
A_μ	3			0	0		U(1)	
$\psi_\mu{}^i$	16	24	32	$-\tfrac{1}{2}$	$\tfrac{1}{2}$	$+$	Q^i	S^i
$\phi_\mu{}^i$	composite			$+\tfrac{1}{2}$	$\tfrac{1}{2}$	$-$	S^i	
T_{ab}^-, T_{abc}^-	6	10	10	1	1			
D	1	1	1	2	0			
χ^i	8	8	8	$\tfrac{3}{2}$	$\tfrac{1}{2}$	$+$		
TOTAL	$24+24$	$32+32$	$40+40$					

The columns $D = 4, 5, 6$ indicate the number of components of the fields in the first column with gauge transformations subtracted. The next columns contain the Weyl weight, the chiral weight (only for $D = 4$, where it means that the transformation under U(1) is $\delta_T \phi = ic\phi\lambda_T$), and the chirality for the fermions for even dimension. Note that for $D = 4$, changing the position of the index changes the chirality, while in 6 dimensions the chirality is generic. Finally, we indicate the symmetry for which it is a gauge field, and possibly other gauge transformations that have been used to reduce its number of degrees of freedom in this counting

The standard Weyl multiplet involves an antisymmetric tensor T, with two indices in 4 dimensions (T_{ab}, you may take it anti-self-dual, but then it is complex) [21], with two or three indices in 5 dimensions (these are dual to each other) [18], or an anti-self-dual real tensor in 6 dimensions [5]. Further there is a real scalar D and a fermion doublet χ^i.

In order to obtain the transformation laws of the standard Weyl multiplet one uses the general procedure outlined in [5].

(1) As a first step one must construct the linear Q-supersymmetry transformations of the matter fields and also the additional transformations for the gauge fields due to the presence of these matter fields. This is done by writing a general ansatz linear in the fields with yet undetermined coefficients. These coefficients will be fixed by demanding that the commutator of two super-symmetry transformations gives a local translation, i.e. $[\delta_Q(\epsilon_1), \delta_Q(\epsilon_2)] = \xi^a(\epsilon_1, \epsilon_2)P_a$ with $\xi^a(\epsilon_1, \epsilon_2)$ given by (1.6). In writing the ansatz one makes use of the Lorentz structure of the fields and of course their spin. Recalling the tricks presented in Sect. 2.3.4, a derivative acting on a gauge field can be replaced with the corresponding (modified) curvature. Similarly, one must also covariantize all the derivatives using the covariant derivative. Both the explicit

expressions of the covariant derivatives and the curvatures will change during the procedure that we outline here. More specifically one will encounter $M_{\mu J}{}^{I}$ terms, which will modify the transformation laws and curvatures as shown in (2.22) and (2.29), respectively.

(2) Next one would like to obtain the bosonic transformation rules of the matter fields. The Lorentz transformations follow straightforwardly from the index structures of the different fields. For the fields that we consider here, the action of the dilatation on a field is of the form

$$\delta_D \phi = w \, \lambda_D \phi \,, \tag{2.87}$$

determined completely by the Weyl weight w of the field. The Weyl weights of the gauge fields are easily found by looking at the algebra. For the matter fields one must use another consequence of the algebra, namely that the Weyl weight of the Q-supersymmetry transformation must be $\frac{1}{2}$ higher than the original field on which it acts. This is equivalent to considering the Weyl weight of the parameter ϵ as $-\frac{1}{2}$ and demand an equality of the Weyl weight of the left-hand side and right-hand side of the Q-supersymmetry transformations that we have just determined. In this one must also use that the Weyl weight of a covariant derivative is 1 and that a curvature has Weyl weight 2 higher than the corresponding gauge field. This is due to the presence of frame fields in their expressions. The determination of the Weyl weights is an important step because they restrict the other transformations severely. Especially for the K-transformation, which lowers the Weyl weight of the field by 1. In the same sense as for ϵ one can consider the parameter λ_K^a as having weight 1. So in order to construct these transformations, one must look whether there are fields present in the multiplet that obey this restriction.

One is then left with the R-symmetries. For the T transformation in $D = 4$ one uses similar rules as for dilatations, using a chiral weight c for every field with the meaning

$$\delta_T \phi = ic\phi\lambda_T \,. \tag{2.88}$$

The complex conjugate of a field has opposite chiral weight. In the same way, the right-projected field, e.g. χ_i has $c = -\frac{1}{2}$ since χ^i has $c = +\frac{1}{2}$. To find the weights, the commutator $[T, Q^i]$ in (1.58) implies that one can count ϵ^i as having chiral weight $\frac{1}{2}$, and thus ϵ_i as weight $-\frac{1}{2}$. The SU(2) transformations are implicit in the position of the indices, see (A.26).

(3) The S-supersymmetry transformation is determined by using the action of the commutator $[K, Q] \sim S$ on the fields. Since for the Weyl multiplet, b_μ is the only one of the independent fields that transforms under K-transformations (as in (2.25)), the S-transformation is easy to get in the following way. First consider all occurrences of b_μ in the transformation law (hidden in covariant derivatives and composite fields such as (2.62)). This should lead to terms proportional to $b_\mu \gamma^\mu \epsilon^i$. Then replace $b_\mu \gamma^\mu \epsilon^i$ by $-2s_D \eta^i$ where s_D are the sign factors in (1.58).

Exercise 2.7 An immediate example of how this procedure works consists in obtaining the first line in (2.25) from the supersymmetry transformation of the gravitino

$$\delta(\epsilon)\psi_\mu = D_\mu\epsilon + \cdots = \left(\tfrac{1}{2}b_\mu + \tfrac{1}{4}\omega_\mu{}^{ab}(e,b)\gamma^{ab}\right)\epsilon + \cdots \qquad (2.89)$$

□

At this point we have determined all the transformations for the fields of the standard Weyl multiplet. However, there are two complications which will further modify these transformations. A first modification applies only to the composite fields as discussed in Sect. 2.4. There we have shown that in order for the constraints to be invariant under the symmetries one must change the transformations of the composite fields. First, it is important that one uses the modified curvatures due to the extra matter terms in, for example, the Q-supersymmetry transformations of the gauge fields.

Further, as shown in Sect. 2.4.2, it is possible to change these constraints by adding additional covariant functions of fields. In some cases, these functions can be chosen in a way that they simplify the transformations of the composite fields. For $D = 4$ and for $D = 6$ (the $(1, 0)$ theory that we treat here) we will be able to choose the constraints invariant under S-supersymmetry, avoiding extra S-transformations for the constrained fields. Observe, however, that this is a choice for convenience, which is even not necessary possible (it is not possible in $D = 5$ [18] or for $(2, 0)$ in $D = 6$ [17]). In principle all constraints of the form (2.68) and (2.69) are equivalent up to field redefinitions.

A second, and more important, complication is the fact that one obtains additional transformations if one determines $\{Q, Q\}$. For example if one calculates this commutator on the frame field one will obtain a general coordinate transformation, as expected, but also an additional Lorentz transformation. These transformations will depend on the fields and thus change the algebra into a *soft algebra*, which has structure functions rather than constants, see [1, Sect. 11.1.3]. This change in the algebra is the start of an iterative process. First one imposes the newly found commutator $\{Q, Q\}$ on all the fields, i.e. one demands that the Q-transformations of all the fields are such that their commutator gives a Lorentz transformation acting on the field. In order to be able to do this one may need to use terms that are nonlinear in the fields, e.g. second order. If these nonlinear Q-transformations are found, one again calculates the commutator on all the fields. If one finds a new transformation as a result, one must impose this again on all the fields and thus change the Q-transformations. This iterative process stops if one finds no new transformations in calculating the commutator and thus obtains closure. A same procedure can be applied to $\{Q, S\}$ and $\{S, S\}$.

2.6.2 $D = 4$

We present here the Q and S supersymmetry transformations in $D = 4$. First for the independent gauge fields:

$$\delta_{Q,S}(\epsilon,\eta)e_\mu{}^a = \tfrac{1}{2}\bar{\epsilon}^i\gamma^a\psi_{\mu i} + \text{h.c.},$$

$$\delta_{Q,S}(\epsilon,\eta)b_\mu = \tfrac{1}{2}\bar{\epsilon}^i\phi_{\mu i} - \tfrac{1}{2}\bar{\eta}^i\psi_{\mu i}$$
$$- \tfrac{3}{8}\bar{\epsilon}^i\gamma_\mu\chi_i + \text{h.c.},$$

$$\delta_{Q,S}(\epsilon,\eta)A_\mu = -\tfrac{1}{2}i\bar{\epsilon}^i\phi_{\mu i} - \tfrac{1}{2}i\bar{\eta}^i\psi_{\mu i}$$
$$- \tfrac{3}{8}i\bar{\epsilon}^i\gamma_\mu\chi_i + \text{h.c.},$$

$$\delta_{Q,S}(\epsilon,\eta)V_{\mu i}{}^j = -\bar{\epsilon}_i\phi_\mu^j - \bar{\eta}_i\psi_\mu^j$$
$$+ \tfrac{3}{4}\bar{\epsilon}_i\gamma_\mu\chi^j - (\text{h.c.}; \ \text{traceless}),$$

$$\delta_{Q,S}(\epsilon,\eta)\psi_\mu^i = \left(\partial_\mu + \tfrac{1}{2}b_\mu + \tfrac{1}{4}\gamma^{ab}\omega_{\mu ab} - \tfrac{1}{2}iA_\mu\right)\epsilon^i - V_{\mu j}{}^i\epsilon^j$$
$$- \tfrac{1}{16}\gamma\cdot T^-\varepsilon^{ij}\gamma_\mu\epsilon_j - \gamma_\mu\eta^i,$$

$$\delta_{Q,S}(\epsilon,\eta)T_{ab}^- = 2\bar{\epsilon}^i\widehat{R}_{ab}(Q)^j\varepsilon_{ij},$$

$$\delta_{Q,S}(\epsilon,\eta)\chi^i = -\tfrac{1}{24}\gamma\cdot\slashed{D}T^-\varepsilon^{ij}\epsilon_j - \tfrac{1}{6}\widehat{R}_j{}^i\cdot\gamma\epsilon^j + \tfrac{1}{6}i\widehat{R}(T)\cdot\gamma\epsilon^i + \tfrac{1}{2}D\,\epsilon^i$$
$$+ \tfrac{1}{12}\gamma\cdot T^-\varepsilon^{ij}\eta_j,$$

$$\delta_{Q,S}(\epsilon,\eta)D = \tfrac{1}{2}\bar{\epsilon}^i\slashed{D}\chi_i + \text{h.c.}, \tag{2.90}$$

where $\widehat{R}_j{}^i$ is the SU(2) curvature, $R(T)$ the U(1) curvature, and $\gamma\cdot R = \gamma^{ab}R_{ab}$. The notation $A^i{}_j - (\text{h.c.}; \ \text{traceless})$ stands for $A^i{}_j - A_j{}^i - \tfrac{1}{2}\delta_j{}^i(A^k{}_k - A_k{}^k)$. For the gauge fields, the first line represents the original gauge transformations, and the second line are the terms that were symbolically represented by $\mathcal{M}$ in (2.26). Observe the covariance of these terms and the transformations of the matter terms.

The dependent fields are defined by the constraints

$$0 = R_{\mu\nu}^a(P),$$

$$0 = \gamma^b\widehat{R}_{ba}(Q)^i + \tfrac{3}{2}\gamma_a\chi^i,$$

$$0 = \widehat{R}_{ac}(M^{bc}) - i\widetilde{\widehat{R}}_a{}^b(T) + \tfrac{1}{4}T_{ca}^-T^{+bc} + \tfrac{3}{2}\delta_a{}^b D. \tag{2.91}$$

The terms that could appear in these constraints are fixed by compatibility with Weyl weights (thus that we do not want to modify the dilatational transformations), and as mentioned above, the coefficients are chosen such that they are invariant under S-supersymmetry.

The expressions of the dependent fields in terms of the physical fields are

$$\omega_\mu{}^{ab} = \omega_\mu{}^{ab}(e, b) + \tfrac{1}{2}\left[\bar{\psi}_\mu^i \gamma^{[a}\psi_i^{b]} + \bar{\psi}_{\mu i}\gamma^{[a}\psi^{b]i} + \bar{\psi}_i^{[a}\gamma_\mu\psi^{b]i}\right],$$

$$f_\mu{}^a = -\tfrac{1}{4}\hat{R}_\mu{}^a + \tfrac{1}{24}e_\mu{}^a\hat{R} + \tfrac{1}{4}i\widetilde{\hat{R}}_\mu{}^a(T) - \tfrac{1}{16}T_{c\mu}^-T^{+ac} - \tfrac{1}{8}e_\mu{}^a D,$$

$$\phi_\mu^i = -\tfrac{1}{2}\gamma^\nu \hat{R}'_{\mu\nu}(Q^i) + \tfrac{1}{12}\gamma_\mu\gamma^{ab}\hat{R}_{ab}(Q^i) + \tfrac{1}{4}\gamma_\mu\chi^i. \tag{2.92}$$

Here appears the dual of the U(1) curvature, and the Ricci tensor (2.71) of

$$\hat{R}'_{\mu\nu}\left(M^{ab}\right) = r_{\mu\nu}(M^{ab}) + \left[\bar{\psi}_{[\mu}^i\gamma_{\nu]}\left(\tfrac{3}{4}\gamma^{ab}\chi_i + \hat{R}^{ab}(Q_i)\right)\right.$$
$$\left. + \tfrac{1}{4}\bar{\psi}_\mu^i\psi_\nu^j\varepsilon_{ij}T^{+ab} + \text{h.c.}\right],$$

$$r_{\mu\nu}(M^{ab}) = 2\partial_{[\mu}\omega_{\nu]}{}^{ab} + 2\omega_{[\mu}{}^a{}_c\omega_{\nu]}{}^{cb} - \bar{\psi}_{[\mu}^i\gamma^{ab}\phi_{\nu]i} - \bar{\psi}_{i[\mu}\gamma^{ab}\phi_{\nu]}^i, \tag{2.93}$$

and

$$\hat{R}'_{\mu\nu}(Q^i) = 2\left(\partial_{[\mu} + \tfrac{1}{2}b_{[\mu} + \tfrac{1}{4}\gamma_{ab}\omega_{[\mu}{}^{ab} - \tfrac{1}{2}iA_{[\mu}\right)\psi_{\nu]}^i + 2V_{[\mu}{}^i{}_j\psi_{\nu]}^j$$
$$- \tfrac{1}{8}\gamma^{ab}T_{ab}^-\varepsilon^{ij}\gamma_{[\mu}\psi_{\nu]j}. \tag{2.94}$$

We present the transformations of the constrained fields in 3 lines: first the original transformations, then the M-transformations determined by the non-invariance of the constraints, and finally terms that represent modified structure functions:

$$\delta_{Q,S}(\epsilon, \eta)\omega_\mu{}^{ab} = \tfrac{1}{2}\bar{\epsilon}^i\gamma^{ab}\phi_{\mu i} + \tfrac{1}{2}\bar{\eta}^i\gamma^{ab}\psi_{\mu i}$$
$$- \tfrac{3}{8}\bar{\epsilon}^i\gamma_\mu\gamma^{ab}\chi_i - \tfrac{1}{2}\bar{\epsilon}^i\gamma_\mu\hat{R}^{ab}(Q)_i$$
$$- \tfrac{1}{4}\bar{\epsilon}^iT^{+ab}\varepsilon_{ij}\psi_\mu^j + \text{h.c.},$$

$$\delta_{Q,S}(\epsilon, \eta)f_\mu{}^a = \tfrac{1}{2}\bar{\eta}^i\gamma^a\phi_{\mu i}$$
$$- \tfrac{3}{16}e_\mu{}^a\bar{\epsilon}^i\mathcal{D}\chi_i + \tfrac{1}{4}\bar{\epsilon}^i\gamma_\mu\mathcal{D}_b\hat{R}^{ba}(Q)_i$$
$$- \tfrac{1}{8}\bar{\epsilon}^i\psi_\mu^j\mathcal{D}_bT^{+ba}\varepsilon_{ij} - \tfrac{3}{16}\bar{\epsilon}^i\gamma^a\psi_{\mu i}D + \text{h.c.},$$

$$\delta_{Q,S}(\epsilon, \eta)\phi_\mu^i = \left(\partial_\mu - \tfrac{1}{2}b_\mu + \tfrac{1}{4}\gamma^{ab}\omega_{\mu ab} - \tfrac{1}{2}iA_\mu\right)\eta^i + V_\mu{}^i{}_j\eta^j - f_\mu^a\gamma_a\epsilon^i$$
$$- \tfrac{1}{32}\mathcal{D}T^-\cdot\gamma\gamma_\mu\varepsilon^{ij}\epsilon_j - \tfrac{1}{8}\hat{R}_j^i\cdot\gamma\gamma_\mu\epsilon^j - \tfrac{1}{8}i\hat{R}(T)\cdot\gamma\gamma_\mu\epsilon^i$$
$$+ \tfrac{3}{8}\left[\left(\bar{\chi}_j\gamma^a\epsilon^j\right)\gamma_a\psi_\mu^i - \left(\bar{\chi}_j\gamma^a\psi_\mu^j\right)\gamma_a\epsilon^i\right]. \tag{2.95}$$

The third line of each of the (2.95) show that the commutator between two supersymmetries is modified as follows:

$$[\delta_Q(\epsilon_1), \delta_Q(\epsilon_2)] = \delta_{\text{cgct}}\left(\xi_3^a\right) + \delta_M\left(\lambda_3^{ab}\right) + \delta_K\left(\lambda_{K3}^a\right) + \delta_S\left(\eta_3^i\right), \qquad (2.96)$$

where the associated parameters are given by the following expressions:

$$\xi_3^a = \tfrac{1}{2}\bar{\epsilon}_2^i\gamma^a\epsilon_{1i} + \text{h.c.},$$

$$\lambda_3^{ab} = \tfrac{1}{4}\bar{\epsilon}_1^i\epsilon_2^j\,T^{+ab}\varepsilon_{ij} + \text{h.c.},$$

$$\Lambda_{K3}^a = \tfrac{1}{8}\bar{\epsilon}_1^i\epsilon_2^j\,\mathcal{D}_b T^{+ba}\varepsilon_{ij} + \tfrac{3}{16}\bar{\epsilon}_2^i\gamma^a\epsilon_{1i}\,D + \text{h.c.},$$

$$\eta_3^i = \tfrac{3}{4}\,\bar{\epsilon}_{[1}^i\epsilon_{2]}^j\,\chi_j. \qquad (2.97)$$

These extra terms become important in applications where they can give rise to central charges if the fields appearing in the structure functions get non-zero vacuum expectation values. We will see that in the presence of vector multiplets, there appear extra terms of a similar nature involving the scalars of the vector multiplet.

2.6.3　D = 5

We report here the Q- and S-supersymmetry transformation laws of the independent fields using again the same splitting in two lines for gauge terms and matter terms[11]

$$\delta_{Q,S}(\epsilon, \eta)e_\mu{}^a = \tfrac{1}{2}\bar{\epsilon}\gamma^a\psi_\mu,$$

$$\delta_{Q,S}(\epsilon, \eta)b_\mu = \tfrac{1}{2}i\bar{\epsilon}\phi_\mu + \tfrac{1}{2}i\bar{\eta}\psi_\mu$$
$$-2\bar{\epsilon}\gamma_\mu\chi,$$

$$\delta_{Q,S}(\epsilon, \eta)V_\mu{}^{ij} = -\tfrac{3}{2}i\bar{\epsilon}^{(i}\phi_\mu^{j)} + \tfrac{3}{2}i\bar{\eta}^{(i}\psi_\mu^{j)}$$
$$+4\bar{\epsilon}^{(i}\gamma_\mu\chi^{j)} + i\bar{\epsilon}^{(i}\gamma^{ab}T_{ab}\psi_\mu^{j)},$$

$$\delta_{Q,S}(\epsilon, \eta)\psi_\mu^i = \partial_\mu\epsilon^i + \tfrac{1}{2}b_\mu\epsilon^i + \tfrac{1}{4}\omega_\mu^{ab}\gamma_{ab}\epsilon^i - V_\mu^{ij}\epsilon_j - i\gamma_\mu\eta^i$$
$$+i\gamma^{ab}T_{ab}\gamma_\mu\epsilon^i,$$

$$\delta_{Q,S}(\epsilon, \eta)T_{ab} = \tfrac{1}{2}i\bar{\epsilon}\gamma_{ab}\chi - \tfrac{3}{32}i\bar{\epsilon}\widehat{R}_{ab}(Q),$$

$$\delta_{Q,S}(\epsilon, \eta)\chi^i = \tfrac{1}{4}\epsilon^i D - \tfrac{1}{64}\gamma^{ab}\widehat{R}_{ab}^{ij}(V)\epsilon_j + \tfrac{1}{8}i\gamma^{ab}\mathcal{D}T_{ab}\epsilon^i - \tfrac{1}{8}i\gamma^a\mathcal{D}^b T_{ab}\epsilon^i$$

---

[11] For $\delta\chi$ the split in two lines is accidental due to the length of the expression.

$$-\tfrac{1}{4}\gamma^{abcd}T_{ab}T_{cd}\epsilon^i + \tfrac{1}{6}T^2\epsilon^i + \tfrac{1}{4}\gamma^{ab}T_{ab}\eta^i \,,$$

$$\delta_{Q,S}(\epsilon,\eta)D = \bar{\epsilon}\slashed{\mathcal{D}}\chi - \tfrac{5}{3}\mathrm{i}\bar{\epsilon}\gamma^{ab}T_{ab}\chi - \mathrm{i}\bar{\eta}\chi \,. \tag{2.98}$$

In 5 dimensions, not much simplifications are possible by taking appropriate C in (2.68) or ρ in (2.69), so that we just took the constraints [18]

$$R^a_{\mu\nu}(P) = 0\,, \qquad \gamma^b\widehat{R}_{ba}(Q)^i = 0\,, \qquad \widehat{R}_{ac}(M^{bc}) = 0\,. \tag{2.99}$$

Following the methods in Sect. 2.4, we find expressions for the dependent gauge fields: gauge fields associated with S and K symmetries, respectively:

$$\omega_\mu{}^{ab} = \omega_\mu{}^{ab}(e,b) - \tfrac{1}{2}\bar{\psi}^{[b}\gamma^{a]}\psi_\mu - \tfrac{1}{4}\bar{\psi}^b\gamma_\mu\psi^a \,,$$

$$\phi^i_\mu = \tfrac{1}{3}\mathrm{i}\gamma^a\,\widehat{R}'_{\mu a}{}^i(Q) - \tfrac{1}{24}\mathrm{i}\gamma_\mu\gamma^{ab}\widehat{R}'_{ab}{}^i(Q)\,,$$

$$\widehat{R}'_{\mu\nu}{}^i(Q) = 2\partial_{[\mu}\psi^i_{\nu]} + \tfrac{1}{2}\omega_{[\mu}{}^{ab}\gamma_{ab}\psi^i_{\nu]} + b_{[\mu}\psi^i_{\nu]} - 2V_{[\mu}{}^{ij}\psi_{\nu]j} + 2\mathrm{i}\gamma\cdot T\gamma_{[\mu}\psi^i_{\nu]}\,,$$

$$f_a{}^a = \tfrac{1}{16}\left(-R(\omega) - \tfrac{1}{3}\bar{\psi}_\rho\gamma^{\rho\mu\nu}\mathcal{D}_\mu\psi_\nu\right.$$

$$\left. + \tfrac{1}{3}\bar{\psi}^i_a\gamma^{abc}\psi^j_b V_{cij} + 16\bar{\psi}_a\gamma^a\chi - 4\mathrm{i}\bar{\psi}^a\psi^b T_{ab} + \tfrac{4}{3}\mathrm{i}\bar{\psi}^b\gamma_{abcd}\psi^a T^{cd}\right)\,. \tag{2.100}$$

The full commutator of two supersymmetry transformations is

$$\left[\delta_Q(\epsilon_1), \delta_Q(\epsilon_2)\right] = \delta_{\mathrm{cgct}}\left(\xi_3^\mu\right) + \delta_M\left(\lambda_3^{ab}\right) + \delta_S(\eta_3) + \delta_U\left(\lambda_3^{ij}\right) + \delta_K\left(\lambda_{\mathrm{K}3}^a\right)\,. \tag{2.101}$$

The covariant general coordinate transformations have been defined in (2.14). The parameters appearing in (2.101) are

$$\xi_3^\mu = \tfrac{1}{2}\bar{\epsilon}_2\gamma_\mu\epsilon_1\,,$$

$$\lambda_3^{ab} = -\mathrm{i}\bar{\epsilon}_2\gamma^{[a}\gamma^{cd}T_{cd}\gamma^{b]}\epsilon_1\,,$$

$$\lambda_3^{ij} = \mathrm{i}\bar{\epsilon}_2^{(i}\gamma^{ab}T_{ab}\epsilon_1^{j)}\,,$$

$$\eta_3^i = -\tfrac{9}{4}\mathrm{i}\,\bar{\epsilon}_2\epsilon_1\chi^i + \tfrac{7}{4}\mathrm{i}\,\bar{\epsilon}_2\gamma_c\epsilon_1\gamma^c\chi^i$$

$$+ \tfrac{1}{4}\mathrm{i}\,\bar{\epsilon}_2^{(i}\gamma_{cd}\epsilon_1^{j)}\left(\gamma^{cd}\chi_j + \tfrac{1}{4}\,\widehat{R}^{cd}{}_j(Q)\right)\,,$$

$$\lambda_{\mathrm{K}3}^a = -\tfrac{1}{2}\bar{\epsilon}_2\gamma^a\epsilon_1 D + \tfrac{1}{96}\bar{\epsilon}_2^i\gamma^{abc}\epsilon_1^j\widehat{R}_{bcij}(V)$$

$$+ \tfrac{1}{12}\mathrm{i}\bar{\epsilon}_2\left(-5\gamma^{abcd}\mathcal{D}_b T_{cd} + 9\mathcal{D}_b T^{ba}\right)\epsilon_1$$

$$+ \bar{\epsilon}_2\left(\gamma^{abcde}T_{bc}T_{de} - 4\gamma^c T_{cd}T^{ad} + \tfrac{2}{3}\gamma^a T^2\right)\epsilon_1\,. \tag{2.102}$$

For the Q, S commutators we find the following algebra:

$$\left[\delta_S(\eta), \delta_Q(\epsilon)\right] = \delta_D\left(\tfrac{1}{2}i\bar{\epsilon}\eta\right) + \delta_M\left(\tfrac{1}{2}i\bar{\epsilon}\gamma^{ab}\eta\right) + \delta_U\left(-\tfrac{3}{2}i\bar{\epsilon}^{(i}\eta^{j)}\right) + \delta_K\left(\lambda_{K3}^a\right),$$

$$\left[\delta_S(\eta_1), \delta_S(\eta_2)\right] = \delta_K\left(\tfrac{1}{2}\bar{\eta}_2\gamma^a\eta_1\right), \tag{2.103}$$

with

$$\lambda_{K3}^a = \tfrac{1}{6}\bar{\epsilon}\left(\gamma^{bc}T_{bc}\gamma_a - \tfrac{1}{2}\gamma_a\gamma^{bc}T_{bc}\right)\eta. \tag{2.104}$$

For practical purposes (see how to calculate transformations of covariant derivatives), it is useful to give the extra parts of the transformation laws of dependent gauge fields, i.e. the parts denoted by $\mathcal{M}_{\mu J}{}^I$ in (2.24). These are (for the gauge field of special conformal transformations we suffice by giving the transformation of the trace, as this is what one often needs)

$$\delta_{Q,S}(\epsilon, \eta)\omega_\mu{}^{ab} = \cdots - \tfrac{1}{2}\bar{\epsilon}\gamma^{[a}\widehat{R}_\mu{}^{b]}(Q) - \tfrac{1}{4}\bar{\epsilon}\gamma_\mu\widehat{R}^{ab}(Q) - 4e_\mu{}^{[a}\bar{\epsilon}\gamma^{b]}\chi,$$

$$\delta_{Q,S}(\epsilon, \eta)\phi_\mu^i = \cdots - \tfrac{1}{12}i\left\{\gamma^{ab}\gamma_\mu - \tfrac{1}{2}\gamma_\mu\gamma^{ab}\right\}\widehat{R}_{ab}{}^i{}_j(V)\epsilon^j +$$

$$+\tfrac{1}{3}\left[\mathcal{D}\gamma^{ab}T_{ab}\gamma_\mu - \mathcal{D}_\mu\gamma^{ab}T_{ab} + \gamma_\mu\gamma^c\mathcal{D}^a T_{ac}\right]\epsilon^i +$$

$$+i\left[-\gamma_{\mu abcd}T^{ab}T^{cd} + 8\gamma_\rho T^{\rho\sigma}T_{\mu\sigma} - 2\gamma_\mu T^2\right]\epsilon^i +$$

$$+\tfrac{1}{3}i\left(8\gamma^b T_{\mu b} - \gamma_\mu\gamma\cdot T\right)\eta^i,$$

$$\delta_S(\eta)f_a{}^a = -5i\bar{\eta}\chi, \tag{2.105}$$

with $\gamma\cdot T = \gamma^{ab}T_{ab}$ and $T^2 = T^{ab}T_{ab}$. Note that there are other terms, which however are proportional to gauge fields, determined by the algebra. For example, the λ_3^{ab} expression in (2.102) implies that the supersymmetry transformation of the spin connection contains a term

$$\delta_Q(\epsilon)\omega_\mu{}^{ab} = \cdots - i\bar{\epsilon}\gamma^{[a}\gamma^{cd}T_{cd}\gamma^{b]}\psi_\mu. \tag{2.106}$$

2.6.4 $D = 6$

The transformation laws of the independent fields are

$$\delta_{Q,S}(\epsilon, \eta)e_\mu{}^a = \tfrac{1}{2}\bar{\epsilon}\gamma^a\psi_\mu,$$

$$\delta_{Q,S}(\epsilon, \eta)b_\mu = -\tfrac{1}{2}\bar{\epsilon}\phi_\mu + \tfrac{1}{2}\bar{\eta}\psi_\mu$$

$$-\tfrac{1}{24}\bar{\epsilon}\gamma_\mu\chi,$$

$$\delta_{Q,S}(\epsilon, \eta) V_\mu{}^{ij} = 2\bar{\epsilon}^{(i}\phi_\mu^{j)} + 2\bar{\eta}^{(i}\psi_\mu^{j)}$$

$$+ \tfrac{1}{6}\bar{\epsilon}^{(i}\gamma_\mu\chi^{j)},$$

$$\delta_{Q,S}(\epsilon, \eta)\psi_\mu^i = \left(\partial_\mu + \tfrac{1}{2}b_\mu + \tfrac{1}{4}\gamma^{ab}\omega_{\mu ab}\right)\epsilon^i + V_\mu{}^i{}_j\epsilon^j$$

$$+ \tfrac{1}{24}\gamma \cdot T^-\gamma_\mu\epsilon^i + \gamma_\mu\eta^i,$$

$$\delta_{Q,S}(\epsilon, \eta)T^-_{abc} = -\tfrac{1}{32}\bar{\epsilon}\gamma^{de}\gamma_{abc}\widehat{R}_{de}(Q) - \tfrac{7}{96}\bar{\epsilon}\gamma_{abc}\chi,$$

$$\delta_{Q,S}(\epsilon, \eta)\chi^i = +\tfrac{1}{8}(\mathcal{D}_\mu\gamma \cdot T^-)\gamma^\mu\epsilon^i - \tfrac{3}{8}\widehat{R}^{ij}(V) \cdot \gamma\epsilon_j + \tfrac{1}{4}D\epsilon^i + \tfrac{1}{2}\gamma \cdot T^-\eta^i,$$

$$\delta_{Q,S}(\epsilon, \eta)D = \bar{\epsilon}^i\mathcal{D}\chi_i - 2\bar{\eta}\chi, \tag{2.107}$$

where $\gamma \cdot T \equiv \gamma^{abc}T_{abc}, \dots$.

The constraints that we took in 6 dimensions are

$$0 = R^a_{\mu\nu}(P),$$

$$0 = \gamma^b\widehat{R}_{ba}(Q)^i + \tfrac{1}{6}\gamma_a\chi^i,$$

$$0 = \widehat{R}_{ac}\left(M^{bc}\right) - T^-_{acd}T^{-bcd} + \tfrac{1}{12}\delta_a{}^bD. \tag{2.108}$$

The last equation contains a sign correction[12] to the equation in [5].

The transformations of the dependent gauge fields $\omega_\mu{}^{ab}$ and ϕ_μ^i contain as covariant terms (terms not determined by the algebra)

$$\delta_Q(\epsilon)\omega_\mu{}^{ab} = \cdots - \tfrac{1}{2}\bar{\epsilon}\gamma^{[a}\widehat{R}_\mu{}^{b]}(Q) - \tfrac{1}{4}\bar{\epsilon}\gamma_\mu\widehat{R}^{ab}(Q) - \tfrac{1}{12}e_\mu{}^{[a}\bar{\epsilon}\gamma^{b]}\chi,$$

$$\delta_Q(\epsilon)\phi_\mu^i = \cdots + \tfrac{1}{32}\left\{\gamma^{ab}\gamma_\mu - \tfrac{1}{2}\gamma_\mu\gamma^{ab}\right\}\widehat{R}_{ab}{}^i{}_j(V)\epsilon^j +$$

$$- \tfrac{1}{96}\left(\mathcal{D}\gamma^{abc}T^-_{abc}\gamma_\mu\right)\epsilon^i. \tag{2.109}$$

In this case the constraints are also S-invariant, and thus there are no extra S-supersymmetry transformations.

The algebra is only modified in the anticommutator of two Q-supersymmetries:

$$[\delta_Q(\epsilon_1), \delta_Q(\epsilon_2)] = \delta_{cgct}\left(\tfrac{1}{2}\bar{\epsilon}_2\gamma_\mu\epsilon_1\right) + \delta_M\left(\tfrac{1}{2}\bar{\epsilon}_2\gamma_c\epsilon_1 T^{-abc}\right) + \delta_S\left(\tfrac{1}{24}\gamma_a\chi^i\bar{\epsilon}_2\gamma^a\epsilon_1\right)$$

$$+ \delta_K\left(-\tfrac{1}{8}\bar{\epsilon}_2\gamma_b\epsilon_1\left(\mathcal{D}_cT^{-abc} + \tfrac{1}{12}\eta^{ab}D\right)\right). \tag{2.110}$$

---

[12]We thank T. Kugo for this correction.

Again, this implies that the transformations of gauge fields contain extra terms with gauge fields as e.g. there is in the transformation of the spin connection $+\frac{1}{2}\bar{\epsilon}\gamma_c\psi_\mu T^{-abc}$.

References

1. D.Z. Freedman, A. Van Proeyen, *Supergravity* (Cambridge University, Cambridge, 2012). http://www.cambridge.org/mw/academic/subjects/physics/theoretical-physics-and-mathematical-physics/supergravity?format=AR
2. B. de Wit, D.Z. Freedman, Combined supersymmetric and gauge-invariant field theories. Phys. Rev. **D12**, 2286 (1975). https://doi.org/10.1103/PhysRevD.12.2286
3. R. Jackiw, Gauge-covariant conformal transformations. Phys. Rev. Lett. **41**, 1635 (1978). https://doi.org/10.1103/PhysRevLett.41.1635
4. P. van Nieuwenhuizen, Constraints in conformal simple supergravity, in *From* SU(3) *to gravity, Festschrift in honor of Y. Ne'eman*, eds. by E. Gotsman, G. Tauber (Cambridge University, Cambridge, 1985), pp. 369-382
5. E. Bergshoeff, E. Sezgin, A. Van Proeyen, Superconformal tensor calculus and matter couplings in six dimensions. Nucl. Phys. **B264**, 653 (1986). https://doi.org/10.1016/0550-3213(86)90503-1, [Erratum: Nucl. Phys. B **598**, 667(2001)]
6. D.Z. Freedman, P. van Nieuwenhuizen, S. Ferrara, Progress toward a theory of supergravity. Phys. Rev. **D13**, 3214–3218 (1976). https://doi.org/10.1103/PhysRevD.13.3214
7. S. Deser, B. Zumino, Consistent Supergravity. Phys. Lett. **62B**, 335 (1976). https://doi.org/10.1016/0370-2693(76)90089-7, [335(1976)]
8. P. Van Nieuwenhuizen, Supergravity. Phys. Rept. **68**, 189–398 (1981). https://doi.org/10.1016/0370-1573(81)90157-5
9. E.S. Fradkin, M.A. Vasiliev, Minimal set of auxiliary fields and S matrix for extended supergravity. Lett. Nuovo Cim. **25**, 79–90 (1979). https://doi.org/10.1007/BF02776267
10. E.S. Fradkin, M.A. Vasiliev, Minimal set of auxiliary fields in SO(2) extended supergravity. Phys. Lett. **B85**, 47–51 (1979)
11. B. de Wit, J.W. van Holten, Multiplets of linearized SO(2) supergravity. Nucl. Phys. **B155**, 530–542 (1979). https://doi.org/10.1016/0550-3213(79)90285-2
12. M. Kaku, P.K. Townsend, P. van Nieuwenhuizen, Properties of conformal supergravity. Phys. Rev. **D17**, 3179–3187 (1978). https://doi.org/10.1103/PhysRevD.17.3179
13. S. Ferrara, B. Zumino, Transformation properties of the supercurrent. Nucl. Phys. **B87**, 207 (1975). https://doi.org/10.1016/0550-3213(75)90063-2
14. M.F. Sohnius, The multiplet of currents for $N = 2$ extended supersymmetry. Phys. Lett. **81B**, 8–10 (1979). https://doi.org/10.1016/0370-2693(79)90703-2
15. E. Bergshoeff, M. de Roo, B. de Wit, Extended conformal supergravity. Nucl. Phys. **B182**, 173 (1981). https://doi.org/10.1016/0550-3213(81)90465-X
16. E.A. Bergshoeff, *Conformal Invariance in Supergravity*, PhD thesis, Leiden University, 1983
17. E. Bergshoeff, E. Sezgin, A. Van Proeyen, (2, 0) tensor multiplets and conformal supergravity in $D = 6$. Class. Quant. Grav. **16**, 3193–3206 (1999). https://doi.org/10.1088/0264-9381/16/10/311, arXiv:hep-th/9904085 [hep-th]
18. E. Bergshoeff, S. Cucu, M. Derix, T. de Wit, R. Halbersma, A. Van Proeyen, Weyl multiplets of $N = 2$ conformal supergravity in five dimensions. J. High Energy Phys. **06**, 051 (2001). https://doi.org/10.1088/1126-6708/2001/06/051, arXiv:hep-th/0104113 [hep-th]
19. D. Butter, S. Hegde, I. Lodato, B. Sahoo, $N = 2$ dilaton Weyl multiplet in 4D supergravity. J. High Energy Phys. **03**, 154 (2018). https://doi.org/10.1007/JHEP03(2018)154, arXiv:1712.05365 [hep-th]

20. W. Siegel, Curved extended superspace from Yang–Mills theory à la strings. Phys. Rev. **D53**, 3324–3336 (1996). https://doi.org/10.1103/PhysRevD.53.3324, arXiv:hep-th/9510150 [hep-th]
21. B. de Wit, J.W. van Holten, A. Van Proeyen, Transformation rules of $N = 2$ supergravity multiplets. Nucl. Phys. **B167**, 186–204 (1980). https://doi.org/10.1016/0550-3213(80)90125-X

Chapter 3
Matter Multiplets

Abstract After the Weyl multiplet is introduced, we can now define matter multiplets whose transformations respect the algebra with structure functions that depend on the fields of the Weyl multiplet. We treat here vector multiplets and hypermultiplets. We define them for $D = 4, 5$ and 6, first for rigid supersymmetry and then for the superconformal theory. In the second part of this chapter we define actions for these multiplets, which will be the basis for the further chapters.

The goal of this chapter is to construct local superconformal actions for the matter multiplets, exploiting our knowledge of the Weyl multiplet. In principle there are many representations of the superconformal algebra that define matter multiplets. The physical theories for $D = 4$ can all be obtained with vector multiplets and hypermultiplets. For $D = 5$ and $D = 6$ tensor multiplets can lead to inequivalent theories. For $D = 5$, this has been included in the treatments of [1–5], to which we will come back to this in Sect. 6.2. One might also prefer formulations in terms of other multiplets to make connections with other descriptions, e.g. in string theory. We will briefly discuss the $D = 4$ tensor multiplet in a superconformal background [6] in Sect. 3.2.5. The action with one tensor multiplet was given in [7] and extended to more multiplets in [8]. On-shell matter couplings using different formalisms have been given in [9–11]. Recently [12], also the (off-shell) coupling of one tensor multiplet to an arbitrary number of vector multiplets has been obtained.

The main focus of this chapter will be on vector and hypermultiplets. Importantly, the latter will be used not only as physical multiplets, but also as compensating multiplets to describe super-Poincaré theories with matter couplings. This is in the spirit of the general strategy outlined in Sect. 1.2.2 that we review in Sect. 3.1.

The remainder of this chapter is split in two parts. In Sect. 3.2 we explain the structure of first vector and then hypermultiplets and their embedding in the superconformal algebra. The construction of actions is postponed to the second part, Sect. 3.3. We explicitly construct the superconformal invariant actions for sets of these multiplets, which will be combined in Chap. 4 by the gauge fixing to Poincaré supergravity. Many parts of this chapter, especially for the case of $D = 4$, have been obtained in the context of the master thesis of De Rydt and Vercnocke [13].

© Springer Nature Switzerland AG 2020

E. Lauria, A. Van Proeyen, $\mathcal{N} = 2$ *Supergravity in $D = 4, 5, 6$ Dimensions*,
Lecture Notes in Physics 966, https://doi.org/10.1007/978-3-030-33757-5_3

3.1 Review of the Strategy

In Sect. 1.2.2, we already outlined the general idea of the superconformal construction for actions with super-Poincaré invariance. At that time, we had not yet
explained the gauging of the conformal algebra. Now we can be more precise. For
this example, we will still restrict to the bosonic case. Consider a scalar field ϕ with
Weyl weight w and no intrinsic special conformal transformations: $k_\mu(\phi) = 0$. Its
superconformal covariant derivative is

$$\mathcal{D}_\mu \phi = \left(\partial_\mu - w b_\mu\right) \phi. \tag{3.1}$$

The transformation of the covariant derivative $\mathcal{D}_a \phi$ can be easily obtained from
the 'easy method' (Sect. 2.3.4). One takes into account (2.25) to find that there is a
K transformation. The transformation law of a covariant derivative determines the
covariant box

$$\Box^C \phi \equiv \eta^{ab} \mathcal{D}_b \mathcal{D}_a \phi = e^{a\mu} \left(\partial_\mu \mathcal{D}_a \phi - (w+1) b_\mu \mathcal{D}_a \phi + \omega_{\mu\, ab} \mathcal{D}^b \phi + 2w f_{\mu a} \phi\right)$$

$$= e^{-1} \left(\partial_\mu - (w+2-D) b_\mu\right) e g^{\mu\nu} \left(\partial_\nu - w b_\nu\right) \phi - \frac{w}{2(D-1)} R \phi. \tag{3.2}$$

We use here the constraint (2.72) (without matter for the pure bosonic case). The
last term is the well-known $R/6$ term in $D = 4$. In fact, choosing $w = \frac{D}{2} - 1$, one
has a conformal invariant scalar action

$$S = \int \mathrm{d}^D x \, e \phi \Box^C \phi. \tag{3.3}$$

Exercise 3.1 Show that $\int \mathrm{d}^D x \, e \mathcal{D}_a \phi \, \mathcal{D}^a \phi$ is not a special conformal invariant,
while $\Box \phi$ is invariant under K iff $w = \frac{D}{2} - 1$. $\Box$

In order to obtain a Poincaré invariant action, we have to break dilatations and
special conformal transformations (as these are not part of the Poincaré algebra).
Considering (2.25), it is clear that the latter can be broken by a gauge choice

$$K - \text{gauge}: \qquad b_\mu = 0. \tag{3.4}$$

One could take as gauge choice for dilatations a fixed value of a scalar ϕ. As a
consequence, the action (3.3) reduces to the Poincaré gravity action: only the frame
field of the 'Weyl multiplet' (which was in the background) remains.

The lesson to learn is: once the gauge for the superfluous symmetries in the
matter action is fixed, without considering any action for the Weyl multiplet, we
find kinetic terms for the gravity sector.

We can schematically summarize this procedure in the following diagram:

Weyl multiplet: $e_\mu{}^a$, b_μ (Background)

$+$

matter field: ϕ

$\downarrow$ gauge fixing K_a, D

Poincaré gravity $e_\mu{}^a$,   (3.5)

namely we introduce, in the background of the Weyl multiplet, the conformally invariant action of a matter field ϕ and we fix the gauge to get the action of Poincaré gravity. In the above scheme, the field ϕ provides the *compensating field* degree of freedom that makes the combined field gauge equivalent to an irreducible multiplet of Poincaré gravity. We remark that, at the classical level, every gauge fixing is equivalent to redefinitions of the fields. In this case, defining (the conformal invariant)

$$\tilde{g}_{\mu\nu} = g_{\mu\nu}\phi^{4/(D-2)} ,$$ (3.6)

and writing the action in terms of $R(\tilde{g})$, the field ϕ disappears from the action

$$S = -\frac{D-2}{4(D-1)} \int d^D x \sqrt{\tilde{g}}\, R(\tilde{g}) .$$ (3.7)

The absence of ϕ from the action above is just a consequence of dilatational invariance

$$\int d^D x \left[\frac{\delta S(\tilde{g}, \phi)}{\delta \tilde{g}_{\mu\nu}(x)} \delta_D \tilde{g}_{\mu\nu}(x) + \frac{\delta S(\tilde{g}, \phi)}{\delta \phi(x)} \delta_D \phi(x) \right] = 0 ,$$ (3.8)

which, together with $\delta_D \tilde{g}_{\mu\nu} = 0$, implies $S(\tilde{g}, \phi) \equiv S(\tilde{g})$.

3.2 Conformal Properties of the Multiplets

Having the Weyl multiplet, the further step now is to introduce other multiplets in the background of the Weyl multiplet. The resulting algebra, which depends for part on the fields of the Weyl multiplet, is fixed for what concerns the superconformal transformations. On the other hand, extra terms with gauge transformations of extra vectors or antisymmetric tensors may still appear in the algebra. As long as the fields of the Weyl multiplet are inert under these transformations (as we will impose by hypothesis), these extra transformations do not modify our previous results.

A first modification of this structure is obtained by the introduction of a gauge vector multiplet. The commutator of the supersymmetries can still be modified by a gauge transformation that depends on fields of this vector multiplet. For this structure to make sense, the algebra of the Weyl multiplet had to close without using an equation of motion. Furthermore, as long as the vector multiplet is well defined off-shell, a matter multiplet (in the background of both the vector and Weyl multiplet) may now be introduced whose algebra closes only modulo equations of motion.

All fields in 'matter multiplets' will now have to obey the same 'soft' algebra defined by the Weyl multiplet. A first step is to define their transformations under the bosonic symmetries. We assume the rules (2.87) and (2.88) under Weyl and chiral transformations, where the weights will be given in Table 3.1. The R-symmetry SU(2) transformation is implicit in the index structure of the fields.

Table 3.1 Fields in some superconformal matter multiplets

field	w	c	#	field	w	#	field	w	#	SU(2)	γ_*
	D = 4			**D = 5**			**D = 6**				
Off-shell vector multiplet											
X	1	1	2	σ	1	1				1	
W_μ	0	0	3	W_μ	0	4	W_μ	0	5	1	
Y_{ij}	2	0	3	Y_{ij}	2	3	Y_{ij}	2	3	3	
Ω_i	3/2	1/2	8	ψ_i	3/2	8	λ_i	3/2	8	2	+
On-shell tensor multiplet											
				$B_{\mu\nu}$	0	3	$B_{\mu\nu}$	0	3	1	
				ϕ	1	1	σ	2	1	1	
				λ^i	3/2	4	ψ^i	5/2	4	2	
On-shell hypermultiplet											
q^X	1	0	4	q^X	3/2	4	q^X	2	4	2	
ζ^A	3/2	−1/2	4	ζ^A	2	4	ζ^A	5/2	4	1	+
Off-shell chiral multiplet											
A	w	w	2							1	
B_{ij}	$w+1$	$w-1$	6							3	
G^-_{ab}	$w+1$	$w-1$	6							1	
C	$w+2$	$w-2$	2							1	
Ψ_i	$w+\frac{1}{2}$	$w-\frac{1}{2}$	8							2	+
Λ_i	$w+\frac{3}{2}$	$w-\frac{3}{2}$	8							2	−
Off-shell linear multiplet											
L_{ij}	2	0	3	L_{ij}	3	3	L_{ij}	4	3	3	
E_a	3	0	3	E_a	4	4	E_a	5	5	1	
G	3	−1	2	N	4	1				1	
φ_i	5/2	1/2	8	φ^i	7/2	8	φ^i	9/2	8	2	−

We indicate for each dimension the Weyl weight (and for $D = 4$ chiral weight), the number of real degrees of freedom, the SU(2) representations, which is the same in any dimension, and the chirality for $D = 4$ and $D = 6$. For each multiplet we give first the bosonic fields, and then the fermionic fields (below the line)

3.2.1 Vector Multiplets

Vector multiplets can first be defined in 6 dimensions, and then reduced to 5 or 4 dimensions.

3.2.1.1 Vector Multiplet in 6 Dimensions (Abelian Case)

Consider the vector multiplet in $D = 6$, which has already been introduced in Sect. 2.3.2. It has been shown in (2.59) that the supersymmetry transformations do not close. The solution to this issue is well-known: the 5 bosonic components of the gauge vector, and the 8 components of the spinor, need an SU(2)-triplet of real scalars, $Y^{(ij)}$. The latter will appear in the transformation law of the fermion.

As an illustrative example, let us show how the transformation laws of the $D = 6$ vector multiplet have been determined with methods that can be used in general. In general, it is useful to first consider the Weyl weights of the fields. One useful principle is that gauge fields (beyond the Weyl multiplet) should have Weyl weight 0, as all transformations beyond the superconformal group must commute with the conformal generators. Equivalently, all the parameters beyond the superconformal group have to be considered[1] as Weyl weight 0.

For the U(1) gauge vector W_μ, whose abelian gauge transformation is $\delta_G W_\mu = \partial_\mu \theta$, the previous argument implies that W_μ has Weyl weight 0.[2] The same argument holds in fact for any gauge field, or gauge two-form, ... Then the associated curvature F_{ab} has Weyl weight 2 (due to the frame fields involved in $F_{ab} = e_a{}^\mu e_b{}^\nu F_{\mu\nu}$). As we have explained, these are the covariant quantities that should appear in the transformations of other matter fields. The supersymmetry parameter ϵ should be considered to be of Weyl weight $-\frac{1}{2}$, identical to its gauge field ψ_μ. Thus the supersymmetry transformation of the gaugino to the field strength of the gauge field determines that the conformal weight of λ is indeed $\frac{3}{2}$.

Exercise 3.2 Determine the same result from the transformation of the gauge field to the gaugino. □

The auxiliary field Y^{ij} can appear in the transformation of the fermion via an extra term $\delta \lambda^i = Y^{ij} \epsilon_j$, hence the auxiliary field should be of Weyl weight 2. In its supersymmetry transformation law can appear a covariant fermionic object of Weyl weight $\frac{5}{2}$. This is consistent with a transformation to the covariant derivative of the

---

[1] In principle parameters do not transform, but the commutators of symmetries can be stated in these terms.

[2] We could straightforwardly have generalized to a non-abelian algebra. We will do this below for $D = 5$ and $D = 4$.

gaugino, in order to cancel (2.59). The full transformation laws are

$$\delta W_\mu = \partial_\mu \theta + \tfrac{1}{2}\bar{\epsilon}\gamma_\mu \lambda \,,$$

$$\delta\lambda^i = \left(\tfrac{3}{2}\lambda_{\mathrm{D}} - \tfrac{1}{4}\gamma^{ab}\lambda_{ab}\right)\lambda^i + \lambda^{ij}\lambda_j - \tfrac{1}{4}\gamma^{ab}\widehat{F}_{ab}\epsilon^i - Y^{ij}\epsilon_j \,,$$

$$\delta Y^{ij} = 2\lambda_{\mathrm{D}}Y^{ij} + 2\lambda^{k(i}Y^{j)}{}_k - \tfrac{1}{2}\bar{\epsilon}^{(i}\slashed{\mathcal{D}}\lambda^{j)} + \bar{\eta}^{(i}\lambda^{j)} \,. \tag{3.9}$$

Starting from the rigid transformations, we replaced F_{ab} by the covariant expression $\widehat{F}_{ab}$ and the derivative of λ^j has been replaced by a covariant derivative.

$$\widehat{F}_{\mu\nu} = F_{\mu\nu} - \bar{\psi}_{[\mu}\gamma_{\nu]} \,,$$

$$\mathcal{D}_\mu \lambda^i = \left(\partial_\mu - \tfrac{3}{2}b_\mu + \tfrac{1}{4}\gamma^{ab}\omega_{\mu ab}\right)\lambda^i - V_\mu{}^{ij}\lambda_j + \tfrac{1}{4}\bar{\psi}^i_\mu\gamma^{ab}\widehat{F}_{ab} + Y^{ij}\psi_{\mu j} \,. \tag{3.10}$$

The consistency with Weyl weights does not leave place for other terms in the Q-transformations. Since the S-supersymmetry parameter η has to be considered as having Weyl weight $\tfrac{1}{2}$, the only S-transformation that can occur consistent with Weyl weights is the last term in (3.9). Its coefficient has to be fixed from calculating the $[\delta_Q(\epsilon), \delta_Q(\eta)]$ commutator on the gaugino or from the method in item (3) in Sect. 2.6.1. One can check that the extra terms from Y^{ij} cancel the non-closure terms (2.59).

Exercise 3.3 Check that all the transformation laws determine (and are consistent with) λ to be a left-chiral spinor, in accordance with Table 1.1. ☐

3.2.1.2 Vector Multiplet in 5 Dimensions

The transformations of the vector multiplet in 5 dimensions can be obtained from dimensional reduction[3] of the transformations for $D = 6$. Note that one component of the $D = 6$ vector is a real scalar σ in $D = 5$.

We will introduce here the vector multiplet in a non-abelian group, based on matrix representations with $[t_I, t_J] = f_{IJ}{}^K t_K$. Note that we will use the index I from now on to enumerate the vector multiplets, and thus the generators of the non-abelian algebra that can be gauged. We hope that this does not lead to confusion with the index I that was used so far to denote all standard gauge transformations as it was done in Chap. 2.

---

[3] The reader can easily find the linearized transformations from those in (3.9) using the rules in Appendix A.4. It may be more difficult to find the nonlinear transformations, since there are redefinitions such as $W_\mu(D = 6) = W_\mu(D = 5) + e^5_\mu \sigma$. It is easier to obtain the nonlinear transformations from directly imposing the supersymmetry algebra in $D = 5$.

The full rules can be found in [14, 5] for a generalization containing also tensor multiplets. For simplicity, we give here the supersymmetry transformations for only vector multiplets:

$$\delta W_\mu^I = \partial_\mu \theta^I - \theta^J W_\mu^K f_{JK}{}^I + \tfrac{1}{2}\bar{\epsilon}\gamma_\mu\psi^I - \tfrac{1}{2}i\sigma^I\bar{\epsilon}\psi_\mu,$$

$$\delta Y^{ijI} = -\tfrac{1}{2}\bar{\epsilon}^{(i}\mathcal{D}\psi^{j)I} + \tfrac{1}{2}i\bar{\epsilon}^{(i}\gamma\cdot T\psi^{j)I} - 4i\sigma^I\bar{\epsilon}^{(i}\chi^{j)},$$

$$+\tfrac{1}{2}i\bar{\epsilon}^{(i} f_{JK}{}^I \sigma^J \psi^{j)K} + \tfrac{1}{2}i\bar{\eta}^{(i}\psi^{j)I},$$

$$\delta\psi^{iI} = -\tfrac{1}{4}\gamma\cdot\widehat{F}^I\epsilon^i - \tfrac{1}{2}i\mathcal{D}\sigma^I\epsilon^i - Y^{ijI}\epsilon_j + \sigma^I\gamma\cdot T\epsilon^i + \sigma^I\eta^i,$$

$$\delta\sigma^I = \tfrac{1}{2}i\bar{\epsilon}\psi^I. \tag{3.11}$$

The (superconformal) covariant derivatives are given by

$$\mathcal{D}_\mu\sigma^I = D_\mu\sigma^I - \tfrac{1}{2}i\bar{\psi}_\mu\psi^I,$$

$$D_\mu\sigma^I = (\partial_\mu - b_\mu)\sigma^I - f_{JK}{}^I W_\mu^K\sigma^J,$$

$$\mathcal{D}_\mu\psi^{iI} = D_\mu\psi^{iI} + \tfrac{1}{4}\gamma\cdot\widehat{F}^I\psi_\mu^i + \tfrac{1}{2}i\mathcal{D}\sigma^I\psi_\mu^i + Y^{ijI}\psi_{\mu j} - \sigma^I\gamma\cdot T\psi_\mu^i,$$

$$+\tfrac{1}{2}f_{JK}{}^I\sigma^J\sigma^K\psi_\mu^i - \sigma^I\phi_\mu^i,$$

$$D_\mu\psi^{iI} = \left(\partial_\mu - \tfrac{3}{2}b_\mu + \tfrac{1}{4}\gamma_{ab}\hat{\omega}_\mu{}^{ab}\right)\psi^{iI} - V_\mu^{ij}\psi_j^I - f_{JK}{}^I W_\mu^K\psi^{iJ}. \tag{3.12}$$

with $\widehat{F}_{\mu\nu}^I$ given by

$$\widehat{F}_{\mu\nu}{}^I = F_{\mu\nu}{}^I - \bar{\psi}_{[\mu}\gamma_{\nu]}\psi^I + \tfrac{1}{2}i\sigma^I\bar{\psi}_{[\mu}\psi_{\nu]}, \qquad F_{\mu\nu}{}^I = 2\partial_{[\mu}W_{\nu]}^I + W_\mu^J W_\nu^K f_{JK}{}^I, \tag{3.13}$$

There is one more aspect in the dimensional reduction (whether the multiplet is abelian or not). Remember that the covariant general coordinate transformations contain a linear combination of all gauge symmetries. That involves also the gauge transformation of the vector. Thus in the commutator of two supersymmetry transformations in $D = 6$ is a term $\bar{\epsilon}_2\gamma^\mu\epsilon_1 W_\mu$. When reduced to 5 dimensions (and below also to 4 dimensions), some components of W_μ are replaced by the scalars σ. This is the origin of a new term in the supersymmetry commutator involving structure functions depending on the scalars, which is implicit in the form of the last term in $\widehat{F}_{\mu\nu}^I$, which is of the form of the last term in (2.5) for gravitini as gauge fields.

3.2.1.3 Vector Multiplet in 4 Dimensions

Further dimensional reduction leads to the vector multiplet in 4 dimensions. As mentioned already in Sect. 1.2.1, it has then a complex scalar, built from the fourth and fifth components of the vector of 6 dimensions. To get the right behaviour of gauge and general coordinate transformations, one has to consider the reduction of the vector with tangent spacetime indices (see [15, 16] and a useful general introduction to dimensional reduction is [17]). In other words, the object from where the scalars originate in the dimensional reduction should be a world scalar, $e_a{}^\mu W_\mu$, which has Weyl weight 1. Therefore the complex[4] scalar X of the $D = 4$ vector multiplets has $w = 1$.

Before giving the supersymmetry transformations, we have to translate the reality of the triplet Y_{ij} in appropriate notation for 4 dimensions. In 6 dimensions the reality is $Y = Y^* = \sigma_2 Y^C \sigma_2$. It is in the form with Y^C that we have to translate it, thus giving rise to

$$Y_{ij} = \varepsilon_{ik}\varepsilon_{j\ell}Y^{k\ell}, \qquad Y^{ij} = (Y_{ij})^*. \tag{3.14}$$

As for $D = 5$, we write the transformations for the non-abelian vector multiplet. The transformations under dilatations and chiral U(1) transformations follow from Table 3.1, with the general rules (2.87) and (2.88). The supersymmetry (Q and S), and the gauge transformations with parameter θ in 4 dimensions are[5]

$$\delta X^I = \tfrac{1}{2}\bar{\epsilon}^i \Omega_i^I - \theta^J X^K f_{JK}{}^I,$$

$$\delta\Omega_i^I = \slashed{D}X^I \epsilon_i + \tfrac{1}{4}\gamma^{ab}\mathcal{F}_{ab}^{I-}\varepsilon_{ij}\epsilon^j + Y_{ij}^I \epsilon^j + X^J \bar{X}^K f_{JK}{}^I \varepsilon_{ij}\epsilon^j$$

$$\qquad +2X^I \eta_i - \theta^J \Omega_i^K f_{JK}{}^I,$$

$$\delta W_\mu^I = \tfrac{1}{2}\varepsilon^{ij}\bar{\epsilon}_i\gamma_\mu\Omega_j^I + \varepsilon^{ij}\bar{\epsilon}_i\psi_{\mu j}X^I + \text{h.c.} + \partial_\mu\theta^I - \theta^J W_\mu^K f_{JK}{}^I,$$

$$\delta Y_{ij}^I = \tfrac{1}{2}\bar{\epsilon}_{(i}\slashed{D}\Omega_{j)}^I + \tfrac{1}{2}\varepsilon_{ik}\varepsilon_{j\ell}\bar{\epsilon}^{(k}\slashed{D}\Omega^{\ell)I} + \varepsilon_{k(i}\left(\bar{\epsilon}_{j)}X^J\Omega^{kK} - \bar{\epsilon}^k \bar{X}^J \Omega_{j)K}\right)f_{JK}{}^I$$

$$\qquad -\theta^I Y_{ij}^K f_{JK}{}^I, \tag{3.15}$$

where

$$\mathcal{F}_{ab}^{I-} \equiv \widehat{F}_{ab}^{I-} - \tfrac{1}{2}\bar{X}^I T_{ab}^-. \tag{3.16}$$

In the latter expression $\widehat{F}_{ab}^{I-}$ denotes the anti-self-dual part of $\widehat{F}_{ab}$, which is covariant with the new structure functions, as dictated by definitions given in Chap. 2 and

---

[4]To be in accordance with common practice here, we denote the complex conjugates of the scalar fields by $\bar{X}$ rather than X^*.

[5]For the translation from $D = 5$, we use $X^I = \tfrac{1}{2}(W_4^I - i\sigma^I)$, and Ω has been defined with the opposite sign as would straightforwardly follow from Appendix A.4: $\psi^{iI} = -\Omega^{iI} - \Omega_j^I \varepsilon^{ji}$.

reported here for convenience

$$\widehat{F}_{\mu\nu}{}^I = F_{\mu\nu}{}^I + \left(-\varepsilon_{ij}\bar{\psi}^i_{[\mu}\gamma_{\nu]}\Omega^{Ij} - \varepsilon_{ij}\bar{\psi}^i_\mu\psi^j_\nu \bar{X}^I + \text{h.c.} \right),$$

$$F_{\mu\nu}^I = \partial_\mu W_\nu^I - \partial_\nu W_\mu{}^I + W_\mu{}^J W_\nu{}^K f_{JK}{}^I. \tag{3.17}$$

Indeed, the second term of the transformation of the vector reflects the presence of the new term in the commutator of two supersymmetries, as already discussed for $D = 5$, and modifies (2.96) to

$$[\delta_Q(\epsilon_1), \delta_Q(\epsilon_2)] = \delta_P\left(\xi_3^a\right) + \delta_M\left(\lambda_3^{ab}\right) + \delta_K\left(\lambda_{K3}^a\right) + \delta_S\left(\eta_3\right)$$

$$+\delta_G\left(\theta_3^I(\epsilon_1, \epsilon_2) = \varepsilon^{ij}\bar{\epsilon}_{2i}\epsilon_{1j}X^I + \text{h.c.} \right), \tag{3.18}$$

where δ_G is the (non-abelian) gauge transformation parameterized by θ^I.

Exercise 3.4 Check that this leads to the form of $\widehat{F}_{\mu\nu}{}^I$ as given in (3.17). □

The covariant derivatives are

$$\mathcal{D}_\mu X^I = D_\mu X^I - \tfrac{1}{2}\bar{\psi}^i_\mu \Omega^I_i,$$

$$D_\mu X^I = \left(\partial_\mu - b_\mu - \mathrm{i}A_\mu\right) X^I + W_\mu^J X^K f_{JK}{}^I,$$

$$\mathcal{D}_\mu \Omega^I_i = D_\mu \Omega^I_i - \slashed{D}X^I \psi_{\mu i} - \tfrac{1}{4}\gamma^{ab}\mathcal{F}^{I-}_{ab}\varepsilon_{ij}\psi^j_\mu$$

$$-Y^I_{ij}\psi^j_\mu - X^J \bar{X}^K f_{JK}{}^I \varepsilon_{ij}\psi^j_\mu - 2X^I \phi_{\mu i}, \tag{3.19}$$

$$D_\mu \Omega^I_i = \left(\partial_\mu + \tfrac{1}{4}\omega_\mu{}^{ab}\gamma_{ab} - \tfrac{3}{2}b_\mu - \tfrac{1}{2}\mathrm{i}A_\mu\right)\Omega^I_i + V_{\mu i}{}^j \Omega_j{}^I + W_\mu^J \Omega^K_i f_{JK}{}^I.$$

As will become clear in the following section, the vector multiplet is a constrained chiral multiplet. This observation becomes relevant when constructing actions for the vector multiplet (Sect. 3.3).

3.2.2 Intermezzo: Chiral Multiplet

A multiplet corresponds to a superfield in superspace. A multiplet or a superfield can be real or chiral, or carry a Lorentz representation, or be in a non-trivial representations of the R-symmetry, ... For the multiplet, this just reflects the property of its 'lowest component'.[6] For example, a chiral multiplet is characterized by the fact that its lowest component transforms 'chirally', i.e. only under the left-handed supersymmetry and not under the right-handed one. In superspace this means that one chiral superspace derivative vanishes on the field. Furthermore multiplets or

---

[6]'Lowest' refers here to the Weyl weight in superconformal language (or to the engineering dimensions, if we do not discuss the superconformal properties).

superfields can be constrained. In this section we explain, first in the context of rigid supersymmetry, how further constraints on a chiral multiplet lead to the vector multiplet, which is smaller. A generalization to the rigid superconformal case follows.

Let us consider a general scalar multiplet, whose 'lowest' component is a complex scalar A. In general, a complex scalar can transform under Q to arbitrary spinors

$$\delta_Q(\epsilon)A = \tfrac{1}{2}\bar{\epsilon}^i\Psi_i + \tfrac{1}{2}\bar{\epsilon}_i\Lambda^i. \tag{3.20}$$

Then the transformations of these arbitrary spinors Ψ_i and Λ^i can have arbitrary expressions containing new fields, as long as it is consistent with the algebra. See e.g. [18, Sect. 14.1.1] for the example of $\mathcal{N} = 1$ chiral multiplets, and in Sect. 2.2 of [19] this is worked out for the chiral multiplet of $\mathcal{N} = 2$, which we consider here.

If $\Lambda^i = 0$, then the lowest component only transforms under left supersymmetry:

$$\delta_Q(\epsilon)A = \tfrac{1}{2}\bar{\epsilon}^i\Psi_i, \tag{3.21}$$

and the multiplet is called *chiral*. Imposing the rigid supersymmetry algebra leads to the following general expressions:

$$
\begin{aligned}
\delta_Q(\epsilon)A &= \tfrac{1}{2}\bar{\epsilon}^i\Psi_i, \\
\delta_Q(\epsilon)\Psi_i &= \slashed{\partial}A\epsilon_i + B_{ij}\epsilon^j + \tfrac{1}{4}\gamma_{ab}G^{-ab}\varepsilon_{ij}\epsilon^j, \\
\delta_Q(\epsilon)B_{ij} &= \tfrac{1}{2}\bar{\epsilon}_{(i}\slashed{\partial}\Psi_{j)} - \tfrac{1}{2}\bar{\epsilon}^k\Lambda_{(i}\varepsilon_{j)k}, \\
\delta_Q(\epsilon)G_{ab}^- &= \tfrac{1}{4}\varepsilon^{ij}\bar{\epsilon}_i\slashed{\partial}\gamma_{ab}\Psi_i + \tfrac{1}{4}\bar{\epsilon}^i\gamma_{ab}\Lambda_i, \\
\delta_Q(\epsilon)\Lambda_i &= -\tfrac{1}{4}\gamma^{ab}G_{ab}^-\overleftarrow{\slashed{\partial}}\,\epsilon_i - \slashed{\partial}B_{ij}\varepsilon^{jk}\epsilon_k + \tfrac{1}{2}C\varepsilon_{ij}\epsilon^j, \\
\delta_Q(\epsilon)C &= -\varepsilon^{ij}\bar{\epsilon}_i\slashed{\partial}\Lambda_j.
\end{aligned}
\tag{3.22}
$$

The reader can count that this is a $16 + 16$ multiplet counted as real components. In fact it is reducible, since one can impose the following *consistent constraints*[7]:

$$
\begin{aligned}
B_{ij} - \varepsilon_{ik}\varepsilon_{j\ell}B^{k\ell} &= 0, \\
\slashed{\partial}\Psi^i - \varepsilon^{ij}\Lambda_j &= 0,
\end{aligned}
$$

---

[7] There is an extension possible that the first of these expressions is not zero [20] but a constant. This leads to magnetic couplings in rigid supersymmetry, and possibilities for partial breaking to $\mathcal{N} = 1$ supersymmetry. Recently [21], it has been shown how to generate these constants dynamically using multiplets with 3-form gauge fields, and in [22] this has been related to deformations in Dirac–Born–Infeld actions. It is not clear how to generalize this to supergravity, and hence we will not further discuss this.

$$\partial_b(G^{+ab} - G^{-ab}) = 0\,,$$

$$C - 2\partial_a\partial^a\bar{A} = 0\,, \tag{3.23}$$

where $B^{k\ell}$ is, as usual, defined by the complex conjugate of $B_{k\ell}$, and similarly G^+ is the complex conjugate of G^-, and thus self-dual as G^- is anti-self-dual. These constraints are *consistent* in the sense that a supersymmetry variation of one of them leads to the other equations, and this is a complete set in that sense.

The third equation is a Bianchi identity that can be solved by interpreting G_{ab} as the field strength of a vector. To conclude, the independent components are then those of the vector multiplet, with the following identifications:

$$X = A\,, \qquad \Omega_i = \Psi_i\,, \qquad F_{ab} = G_{ab}\,, \qquad Y_{ij} = B_{ij}\,. \tag{3.24}$$

Indeed the linear part of (3.15) corresponds to (3.22). We have thus identified the vector multiplet as a *constrained* chiral multiplet.

To define the chiral multiplet in the conformal algebra, one first allows an arbitrary Weyl weight for A, say that this is w. Then consistency with Weyl weights imposes that a general S-supersymmetry transformation for Ψ_i should be proportional to A. Imposing the $\{Q, S\}$ anticommutator immediately shows that the chiral U(1) weight of A should be related to its Weyl weight. In fact, to avoid the ϵ_i terms in this anticommutator, one should impose that under dilatations and U(1),

$$\delta_{D,T}(\lambda_D, \lambda_T)A = w\,(\lambda_D + i\lambda_T)\,A\,. \tag{3.25}$$

The same transformations for the other fields can be obtained by requiring compatibility with Q-transformations, to obtain

$$\delta_{D,T}(\lambda_D, \lambda_T)\Psi_i = \left(\left(w + \tfrac{1}{2}\right)\lambda_D + i\left(w - \tfrac{1}{2}\right)\lambda_T\right)\Psi_i\,,$$

$$\delta_{D,T}(\lambda_D, \lambda_T)B_{ij} = ((w + 1)\lambda_D + i(w - 1)\lambda_T)\,B_{ij}\,,$$

$$\delta_{D,T}(\lambda_D, \lambda_T)G_{ab}^- = ((w + 1)\lambda_D + i(w - 1)\lambda_T)\,G_{ab}^-\,,$$

$$\delta_{D,T}(\lambda_D, \lambda_T)\Lambda_i = \left(\left(w + \tfrac{3}{2}\right)\lambda_D + i\left(w - \tfrac{3}{2}\right)\lambda_T\right)\Lambda_i\,,$$

$$\delta_{D,T}(\lambda_D, \lambda_T)C = ((w + 2)\lambda_D + i(w - 2)\lambda_T)\,C\,. \tag{3.26}$$

To complete the superconformal multiplet, one has to add S-transformations, and there are nonlinear transformations involving the matter fields of the Weyl multiplet χ_i and T_{ab}, necessary in order to represent the anticommutators (2.96). The full result was found in [23]:

$$\delta_{Q,S}(\epsilon, \eta)A = \tfrac{1}{2}\bar{\epsilon}^i\Psi_i\,,$$

$$\delta_{Q,S}(\epsilon, \eta)\Psi_i = \slashed{D}A\epsilon_i + B_{ij}\epsilon^j + \tfrac{1}{4}\gamma\cdot G^-\varepsilon_{ij}\epsilon^j + 2wA\eta_i\,,$$

$$\delta_{Q,S}(\epsilon,\eta)B_{ij} = \tfrac{1}{2}\bar{\epsilon}_{(i}\mathcal{D}\Psi_{j)} - \tfrac{1}{2}\bar{\epsilon}^k\Lambda_{(i}\varepsilon_{j)k} + (1-w)\bar{\eta}_{(i}\Psi_{j)},$$

$$\delta_{Q,S}(\epsilon,\eta)G^-_{ab} = \tfrac{1}{4}\varepsilon^{ij}\bar{\epsilon}_i\mathcal{D}\gamma_{ab}\Psi_j + \tfrac{1}{4}\bar{\epsilon}^i\gamma_{ab}\Lambda_i - \tfrac{1}{2}\varepsilon^{ij}(1+w)\bar{\eta}_i\gamma_{ab}\Psi_j,$$

$$\delta_{Q,S}(\epsilon,\eta)\Lambda_i = -\tfrac{1}{4}\gamma\cdot G^-\overleftarrow{\mathcal{D}}\epsilon_i - \mathcal{D}B_{ij}\epsilon_k\varepsilon^{jk} + \tfrac{1}{2}C\epsilon^j\varepsilon_{ij}$$

$$-\tfrac{1}{8}(\mathcal{D}A)T\cdot\gamma)\epsilon_i - \tfrac{1}{8}wA(\mathcal{D}T)\cdot\gamma\epsilon_i - \tfrac{3}{4}(\bar{\chi}_{[i}\gamma_a\Psi_{j]})\gamma^a\epsilon_k\varepsilon^{jk}$$

$$-2(1+w)B_{ij}\varepsilon^{jk}\eta_k + \tfrac{1}{2}(1-w)\gamma\cdot G^-\eta_i,$$

$$\delta_{Q,S}(\epsilon,\eta)C = -\varepsilon^{ij}\bar{\epsilon}_i\mathcal{D}\Lambda_j - 6\bar{\epsilon}_i\chi_j B_{k\ell}\varepsilon^{ik}\varepsilon^{j\ell}$$

$$+\tfrac{1}{8}(w-1)\bar{\epsilon}_i\gamma\cdot T\overleftarrow{\mathcal{D}}\Psi_j\varepsilon^{ij} + \tfrac{1}{8}\bar{\epsilon}_i\gamma\cdot T\mathcal{D}\Psi_j\varepsilon^{ij} + 2w\varepsilon^{ij}\bar{\eta}_i\Lambda_j.$$

$$(3.27)$$

This time, the set of consistent constraints is[8]

$$0 = B_{ij} - \varepsilon_{ik}\varepsilon_{jl}B^{kl},$$

$$0 = \mathcal{D}\Psi^i - \varepsilon^{ij}\Lambda_j,$$

$$0 = \mathcal{D}^a\left(G^+_{ab} - G^-_{ab} + \tfrac{1}{2}AT_{ab} - \tfrac{1}{2}\bar{A}T_{ab}\right) - \tfrac{3}{4}\left(\varepsilon^{ij}\bar{\chi}_i\gamma_b\Psi_j - \text{h.c.}\right),$$

$$0 = -2\Box\bar{A} - \tfrac{1}{2}G^+_{\mu\nu}T^{\mu\nu} - 3\bar{\chi}_i\Psi^i - C.$$

$$(3.28)$$

Interestingly, the constraints above are consistent only for a specific choice of w. For example, the first constraint is a reality condition, and it is easy to check that this is only consistent if the chiral weight of B_{ij} is zero. This fixes $w = 1$, which in turn is the appropriate value also to interpret G_{ab} as a covariant field strength. Note that the Bianchi identity in the third line of (3.28) shows the shift between the pure covariant field strength and the G. Compare this with (3.16).

The chiral multiplet plays an important role in the construction of the actions in rigid supersymmetry, as its highest component C is a scalar transforming to a total derivative. That action corresponds in superspace to take the full chiral superspace integral of the chiral superfield. However, in local supersymmetry, as in the superconformal transformations in (3.27), the transformation of C is not a pure derivative. Therefore in order to have an invariant action, one has to include more terms, i.e. something of the form

$$I = \int d^4x\, e\, C + \cdots + \text{h.c.}. \qquad (3.29)$$

The $+\cdots$ in (3.29) are terms that should be such that the transformation of the integrand is a total derivative.

---

[8]For rigid supersymmetry, an imaginary constant in B_{ij} would be possible, describing magnetic charges.

But we can first make a few general observations. The integrand should be invariant under all superconformal transformations. Let us start with the Weyl transformations. The Weyl weight of the determinant of the frame field is -4, so C should have Weyl weight 4. It should also be invariant under T-transformations, which means that the chiral weight should be zero. We see from (3.26) that these two requirements are consistent with a requirement that the chiral multiplet should have Weyl weight 2. Note that this implies that it will not be a constrained chiral (i.e. vector) multiplet. We found above that these have Weyl weight 1. But if we start from a vector multiplet, any holomorphic function of X still transforms only under 1 chirality of Q. Hence any $F(X)$ is a chiral multiplet. If we take a homogeneous function of second degree in X, this gives us a chiral multiplet with $w = 2$ on which we can use the action formula.

To determine the full expression in (3.29) one considers other terms that have Weyl weight 4 and chiral weight 0, and imposes the condition of invariance of the action. In practice, imposing S-supersymmetry is easiest to determine all the coefficients of these terms. For local superconformal symmetry the result is [23]

$$
e^{-1}\mathcal{L} = C - \bar{\psi}_i \cdot \gamma \Lambda_j \varepsilon^{ij} + \frac{1}{8}\bar{\psi}_{\mu i}\gamma \cdot T^+ \gamma^\mu \Psi_j \varepsilon^{ij} - \frac{1}{4}A T^+_{ab} T^{+ab}
$$

$$
- \frac{1}{2}\bar{\psi}_{\mu i}\gamma^{\mu\nu}\psi_{vj}B_{kl}\varepsilon^{ik}\varepsilon^{jl} + \bar{\psi}_{\mu i}\psi_{vj}\varepsilon^{ij}\left(G^{-\mu\nu} - AT^{+\mu\nu}\right)
$$

$$
+ \frac{1}{2}i\varepsilon^{ij}\varepsilon^{kl}e^{-1}\varepsilon^{\mu\nu\rho\sigma}\bar{\psi}_{\mu i}\psi_{vj}\bar{\psi}_{\rho k}\left(\gamma_\sigma\Psi_\ell + \psi_{\sigma\ell}A\right) + \text{h.c.}. \quad (3.30)
$$

This is called the *chiral density formula*.

3.2.3   Rigid Hypermultiplets

Hypermultiplets are the analogues of the chiral multiplets of $\mathcal{N} = 1$ supersymmetry. They contain four scalars and two spin- $1/2$ fields. In supergravity, they are defined in the background of the Weyl multiplet and possibly also in the background of the vector multiplet (i.e. they can transform non-trivially under the gauge transformations of the vector multiplets). One can further introduce auxiliary fields to close the algebra for the simplest quaternionic manifolds. The methods of harmonic or projective superspace mentioned in the introduction [24–30] are also equivalent to introducing an infinite number of auxiliary fields. However, we do not need auxiliary fields any more at this point because the hypermultiplets are at the end of the hierarchy line.[9]

---

[9]We are not going to introduce any further multiplet in the background of the hypermultiplets, as these do not introduce new gauge symmetries. This is to be confronted to when we considered the

The closure of the supersymmetry algebra will impose equations that we will interpret as equations of motion, even though we have not defined an action yet. Later we will see how an action can be constructed that gives precisely these equations as Euler–Lagrange equations.

Although our interest is the local case, the present section is mostly devoted to rigid super(conformal) symmetry. This choice has been made since the rigid case provides already simpler and explicative examples of the story. We remark that since dimensional reduction for scalars and spin-1/2 fermions leads to the same type of particles in lower dimension, the properties of the hypermultiplets do not depend on whether we consider $D = 6$, $D = 5$, or $D = 4$ (or even $D = 3$). There is a technical difference since the four on-shell (or eight off-shell) degrees of freedom are captured in symplectic Weyl, symplectic or Majorana spinors, respectively. In practice, we mostly report formulae in $D = 5$. These can be translated to $D = 6$ and $D = 4$ by the rules in Appendix A.4.

Before starting the mathematical formulation, we still want to point out how massive hypermultiplets can be described in this context, since the readers will mainly see equations of motion that describe only massless hypermultiplets. This is of course also related to the fact that we are mainly interested in conformal theories. Massive hypermultiplets in rigid supersymmetry are obtained in this setting by adding a coupling to a vector multiplet that has just a first scalar component equal to the mass, and all other components zero. The reader can glimpse at (3.88) for $D = 5$ with σ^I equal to a mass, or to (3.93) and (3.95) for X^I providing the mass to see that with a suitable choice of the Killing vectors these are massive field equations. In supergravity this will be natural for the σ^I or X^I referring to the compensating multiplet.

3.2.3.1 Rigid Supersymmetry

We consider a set of n_H hypermultiplets. The real scalars are denoted as q^X, with $X = 1, \ldots, 4n_H$, and the fermions are indicated by $\zeta^{\mathcal{A}}$, where the indices $\mathcal{A} = 1, \ldots, 2n_H$ will indicate a fundamental representation of $\mathrm{Sp}(2n_H)$. Imposing the supersymmetry transformations on the bosons lead to the identification of a hypercomplex manifold[10] parameterized by these bosons q^X. The structure is determined by frame fields $f^{i\mathcal{A}}{}_X$, connections $\omega_{X\mathcal{A}}{}^{\mathcal{B}}$ and Γ^Z_{XY} (the latter symmetric in its lower indices) such that

$$f^{i\mathcal{A}}{}_Y f^X{}_{i\mathcal{A}} = \delta^X_Y , \qquad f^{i\mathcal{A}}{}_X f^X{}_{j\mathcal{B}} = \delta^i_j \delta^{\mathcal{A}}_{\mathcal{B}} . \tag{3.31}$$

---

vector multiplets. The construction of the latter had to take into account that the multiplets can be used for various possible actions (including hypermultiplets or not).

[10] In supergravity the scalars span a quaternionic manifold, see Sect. 5.6.

and

$$\nabla_Y f^X{}_{i\mathcal{A}} \equiv \partial_Y f^X{}_{i\mathcal{A}} - \omega_{Y\mathcal{A}}{}^{\mathcal{B}}(q) f^X{}_{i\mathcal{B}} + \Gamma^X_{YZ}(q) f^Z{}_{i\mathcal{A}} = 0,$$
$$\nabla_Y f^{i\mathcal{A}}{}_X \equiv \partial_Y f^{i\mathcal{A}}{}_X + f^{i\mathcal{B}}{}_X \omega_{Y\mathcal{B}}{}^{\mathcal{A}}(q) - \Gamma^Z_{YX}(q) f^{i\mathcal{A}}{}_Z = 0, \qquad (3.32)$$

are satisfied. The frame field satisfies a reality condition, for which we will also introduce indices $\bar{\mathcal{A}}$:

$$\left(f^{i\mathcal{A}}{}_X\right)^* = f^{j\mathcal{B}}{}_X \varepsilon_{ji} \rho_{\mathcal{B}\bar{\mathcal{A}}}, \qquad \left(f^X{}_{i\mathcal{A}}\right)^* = \varepsilon^{ij} \rho^{\bar{\mathcal{A}}\mathcal{B}} f^X{}_{j\mathcal{B}}, \qquad (3.33)$$

in terms of a non-degenerate covariantly constant tensor $\rho_{\mathcal{A}\bar{\mathcal{B}}}$ that satisfies

$$\rho_{\mathcal{A}\bar{\mathcal{B}}} \rho^{\bar{\mathcal{B}}C} = -\delta^C_{\mathcal{A}}, \qquad \rho^{\bar{\mathcal{A}}\mathcal{B}} = \left(\rho_{\mathcal{A}\bar{\mathcal{B}}}\right)^*. \qquad (3.34)$$

By field redefinitions, we could bring it in the standard antisymmetric form

$$\rho_{\mathcal{A}\bar{\mathcal{B}}} = \begin{pmatrix} 0 & \mathbb{1}_{n_H} \\ -\mathbb{1}_{n_H} & 0 \end{pmatrix} = \rho^{\bar{\mathcal{A}}\mathcal{B}}. \qquad (3.35)$$

We will not impose this basis choice in general. In Sect. 3.3.4 we will show how such a basis could be implemented.

The complex conjugate of $\omega_{X\mathcal{A}}{}^{\mathcal{B}}$ is

$$\left(\omega_{X\mathcal{A}}{}^{\mathcal{B}}\right)^* \equiv \bar{\omega}_{X}{}^{\bar{\mathcal{A}}}{}_{\bar{\mathcal{B}}} = -\rho^{\bar{\mathcal{A}}C} \omega_{XC}{}^{\mathcal{D}} \rho_{\mathcal{D}\bar{\mathcal{B}}}. \qquad (3.36)$$

The above conditions lead to the identification of almost quaternionic structures

$$2 f^{i\mathcal{A}}{}_X f^Y{}_{j\mathcal{A}} = \delta^Y_X \delta^i_j + J_X{}^Y{}_j{}^i, \qquad J_X{}^Y{}_j{}^i = \boldsymbol{\tau}_j{}^i \cdot \mathbf{J}_X{}^Y,$$
$$\mathbf{J}_X{}^Y = \left(\mathbf{J}_X{}^Y\right)^* = -f^{i\mathcal{A}}{}_X f^Y{}_{j\mathcal{A}} \boldsymbol{\tau}_i{}^j. \qquad (3.37)$$

We use here the 3-vectors notation and $\boldsymbol{\tau}_i{}^j = i\boldsymbol{\sigma}_i{}^j$ in terms of the three Pauli-matrices $\boldsymbol{\sigma}_i{}^j$ as in (1.52), (1.54). Related formulas are given in Appendix A.2.2. The three matrices $\mathbf{J}$ satisfy the quaternionic algebra, i.e. for any vectors $\mathbf{A}$, $\mathbf{B}$

$$\mathbf{A} \cdot \mathbf{J}_X{}^Z \mathbf{B} \cdot \mathbf{J}_Z{}^Y = -\delta_X{}^Y \mathbf{A} \cdot \mathbf{B} + (\mathbf{A} \times \mathbf{B}) \cdot \mathbf{J}_X{}^Y. \qquad (3.38)$$

In passing, we note that we can solve (3.32) for $\omega_{X\mathcal{A}}{}^{\mathcal{B}}$, such that the independent connection is $\Gamma_{XY}{}^Z$. The latter is the unique connection on the scalar manifold respect to which

$$\nabla_Z \mathbf{J}_X{}^Y \equiv \partial_Z \mathbf{J}_X{}^Y - \Gamma_{ZX}{}^U \mathbf{J}_U{}^Y + \Gamma_{ZU}{}^Y \mathbf{J}_X{}^U = 0. \qquad (3.39)$$

This last condition promotes $\mathbf{J}_X{}^Y$ to be quaternionic structures.

The integrability condition of (3.32) relates the curvatures defined by the two connections:

$$R_{XY}{}^W{}_Z \equiv 2\partial_{[X}\Gamma^W_{Y]Z} + 2\Gamma^W_{V[X}\Gamma^V_{Y]Z},$$

$$\mathcal{R}_{XY\mathcal{B}}{}^{\mathcal{A}} \equiv 2\partial_{[X}\omega_{Y]\mathcal{B}}{}^{\mathcal{A}} + 2\omega_{[X|\mathcal{C}|}{}^{\mathcal{A}}\omega_{Y]\mathcal{B}}{}^{\mathcal{C}},$$

$$R_{XY}{}^W{}_Z = f^W{}_{i\mathcal{A}}f^{i\mathcal{B}}{}_Z\mathcal{R}_{XY\mathcal{B}}{}^{\mathcal{A}}, \qquad \mathcal{R}_{XY\mathcal{B}}{}^{\mathcal{A}} = \tfrac{1}{2}f^W{}_{i\mathcal{B}}f^{i\mathcal{A}}{}_Z R_{XY}{}^Z{}_W.$$

$$(3.40)$$

In order to work with these tensors, it can be useful to introduce also tensors L that are orthogonal to the complex structures:

$$L_Y{}^Z{}_{\mathcal{A}}{}^{\mathcal{B}} \equiv f^Z{}_{i\mathcal{A}}f^{i\mathcal{B}}{}_Y, \quad \mathbf{J}_Z{}^Y L_Y{}^Z{}_{\mathcal{A}}{}^{\mathcal{B}} = 0,$$

$$L_X{}^Y{}_{\mathcal{A}}{}^{\mathcal{B}}L_Y{}^Z{}_{\mathcal{C}}{}^{\mathcal{D}} = L_X{}^Z{}_{\mathcal{C}}{}^{\mathcal{B}}\delta_{\mathcal{A}}{}^{\mathcal{D}},$$

$$L_X{}^X{}_{\mathcal{A}}{}^{\mathcal{B}} = 2\delta_{\mathcal{A}}{}^{\mathcal{B}}, \quad L_X{}^Y{}_{\mathcal{A}}{}^{\mathcal{B}}L_Y{}^X{}_{\mathcal{C}}{}^{\mathcal{D}} = 2\delta_{\mathcal{C}}{}^{\mathcal{B}}\delta_{\mathcal{A}}{}^{\mathcal{D}}.$$

$$(3.41)$$

If the affine connections is the Levi-Civita connection of a metric, then the curvatures satisfy the cyclicity properties $R_{(XY}{}^W{}_{Z)} = 0$, and one can show that

$$f^X{}_{iC}f^Y{}_{j\mathcal{D}}\mathcal{R}_{XY\mathcal{B}}{}^{\mathcal{A}} = -\tfrac{1}{2}\varepsilon_{ij}W_{C\mathcal{D}\mathcal{B}}{}^{\mathcal{A}}, \qquad W_{\mathcal{A}\mathcal{B}C}{}^{\mathcal{D}} \equiv -\varepsilon^{ij}f^X{}_{i\mathcal{A}}f^Y{}_{j\mathcal{B}}\mathcal{R}_{XYC}{}^{\mathcal{D}}.$$

$$(3.42)$$

The tensor $W_{\mathcal{A}\mathcal{B}C}{}^{\mathcal{D}}$ is symmetric in its lower indices, and the other curvatures can be expressed in function of this one as

$$R_{XY}{}^W{}_Z = L_Z{}^W{}_{\mathcal{D}}{}^C \mathcal{R}_{XYC}{}^{\mathcal{D}} = -\varepsilon_{ij}\tfrac{1}{2}L_Z{}^W{}_C{}^{\mathcal{D}}f^{i\mathcal{A}}{}_X f^{j\mathcal{B}}{}_Y W_{\mathcal{A}\mathcal{B}C}{}^{\mathcal{D}}. \qquad (3.43)$$

The Bianchi identity on $\mathcal{R}_{XY\mathcal{A}}{}^{\mathcal{B}}$ implies also a symmetry of the covariant derivative of W:

$$f^X{}_{i\mathcal{A}}\nabla_X W_{\mathcal{B}C\mathcal{D}}{}^{\mathcal{E}} = f^X{}_{i(\mathcal{A}|}\nabla_X W_{|\mathcal{B}C\mathcal{D})}{}^{\mathcal{E}}. \qquad (3.44)$$

When a metric will be defined on the manifold, the W-tensor will become symmetric in the 4 indices. As a consequence, the manifold will be *Ricci flat*:

$$R_{YZ} = R_{XY}{}^X{}_Z = 0. \qquad (3.45)$$

3.2.3.2 Reparameterizations and Covariant Quantities

The hypermultiplet is defined in terms of the scalars q^X, which form a parameterization of a $4n_H$-dimensional manifold, and the fermions $\zeta^{\mathcal{A}}$, which are a parameterization of a $2n_H$-dimensional manifold of fermions. Both these basic parameterizations can be changed [14]. There are thus two kinds of reparameterizations. The first ones are the target space diffeomorphisms, $q^X \rightarrow \tilde{q}^X(q)$, under which $f^X{}_{i\mathcal{A}}$ transforms as a vector, $\omega_{X\mathcal{A}}{}^{\mathcal{B}}$ as a one-form, and $\Gamma_{XY}{}^Z$ as a connection. The second set are the reparameterizations, which act on the tangent space indices $\mathcal{A}$, $\mathcal{B}$ etc. On the fermions, they act as

$$\zeta^{\mathcal{A}} \rightarrow \tilde{\zeta}^{\mathcal{A}}(q) = \zeta^{\mathcal{B}} U_{\mathcal{B}}{}^{\mathcal{A}}(q),\tag{3.46}$$

where $U_{\mathcal{A}}{}^{\mathcal{B}}(q)$ is an invertible matrix, and the reality conditions impose $U^* = \rho^{-1}U\rho$, defining $G\ell(r, \mathbb{H})$. In general, the right-hand side of (3.46) depends on the $\zeta^{\mathcal{A}}$ and on the scalars. Thus the new basis $\tilde{\zeta}^{\mathcal{A}}$ is a basis where the fermions depend on the scalars q^X. In this sense, the hypermultiplet is written in a special basis where q^X and $\zeta^{\mathcal{A}}$ are independent fields. We will develop a covariant formalism which also takes into account these reparameterizations.

The supersymmetry transformations in $D = 5$ are

$$\delta q^X = -i\bar{\epsilon}^i \zeta^{\mathcal{A}} f^X{}_{i\mathcal{A}},$$
$$\delta \zeta^{\mathcal{A}} = \tfrac{1}{2}i f^{i\mathcal{A}}{}_X \slashed{\partial} q^X \epsilon_i - \zeta^{\mathcal{B}} \omega_{X\mathcal{B}}{}^{\mathcal{A}} \delta q^X.\tag{3.47}$$

They are covariant under (3.46) if we transform $f^{i\mathcal{A}}{}_X(q)$ as a vector and $\omega_{X\mathcal{A}}{}^{\mathcal{B}}$ as a connection,

$$\omega_{X\mathcal{A}}{}^{\mathcal{B}} \rightarrow \tilde{\omega}_{X\mathcal{A}}{}^{\mathcal{B}} = \left[\left(\partial_X U^{-1} \right) U + U^{-1} \omega_X U \right]_{\mathcal{A}}{}^{\mathcal{B}}.\tag{3.48}$$

These considerations lead us to define the covariant variation of vectors (see [18, Appendix 14B]) with indices in the tangent space, as $\zeta^{\mathcal{A}}$, or a quantity Δ^X with coordinate indices:

$$\widehat{\delta\zeta}^{\mathcal{A}} \equiv \delta\zeta^{\mathcal{A}} + \zeta^{\mathcal{B}} \omega_{X\mathcal{B}}{}^{\mathcal{A}} \delta q^X, \qquad \widehat{\delta\Delta}^X \equiv \delta\Delta^X + \Delta^Y \Gamma_{YZ}{}^X \delta q^Z,\tag{3.49}$$

for any transformation δ (as e.g. supersymmetry, conformal transformations,...).

Two models related by either target space diffeomorphisms or fermion reparameterizations of the form (3.46) are equivalent; they are different coordinate descriptions of the same system. We usually work in a basis where the fermions and the bosons are independent, i.e. $\partial_X \zeta^{\mathcal{A}} = 0$. But in a covariant formalism, after a transformation (3.46), this is not anymore valid. This shows that the expression $\partial_X \zeta^{\mathcal{A}}$ has no basis-independent meaning. It makes only sense if one compares a transformed basis, like the $\tilde{\zeta}^A$ with the original basis where $\partial_X \zeta^{\mathcal{A}} = 0$. But in the same way also the expression $\zeta^{\mathcal{B}} \omega_{X\mathcal{B}}{}^{\mathcal{A}}$ makes only sense if one compares different bases, as the connection has no absolute value. The only object that has a coordinate-invariant meaning is the covariant derivative

$$\nabla_X \zeta^{\mathcal{A}} \equiv \partial_X \zeta^{\mathcal{A}} + \zeta^{\mathcal{B}} \omega_{X\mathcal{B}}{}^{\mathcal{A}}. \tag{3.50}$$

In the basis where the fermions $\zeta^{\mathcal{A}}$ are considered independent of the bosons, i.e. $\partial_X \zeta^{\mathcal{A}} = 0$, which is the basis used to write down the transformation rules (3.47), only the second term in the covariant derivative above remains, and thus (3.49) becomes

$$\widehat{\delta}\zeta^{\mathcal{A}} = \delta\zeta^{\mathcal{A}} + \nabla_X \zeta^{\mathcal{A}} \delta q^X. \tag{3.51}$$

We will always consider independent bosons and fermions when we write variations.

On any covariant coordinate quantity that depends only on the coordinates q^X, covariant transformations act by covariant derivatives, e.g. for some vectors $V^X(q)$, $W_{\mathcal{A}}$ or $W^{\mathcal{A}}$:

$$\widehat{\delta}V^X(q) = \delta q^Y \nabla_Y V^X(q) = \delta q^Y \left(\partial_Y V^X(q) + \Gamma^X_{YZ} V^Z(q) \right),$$

$$\widehat{\delta}W^{\mathcal{A}}(q) = \delta q^Y \nabla_Y W^{\mathcal{A}}(q) = \delta q^Y \left(\partial_Y W^{\mathcal{A}}(q) + W^{\mathcal{B}}(q) \omega_{Y\mathcal{B}}{}^{\mathcal{A}} \right),$$

$$\widehat{\delta}W_{\mathcal{A}}(q) = \delta q^Y \nabla_Y W_{\mathcal{A}}(q) = \delta q^Y \left(\partial_Y W^{\mathcal{A}}(q) - \omega_{Y\mathcal{A}}{}^{\mathcal{B}} W_{\mathcal{B}}(q) \right).$$

$$\tag{3.52}$$

In particular, $\widehat{\delta}$ of any covariantly constant object (like the frame fields $f^{i\mathcal{A}}{}_X$) is zero.

Note that we can exploit covariant transformations to calculate any transformation on e.g. a quantity $W_{\mathcal{A}}(q)\zeta^{\mathcal{A}}$:

$$\delta \left(W_{\mathcal{A}}(q)\zeta^{\mathcal{A}} \right) = \widehat{\delta} \left(W_{\mathcal{A}}(q)\zeta^{\mathcal{A}} \right) = \delta q^X \nabla_X W_{\mathcal{A}} \zeta^{\mathcal{A}} + W_{\mathcal{A}} \widehat{\delta}\zeta^{\mathcal{A}}. \tag{3.53}$$

Coordinates are not covariant, but their derivatives are, and e.g. the Laplacian[11]

$$\Box q^X = \nabla^\mu \partial_\mu q^X = \partial^\mu \partial_\mu q^X + \Gamma_{YZ}{}^X \left(\partial_\mu q^Y \right) \left(\partial^\mu q^Z \right), \tag{3.54}$$

is covariant for target space transformations.

---

[11]In the local (gravity) theory, the first term should be $(\sqrt{g})^{-1}\partial_\mu \sqrt{g} g^{\mu\nu}\partial_\nu$.

Another interesting relation is that the commutator of $\widehat{\delta}$ and ∇ gives rise to curvature terms:

$$\widehat{\delta}\nabla_\mu V^X = \nabla_\mu \widehat{\delta} V^X + R_{ZW}{}^X{}_Y V^Y \left(\delta q^Z\right)\left(\partial_\mu q^W\right) . \tag{3.55}$$

Similarly the commutator gets adapted by curvature terms:

$$[\delta(\epsilon_1), \delta(\epsilon_2)] V^X = \delta(\epsilon_3) V^X \rightarrow$$

$$\left[\widehat{\delta}(\epsilon_1), \widehat{\delta}(\epsilon_2)\right] V^X = \widehat{\delta}(\epsilon_3) V^X + R_{ZW}{}^X{}_Y V^Y \left(\delta(\epsilon_1) q^Z\right)\left(\delta(\epsilon_2) q^W\right) , \tag{3.56}$$

where ϵ_3 is the function of ϵ_1 and ϵ_2 determined by the structure functions. With these methods, it is easy to compute the commutator of two covariant derivatives. E.g. in $D = 5$ with (3.47) for the fermions

$$\left[\widehat{\delta}(\epsilon_1), \widehat{\delta}(\epsilon_2)\right] \zeta^{\mathcal{A}} = \tfrac{1}{2} \gamma^\mu \epsilon_{2i} f^{i\mathcal{A}}{}_X \bar{\epsilon}_1^j \nabla_\mu \zeta^{\mathcal{B}} f^X{}_{j\mathcal{B}} - (1 \leftrightarrow 2)$$

$$= \tfrac{1}{4} \gamma^\mu \left[(\bar{\epsilon}_2 \epsilon_1) + \gamma^\nu (\bar{\epsilon}_2 \gamma_\nu \epsilon_1)\right] \nabla_\mu \zeta^{\mathcal{A}}$$

$$= \tfrac{1}{2} \nabla_\mu \zeta^{\mathcal{A}} (\bar{\epsilon}_2 \gamma^\mu \epsilon_1) + \tfrac{1}{4} \left[(\bar{\epsilon}_2 \epsilon_1) - \gamma^\nu (\bar{\epsilon}_2 \gamma_\nu \epsilon_1)\right] \slashed{\nabla} \zeta^{\mathcal{A}} , \tag{3.57}$$

with the definition

$$\nabla_\mu \zeta^{\mathcal{A}} \equiv \partial_\mu \zeta^{\mathcal{A}} + \left(\partial_\mu q^X\right) \zeta^{\mathcal{B}} \omega_{X\mathcal{B}}{}^{\mathcal{A}}. \tag{3.58}$$

Indices i, j are raised and contracted as in Appendix A.3.2. This result shows that the algebra does not close: we will interpret the extra parts as equations of motions of a putative action, see Sect. 3.3.3.

3.2.3.3 Non-closure Relations for Fermions and Bosons

From the result (3.57), using (3.56), we can obtain the following commutator of transformations:

$$[\delta(\epsilon_1), \delta(\epsilon_2)] \zeta^{\mathcal{A}} = \tfrac{1}{2} \partial_\mu \zeta^{\mathcal{A}} (\bar{\epsilon}_2 \gamma^\mu \epsilon_1) + \tfrac{1}{4} \left[(\bar{\epsilon}_2 \epsilon_1) - \gamma^\nu (\bar{\epsilon}_2 \gamma_\nu \epsilon_1)\right] \slashed{\nabla} \zeta^{\mathcal{A}}$$

$$+ \zeta^{\mathcal{B}} R_{XY\mathcal{B}}{}^{\mathcal{A}} \bar{\epsilon}_1^i \zeta^C f^X{}_{iC} \bar{\epsilon}_2^j \zeta^D f^Y{}_{jD} . \tag{3.59}$$

With (3.42) and a Fierz transformation, we obtain that the non-closure terms (the last term on the first line and the second line) are

$$+ \tfrac{1}{4} \left[(\bar{\epsilon}_2 \epsilon_1) - \gamma^\nu (\bar{\epsilon}_2 \gamma_\nu \epsilon_1)\right] \slashed{\nabla} \zeta^{\mathcal{A}} + \tfrac{1}{8} W_{CD\mathcal{B}}{}^{\mathcal{A}} \zeta^{\mathcal{B}} \bar{\zeta}^C \left[(\bar{\epsilon}_2 \epsilon_1) + \gamma^\nu (\bar{\epsilon}_2 \gamma_\nu \epsilon_1)\right] \zeta^{\mathcal{D}} . \tag{3.60}$$

Using some $D = 5$ Fierz identities:

$$5\zeta^{(\mathcal{B}}\bar{\zeta}^C\zeta^{D)} = -\gamma^\mu\zeta^{(\mathcal{B}}\bar{\zeta}^C\gamma_\mu\zeta^{D)} ,$$
$$\zeta^{(\mathcal{B}}\bar{\zeta}^C\gamma^\nu\zeta^{D)} = -\gamma^\nu\zeta^{(\mathcal{B}}\bar{\zeta}^C\zeta^{D)} , \tag{3.61}$$

we find

$$[\delta(\epsilon_1), \delta(\epsilon_2)]\zeta^{\mathcal{A}} = \xi^\mu\partial_\mu\zeta^{\mathcal{A}} + \tfrac{1}{4}\left[\left(\bar{\epsilon}_2^i\epsilon_1^j\right) - \gamma^\nu\left(\bar{\epsilon}_2^i\gamma_\nu\epsilon_1^j\right)\right]\varepsilon_{ji}\mathrm{i}\Gamma^{\mathcal{A}} , \tag{3.62}$$

i.e., the non-closure terms are proportional to[12]

$$\mathrm{i}\Gamma^{\mathcal{A}} \equiv \slashed{\nabla}\zeta^{\mathcal{A}} + \frac{1}{2}W_{\mathcal{B}CD}{}^{\mathcal{A}}\zeta^{\mathcal{B}}\bar{\zeta}^C\zeta^{D} . \tag{3.63}$$

The expression above must be interpreted as an equation of motion for the fermions. The supersymmetry transformation of (3.63) gives then also an equation of motion for the scalar fields:

$$\widehat{\delta}(\epsilon)\Gamma^{\mathcal{A}} = \frac{1}{2}f^{i\mathcal{A}}{}_X\epsilon_i\Delta^X , \tag{3.64}$$

where

$$\Delta^X = \Box q^X - \frac{1}{2}\bar{\zeta}^{\mathcal{B}}\gamma_a\zeta^{D}\partial^a q^Y f^{iC}{}_Y f^X{}_{i\mathcal{A}}W_{\mathcal{B}CD}{}^{\mathcal{A}}$$

$$- \frac{1}{4}\nabla_Y W_{\mathcal{B}CD}{}^{\mathcal{A}}\bar{\zeta}^{\mathcal{E}}\zeta^{D}\bar{\zeta}^C\zeta^{\mathcal{B}}f^{iY}\varepsilon f^X{}_{i\mathcal{A}} . \tag{3.65}$$

The equations of motion given by (3.63) and (3.65) form a multiplet, since (3.64) has the counterpart

$$\widehat{\delta}(\epsilon)\Delta^X = \bar{\epsilon}^i\slashed{\nabla}\Gamma^{\mathcal{A}}f^X{}_{i\mathcal{A}} - \bar{\epsilon}^i\Gamma^{\mathcal{B}}\bar{\zeta}^C\zeta^{D}f^X{}_{i\mathcal{A}}W_{\mathcal{B}CD}{}^{\mathcal{A}} , \tag{3.66}$$

where the covariant derivative of $\Gamma^{\mathcal{A}}$ is defined similar to (3.58). As announced before, we thus have already physical equations despite the absence of an action.

3.2.3.4 Rigid Superconformal

To allow the generalization to superconformal couplings, the essential question is whether the manifold has dilatational symmetry. This means, according to (1.30),

---

[12]We inserted a factor i in order that Γ^A is symplectic Majorana.

that there is a 'closed homothetic Killing vector' [31] (see also [18, Sect. 15.7]). The dilatations act as[13]

$$\delta_D(\lambda_D)q^X = \lambda_D k_D{}^X(q),$$ (3.67)

where $k_D{}^X$ satisfies (we generalize here already to D dimensions, as the modifications involve only a normalization factor)

$$\nabla_Y k_D{}^X \equiv \partial_Y k_D{}^X + \Gamma^X_{YZ} k_D{}^Z = \frac{D-2}{2}\delta_Y{}^X.$$ (3.68)

On a flat manifold, the fields q^X have thus Weyl weight $(D-2)/2$. The presence of this vector allows one to extend the transformations of rigid supersymmetry to the superconformal group [31, 32, 14], with e.g. transformations under the SU(2) R-symmetry group:

$$\delta_{SU(2)}(\lambda)q^X = \mp 2\lambda \cdot \mathbf{k}^X, \qquad \mathbf{k}^X \equiv \frac{1}{D-2}k_D{}^Y \mathbf{J}_Y{}^X.$$ (3.69)

Note the sign difference between $D = 4$, upper sign, and $D = 5, 6$, lower sign, as in (A.24).

In general, one can introduce the sections

$$A^{i\mathcal{A}} = k_D^X f^{i\mathcal{A}}{}_X,$$ (3.70)

and in terms of these

$$\mathbf{k}^X = -\frac{1}{D-2}A^{i\mathcal{A}}\boldsymbol{\tau}_i{}^j f^X{}_{j\mathcal{A}}.$$ (3.71)

Using the rules of covariant transformations (and in particular that $\nabla_Y f^{i\mathcal{A}}{}_X$ implies $\widehat{\delta} f^{i\mathcal{A}}{}_X = 0$), the $A^{i\mathcal{A}}$ transform as

$$\widehat{\delta}A^{i\mathcal{A}} = f^{i\mathcal{A}}{}_X \nabla_Y k_D^X \delta q^Y = \frac{D-2}{2}f^{i\mathcal{A}}{}_X \delta q^X$$

$$= \frac{D-2}{2}\left(-i\bar{\epsilon}^i \zeta^{\mathcal{A}} + \lambda_D A^{i\mathcal{A}}\right) + A^{j\mathcal{A}}\lambda_j{}^i,$$ (3.72)

---

[13]Note that we give here only the intrinsic part of the dilatations, i.e. the λ_D term in (1.24), and not the 'orbital' part included in the general coordinate transformation $\xi^\mu(x)$. Similarly for special conformal transformations, we will write here only the intrinsic part represented as $(k_\mu \phi)$ in that equation and also the 'orbital' S-supersymmetry part (1.60) is not mentioned explicitly.

using (3.68) and   (3.69). Note that the supersymmetry transformation in this equation is written for the symplectic spinors of $D = 5, 6$. Below, we will write them for $D = 4$.

We can then derive the other (super)conformal transformations using the algebra. The intrinsic special conformal transformations on q^X and $\zeta^{\mathcal{A}}$ vanish. They have only the 'orbital' parts as follows from (1.24). The latter imply e.g. that $\delta_K(\lambda_K)\partial q^X \neq 0$. The algebra gives then for the intrinsic S-supersymmetry

$$\delta_S(\eta^i)\zeta^{\mathcal{A}} = -A^{i\mathcal{A}}\eta_i \,. \tag{3.73}$$

The (intrinsic) bosonic conformal symmetries act as

$$\widehat{\delta}_D\zeta^{\mathcal{A}} = \frac{D-1}{2}\lambda_D\zeta^{\mathcal{A}} \,, \qquad \widehat{\delta}_{SU(2)}\zeta^{\mathcal{A}} = 0 \,. \tag{3.74}$$

The fermions are inert under $SU(2)$ R-symmetries group.

3.2.3.5   Isometries and Coupling to Vector Multiplets

So far we considered the hypermultiplet with ungauged isometries. A more general situation includes couplings to vector multiplets and in this case one has to define the hypermultiplet in the algebra including the vector multiplet with its gauge transformations. Let us consider general isometries (not necessarily gauged) of the hypermultiplet:

$$\delta_G(\theta)q^X = \theta^I k_I{}^X(q) \,, \tag{3.75}$$

where θ^I are constant parameters and the $k_I{}^X(q)$ represent the transformations. The index I identifies the different generators of the isometry group. Then a subgroup of these could be gauged, identified by an embedding tensor [33–36] projecting from all the symmetries to those that are gauged.[14] When we have a metric, $k_I{}^X(q)$ should be Killing vectors in order to define symmetries of the action. As we have not discussed a metric yet, we could define here some generalization of symmetries, but we just refer the interested reader to [14]. The transformations (3.75) constitute an algebra with structure constants $f_{IJ}{}^K$,

$$k_I{}^Y \partial_Y k_J{}^X - k_J{}^Y \partial_Y k_I{}^X = f_{IJ}{}^K k_K{}^X \,. \tag{3.76}$$

---

[14]However, we will here soon gauge the symmetries, and thus restrict the index I to the gauged symmetries.

We consider symmetries that respect the hypercomplex structure. This is the requirement that $k_I{}^X(q)$ is tri-holomorphic:

$$\left(\nabla_X k_I^Y\right) \mathbf{J}_Y{}^Z = \mathbf{J}_X{}^Y \left(\nabla_Y k_I^Z\right). \tag{3.77}$$

Extracting affine connections from this equation, it can be written as

$$\left(\mathcal{L}_{k_I} \mathbf{J}\right) x{}^Y \equiv k_I{}^Z \partial_Z \mathbf{J}_X{}^Y - \partial_Z k_I{}^Y \mathbf{J}_X{}^Z + \partial_X k_I{}^Z \mathbf{J}_Z{}^Y = 0. \tag{3.78}$$

This is the Lie derivative of the complex structure in the direction of the vector k_I.

Multiplying (3.77) with $f^X{}_{i\mathcal{A}} f^{j\mathcal{B}}{}_Y$ proves that $f^Y{}_{i\mathcal{A}} \nabla_Y k_I{}^X f^{j\mathcal{B}}{}_X$ should be proportional to δ_i^j. This leads to the definition of the matrices

$$t_{I\mathcal{A}}{}^{\mathcal{B}} = \frac{1}{2} f^Y{}_{i\mathcal{A}} \nabla_Y k_I{}^X f^{i\mathcal{B}}{}_X, \qquad f^Y{}_{i\mathcal{A}} \nabla_Y k_I{}^X f^{j\mathcal{B}}{}_X = \delta_i^j t_{I\mathcal{A}}{}^{\mathcal{B}}. \tag{3.79}$$

These matrices satisfy a reality and an almost covariant constancy equation[15]

$$\left(t_{I\mathcal{A}}{}^{\mathcal{B}}\right)^* = -\rho^{\bar{\mathcal{A}}C} t_{IC}{}^{\mathcal{D}} \rho_{\mathcal{D}\bar{\mathcal{B}}} = -t_I{}^{\bar{\mathcal{A}}}{}_{\bar{\mathcal{B}}}, \qquad \nabla_X t_{I\mathcal{A}}{}^{\mathcal{B}} = k_I^Y R_{XYA}{}^B, \tag{3.80}$$

as well as the commutation relations

$$[t_I, t_J]_{\mathcal{B}}{}^{\mathcal{A}} = f_{IJ}{}^K t_{K\mathcal{B}}{}^{\mathcal{A}} - k_I^X k_J^Y R_{XY\mathcal{B}}{}^{\mathcal{A}}, \tag{3.81}$$

which are consistent with (3.56).

The transformation of the fermions under the gauge group follows from the requirement that the commutator of supersymmetry and Killing symmetries vanishes. It is given by the above-defined matrices:

$$\widehat{\delta}_G(\theta) \zeta^{\mathcal{A}} = \theta^I t_{I\mathcal{B}}{}^{\mathcal{A}}(q) \zeta^{\mathcal{B}}. \tag{3.82}$$

For the coupling of the hypermultiplet to the vector gauge multiplets in the presence of the superconformal algebra, these isometries should be consistent with the conformal structure. The requirement that dilatations commute with the isometries is the equation

$$0 = k_D{}^Y \partial_Y k_I{}^X - k_I{}^Y \partial_Y k_D{}^X = k_D{}^Y \nabla_Y k_I{}^X - \frac{D-2}{2} k_I{}^X. \tag{3.83}$$

---

[15]Note that we defined $t_I{}^{\bar{\mathcal{A}}}{}_{\bar{\mathcal{B}}}$ using the common NW–SE convention for raising and lowering indices, and that the equation implies in this sense that t_I is imaginary.

This implies that the dilatations also commute with the SU(2) transformations generated by $\mathbf{k}^X$, defined in (3.69). This equation can also be written as

$$A^{i\mathcal{B}}t_{I\mathcal{B}}{}^{\mathcal{A}} = \frac{D-2}{2} f^{i\mathcal{A}}{}_X k_I{}^X . \tag{3.84}$$

One can also obtain the covariant transformation of $A^{i\mathcal{A}}$ (as for the other transformations in (3.72)), using (3.77) and (3.83)

$$\widehat{\delta}_G(\theta) A^{i\mathcal{A}} = A^{i\mathcal{B}} \theta^I t_{I\mathcal{B}}{}^{\mathcal{A}} . \tag{3.85}$$

3.2.3.6 Non-closure Relations in $D = 5$

We now have all the ingredients to understand the case when the isometry with index I is coupled to the gauge symmetry of the vector multiplet (label by index I)—see Sect. 3.2.1.2. The full form of (3.47) is now

$$\delta_Q(\epsilon) q^X = -i\bar{\epsilon}^i \zeta^{\mathcal{A}} f^X{}_{i\mathcal{A}} ,$$
$$\widehat{\delta}_Q(\epsilon) \zeta^{\mathcal{A}} = \tfrac{1}{2} i \not{D} q^X f^{i\mathcal{A}}{}_X \epsilon_i + \tfrac{1}{2} \sigma^I k_I{}^X f^{i\mathcal{A}}{}_X \epsilon_i , \tag{3.86}$$

with covariant derivatives defined as follows:

$$D_\mu q^X = \partial_\mu q^X - W_\mu^I k_I{}^X ,$$
$$\nabla_\mu \zeta^{\mathcal{A}} \equiv \partial_\mu \zeta^{\mathcal{A}} + \left(\partial_\mu q^X \right) \zeta^{\mathcal{B}} \omega_{X\mathcal{B}}{}^{\mathcal{A}} - W_\mu^I \zeta^{\mathcal{B}} t_{I\mathcal{B}}{}^{\mathcal{A}} . \tag{3.87}$$

Due to the gaugings, there are extra terms in the supersymmetry transformation of the fermions and the non-closure functions (3.63) and (3.65) are now modified to [14]

$$i\Gamma^{\mathcal{A}} = \not{\nabla} \zeta^{\mathcal{A}} + \frac{1}{2} W_{\mathcal{BCD}}{}^{\mathcal{A}} \zeta^{\mathcal{B}} \bar{\zeta}^C \zeta^{\mathcal{D}} - i k_I{}^X f^{i\mathcal{A}}{}_X \psi_i{}^I + i \zeta^{\mathcal{B}} \sigma^I t_{I\mathcal{B}}{}^{\mathcal{A}} ,$$

$$\Delta^X = \Box q^X - \frac{1}{2} \bar{\zeta}^{\mathcal{A}} \gamma_a \zeta^{\mathcal{B}} D^a q^Y W_Y{}^X{}_{\mathcal{AB}} - \frac{1}{4} f^X{}_{i\mathcal{A}} \varepsilon^{ij} f^Y{}_{j\mathcal{E}} \nabla_Y W_{\mathcal{BCD}}{}^{\mathcal{A}} \bar{\zeta}^{\mathcal{E}} \zeta^{\mathcal{D}} \bar{\zeta}^C \zeta^{\mathcal{B}}$$

$$\quad - k_I^Y J_Y{}^X \cdot \mathbf{Y}^I + \sigma^I \sigma^J k_J^Y \nabla_Y k_I^X$$

$$\quad + 2i \bar{\psi}^{iI} \zeta^{\mathcal{B}} t_{I\mathcal{B}}{}^{\mathcal{A}} f^X{}_{i\mathcal{A}} - \tfrac{1}{2} \sigma^I k_I^Y W_Y{}^X{}_{\mathcal{AB}} \bar{\zeta}^{\mathcal{A}} \zeta^{\mathcal{B}} , \tag{3.88}$$

where $\Box q^X$ is now also covariant for gauge transformations:

$$\Box q^X = \partial_a D^a q^X - D_a q^Y \nabla_Y k_I^X W^{aI} + D_a q^Y D^a q^Z \Gamma_{YZ}^X , \tag{3.89}$$

and we introduced the notation, using (3.41),

$$W_X{}^Y{}_{\mathcal{AB}} = L_X{}^Y{}_{\mathcal{C}}{}^{\mathcal{D}} W_{\mathcal{ABC}}{}^{\mathcal{D}} .$$

(3.90)

3.2.3.7   Rigid Superconformal Case in $D = 4$

To formulate the results in 4 dimensions, we consider the same bosonic fields q^X. The fermionic formulae have to be translated using the rules explained in Appendix A.4. This leads again to $2n_H$ spinors, whose left-handed part is $\zeta^{\mathcal{A}}$, with $\mathcal{A} = 1, \ldots, 2n_H$ and the left-handed ones (C-conjugates of the former) are $\zeta_{\bar{\mathcal{A}}}$. Thus, in absence of an SU(2) index on these spinors, the chirality is indicated by the fact that it has the index $\mathcal{A}$ up or down. One can start again by allowing arbitrary transformations for the scalars, and transformations of the spinors to derivatives of the scalars and deduce again the conditions on quantities that appear in these transformations. We would arrive again at (3.31) and (3.32). But as we have already done all the work for $D = 5$ (for which in fact the formalism is easier) we can also translate the results from what we already know.

This leads in 4 dimensions to the transformations [18, (20.33)]

$$\delta_Q(\epsilon)q^X = -i f^X{}_{i\mathcal{A}}\bar{\epsilon}^i \zeta^{\mathcal{A}} + i f^{Xi\bar{\mathcal{A}}}\bar{\epsilon}_i \zeta_{\bar{\mathcal{A}}} ,$$

$$\hat{\delta}_Q(\epsilon)\zeta^{\mathcal{A}} = \tfrac{1}{2} i f^{i\mathcal{A}}{}_X \not{D}q^X \epsilon_i + i \bar{X}^I k_I{}^X f^{i\mathcal{A}}{}_X \varepsilon_{ij}\epsilon^j ,$$

$$\delta_Q(\epsilon)\zeta_{\bar{\mathcal{A}}} = -\tfrac{1}{2} i f_{i\bar{\mathcal{A}}X} \not{D}q^X \epsilon^i - i X^I k_I{}^X f_{i\bar{\mathcal{A}}X} \varepsilon^{ij}\epsilon_j ,$$

(3.91)

where the complex conjugates of the frame fields are denoted as $f^{Xi\bar{\mathcal{A}}} = (f^X{}_{i\mathcal{A}})^*$ and $f_{i\bar{\mathcal{A}}X} = (f^{i\mathcal{A}}{}_X)^*$, see e.g. (A.31). $D_\mu q^X$ is given in (3.87).

The non-closure of the supersymmetries on the fermions is obtained in Appendix A.4 as an example of the translation rules from $D = 5$ to $D = 4$. The result is

$$\left[\delta_Q(\epsilon_1), \delta_Q(\epsilon_2)\right]\zeta^{\mathcal{A}} = \xi^\mu \partial_\mu \zeta^{\mathcal{A}} - \tfrac{1}{2}\varepsilon^{ij}\rho^{\bar{\mathcal{B}}\mathcal{A}}\Gamma_{\bar{\mathcal{B}}}\bar{\epsilon}_{1i}\epsilon_{2j} - \tfrac{1}{2}\gamma_\mu \bar{\epsilon}^i_{[1}\gamma^\mu \epsilon_{2]i}\Gamma^{\mathcal{A}} ,$$

(3.92)

with ξ^μ as in (1.6). The non-closure functions are

$$\Gamma^{\mathcal{A}} \equiv -\not{\nabla}\zeta^{\mathcal{A}} + \tfrac{1}{2}W_{\mathcal{BC}}{}^{\bar{\mathcal{D}}\mathcal{A}}\zeta_{\bar{\mathcal{D}}}\bar{\zeta}^{\mathcal{B}}\zeta^{\mathcal{C}} + 2\bar{X}^I t_{I}{}^{\bar{\mathcal{B}}\mathcal{A}}\zeta_{\bar{\mathcal{B}}} + i f^{i\mathcal{A}}{}_X k_I{}^X \varepsilon_{ij}\Omega^{Ij} ,$$

$$\Gamma_{\bar{\mathcal{A}}} \equiv -\not{\nabla}\zeta_{\bar{\mathcal{A}}} + \tfrac{1}{2}W^{\bar{\mathcal{B}}\bar{\mathcal{C}}}{}_{\mathcal{D}\bar{\mathcal{A}}}\bar{\zeta}_{\bar{\mathcal{B}}}\zeta_{\bar{\mathcal{C}}}\zeta^{\mathcal{D}} + 2X^I t_{I\mathcal{B}\bar{\mathcal{A}}}\zeta^{\mathcal{B}} + i k_I{}^X f^{i\mathcal{B}}{}_X \Omega^I_i \rho_{\mathcal{B}\bar{\mathcal{A}}} ,$$

(3.93)

where $W_{\mathcal{BC}}{}^{\bar{\mathcal{D}}\mathcal{A}} = \rho^{\bar{\mathcal{D}}\mathcal{E}}W_{\mathcal{BC}\mathcal{E}}{}^{\mathcal{A}}$ and $W^{\bar{\mathcal{B}}\bar{\mathcal{C}}}{}_{\mathcal{D}\bar{\mathcal{A}}}$ is its complex conjugate. We will raise or lower indices changing the holomorphicity with the tensors $\rho_{\mathcal{A}\bar{\mathcal{B}}}$ in NE–SW convention, e.g.

$$t_I{}^{\bar{\mathcal{A}}\mathcal{B}} = \rho^{\bar{\mathcal{A}}\mathcal{A}} t_{I\mathcal{A}}{}^{\mathcal{B}}, \qquad t_{I\mathcal{A}\bar{\mathcal{B}}} = t_{I\mathcal{A}}{}^{\mathcal{B}} \rho_{\mathcal{B}\bar{\mathcal{B}}} = \left(t_I{}^{\bar{\mathcal{A}}\mathcal{B}}\right)^*. \tag{3.94}$$

These fermionic non-closure functions transform in real bosonic quantities Δ^X as in (3.64)[16]:

$$\delta(\epsilon)\Gamma^{\mathcal{A}} = -\tfrac{1}{2}\mathrm{i} f^{i\mathcal{A}}{}_X \epsilon_i \Delta^X,$$

$$\Delta^X = \Box q^X + 2\left(X^I \bar{X}^J + X^J \bar{X}^I\right) k_I^Y \nabla_Y k_J{}^X - 2k_I{}^Y \mathbf{J}_Y{}^X \cdot \mathbf{Y}^I$$

$$+ X^I k_I{}^Y W_Y{}^X{}_{\mathcal{A}\mathcal{B}} \bar{\zeta}^{\mathcal{A}} \zeta^{\mathcal{B}} + \bar{X}^I k_I{}^Y W_Y{}^{X\bar{\mathcal{A}}\bar{\mathcal{B}}} \bar{\zeta}_{\bar{\mathcal{A}}} \zeta_{\bar{\mathcal{A}}} + \bar{\zeta}^A \gamma_a \zeta_{\bar{\mathcal{B}}} D_a q^Y W_Y{}^X{}_{\mathcal{A}}{}^{\bar{\mathcal{B}}}$$

$$+ \tfrac{1}{2} f^X{}_{i\mathcal{A}} \varepsilon^{ij} f^Y{}_{j\mathcal{B}} \nabla_Y W^{\bar{D}\bar{\mathcal{E}}}{}_C{}^A \bar{\zeta}^{\mathcal{B}} \zeta^C \bar{\zeta}_{\bar{D}} \zeta_{\bar{\mathcal{E}}}$$

$$- 2\mathrm{i} f^X{}_{i\mathcal{A}} \bar{\Omega}^{Ii} \zeta_{\bar{\mathcal{B}}} t_I{}^{\bar{\mathcal{B}}\mathcal{A}} + 2\mathrm{i} f^{Xi\bar{\mathcal{A}}} \bar{\Omega}_i^I \zeta^{\mathcal{B}} t_{I\mathcal{B}\bar{\mathcal{A}}}. \tag{3.95}$$

Finally, for the remaining U(1) factor in the R-symmetry group we find

$$\widehat{\delta}_{\mathrm{U}(1)} q^X = 0,$$

$$\widehat{\delta}_{\mathrm{U}(1)} \zeta^{\mathcal{A}} = \tfrac{1}{2} \mathrm{i} \lambda_T \zeta^{\mathcal{A}}. \tag{3.96}$$

3.2.4　Hypermultiplets in Superconformal Gravity

The previous results (Sect. 3.2.3) for rigid hypermultiplets can be generalized to local superconformal invariant theories by properly 'covariantizing' the previous expressions with respect to the superconformal algebra.

3.2.4.1　Case $D = 5$

The supersymmetry rules for the hypermultiplet coupled to the $D = 5$ standard Weyl multiplet and the gauge symmetry of the vector multiplet were found to be [14][17]

$$\delta q^X = -\mathrm{i} \bar{\epsilon}^i \zeta^{\mathcal{A}} f^X{}_{i\mathcal{A}},$$

$$\widehat{\delta}\zeta^{\mathcal{A}} = \tfrac{1}{2} \mathrm{i} \slashed{D} q^X f^{i\mathcal{A}}{}_X \epsilon_i + \tfrac{1}{2} \sigma^I k_I{}^X f^{i\mathcal{A}}{}_X \epsilon_i - A^{i\mathcal{A}} \eta_i. \tag{3.97}$$

---

[16]Note that we use here the translation between Y^{ij} and $\mathbf{Y}$ from (A.21), which will be used a lot further on.

[17]A few changes of notation can be found in (C.3).

The new ingredients with respect to (3.86) are the 'matter terms' of the Weyl multiplets and the S-supersymmetry. These transformations and the conformal and R-symmetry transformations determine the superconformal covariant derivatives

$$\mathcal{D}_\mu q^X = D_\mu q^X + i\bar{\psi}^i_\mu \zeta^{\mathcal{A}} f^X{}_{i\mathcal{A}},$$

$$D_\mu q^X = \partial_\mu q^X - b_\mu k^X_D - 2\mathbf{V}_\mu \cdot \mathbf{k}^X - W^I_\mu k_I{}^X,$$

$$\widehat{\mathcal{D}}_\mu \zeta^{\mathcal{A}} = \widehat{D}_\mu \zeta^{\mathcal{A}} - \tfrac{1}{2} i \not{D} q^X f^{i\mathcal{A}}{}_X \psi_{\mu i} - \tfrac{1}{3}\gamma \cdot T k_D{}^X f^{i\mathcal{A}}{}_X \psi_{\mu i} - \tfrac{1}{2}\sigma^I k^X_I f^{i\mathcal{A}}{}_X \psi_{\mu i}$$

$$+ A^{i\mathcal{A}} \phi_{\mu i}, \tag{3.98}$$

$$\widehat{D}_\mu \zeta^{\mathcal{A}} = \partial_\mu \zeta^{\mathcal{A}} + \tfrac{1}{4}\omega_\mu{}^{bc}\gamma_{bc}\zeta^{\mathcal{A}} - 2b_\mu \zeta^{\mathcal{A}} - W^I_\mu \zeta^{\mathcal{B}} t_{I\mathcal{B}}{}^{\mathcal{A}} + \partial_\mu q^X \omega_{X\mathcal{B}}{}^{\mathcal{A}}\zeta^{\mathcal{B}}.$$

The equations of motion for $\zeta^{\mathcal{A}}$ can be obtained by imposing the closure of the superconformal algebra

$$i\Gamma^{\mathcal{A}} \equiv \not{\mathcal{D}}\zeta^{\mathcal{A}} + \frac{1}{2}W_{\mathcal{BCD}}{}^{\mathcal{A}}\zeta^{\mathcal{B}}\bar{\zeta}^{\mathcal{C}}\zeta^{\mathcal{D}} + 2i\gamma^{ab}T_{ab}\zeta^{\mathcal{A}}$$

$$- ik_I{}^X f^{i\mathcal{A}}{}_X \psi^I_i + i\zeta^{\mathcal{B}}\sigma^I t_{I\mathcal{B}}{}^{\mathcal{A}} + \tfrac{8}{3}ik_D{}^X f^{i\mathcal{A}}{}_X \chi_i. \tag{3.99}$$

3.2.4.2 Case $D = 4$

The covariant supersymmetry transformations are those from (3.91) with only a replacement of D_μ by the fully covariant $\mathcal{D}_\mu$, which are

$$\mathcal{D}_\mu q^X = D_\mu q^X + i\bar{\psi}^i_\mu \zeta^{\mathcal{A}} f^X{}_{i\mathcal{A}} - i\varepsilon^{ij}\rho^{\bar{\mathcal{A}}\mathcal{B}}\bar{\psi}_{\mu i}\zeta_{\bar{\mathcal{A}}} f^X{}_{j\mathcal{B}},$$

$$D_\mu q^X = \partial_\mu q^X - b_\mu k_D{}^X + 2\mathbf{V}_\mu \cdot \mathbf{k}^X - W_\mu{}^I k_I{}^X, \tag{3.100}$$

$$\widehat{\mathcal{D}}_\mu \zeta^{\mathcal{A}} = \widehat{D}_\mu \zeta^{\mathcal{A}} - \tfrac{1}{2}i f^{i\mathcal{A}}{}_X \not{D} q^X \psi_{\mu i} - i\bar{X}^I k_I{}^X f^{i\mathcal{A}}{}_X \varepsilon_{ij}\psi^j_\mu - iA^{i\mathcal{A}}\phi_{\mu i},$$

$$\widehat{D}_\mu \zeta^{\mathcal{A}} = \left(\partial_\mu + \tfrac{1}{4}\omega_\mu{}^{ab}\gamma_{ab} - \tfrac{3}{2}b_\mu + \tfrac{1}{2}iA_\mu\right)\zeta^{\mathcal{A}} - W^I_\mu t_{I\mathcal{B}}{}^{\mathcal{A}}\zeta^{\mathcal{B}} + \partial_\mu q^X \omega_{X\mathcal{B}}{}^{\mathcal{A}}\zeta^{\mathcal{B}}.$$

Note that the hatted covariant derivatives are covariant for target space transformations as well and that $\partial_\mu q^X$ in the last term should not be covariantized to obtain this covariant expression $\widehat{D}_\mu \zeta^{\mathcal{A}}$. Because of central-charge like terms, the algebra does not close on the spinors. The new non-closure functions Γ^A will be used to derive the action for the hypermultiplet, as we will explain in Sect. 3.3.3.

In terms of $A^{i\mathcal{A}}$ (3.70), the covariant transformations are

$$\widehat{\delta}A^{i\mathcal{A}} \equiv \delta A^{i\mathcal{A}} + A^{i\mathcal{B}}\omega_{X\mathcal{B}}{}^{\mathcal{A}}\delta q^X = -i\bar{\epsilon}^i \zeta^{\mathcal{A}} + i\bar{\epsilon}_j \zeta_{\bar{\mathcal{B}}}\varepsilon^{ji}\rho^{\mathcal{B}\mathcal{A}},$$

$$\widehat{\delta}\zeta^{\mathcal{A}} = \tfrac{1}{2}i\not{\mathcal{D}}A^{i\mathcal{A}}\epsilon_i + i\bar{X}^I k_I{}^X f^{i\mathcal{A}}{}_X \varepsilon_{ij}\epsilon^j + iA^{i\mathcal{A}}\eta_i, \tag{3.101}$$

where we used

$$\widehat{\mathcal{D}}_\mu A^{i\mathcal{A}} = f^{i\mathcal{A}}{}_X \nabla_Y k_D{}^X \mathcal{D}_\mu q^Y = f^{i\mathcal{A}}{}_X \mathcal{D}_\mu q^X$$
$$= f^{i\mathcal{A}}{}_X \partial_\mu q^X - b_\mu A^{i\mathcal{A}} - A^{j\mathcal{A}} V_{\mu j}{}^i - W_\mu^I A^{i\mathcal{B}} t_{I\mathcal{B}}{}^{\mathcal{A}}$$
$$+ i\bar\psi_\mu^i \zeta^{\mathcal{A}} - i\bar\psi_{\mu j} \zeta_{\bar{\mathcal{B}}} \varepsilon^{ji} \rho^{\bar{\mathcal{B}}\mathcal{A}} . \tag{3.102}$$

Note that the $\widehat{\delta}$ used in (3.101) has no SU(2) connection, similar as in (3.32).

3.2.5   Tensor Multiplet in $D = 4$ Local Superconformal Case

The tensor multiplet in $D = 4$ dimensions was obtained in [7]. It is in fact the multiplet of the constraints (3.28). We can name these constraints, respectively, as L_{ij}, φ^i, E_b (satisfying a differential constraint) and G. These transform in each other and thus form a multiplet. It starts from an SU(2) triplet L_{ij} (hence satisfying the reality property as in (A.21)). The constrained E_a implies that the multiplet has a gauge tensor $E_{\mu\nu}$ (3 degrees of freedom) and a complex auxiliary G, to balance the 8 fermionic degrees of freedom in φ_i. The transformation rules in the background of conformal supergravity are[18]

$$\delta L_{ij} = \bar\epsilon_{(i}\varphi_{j)} + \varepsilon_{ik}\varepsilon_{j\ell} \bar\epsilon^{(k}\varphi^{\ell)} + 2\lambda_D L_{ij} ,$$

$$\delta\varphi^i = \tfrac{1}{2}\mathcal{D}L^{ij} \epsilon_j + \tfrac{1}{2}\varepsilon^{ij} \not{E} \epsilon_j - \tfrac{1}{2}G \epsilon^i + 2L^{ij} \eta_j + \left(\tfrac{5}{2}\lambda_D + \tfrac{1}{2}i\lambda_T\right) \varphi^i ,$$

$$\delta G = -\bar\epsilon_i \mathcal{D}\varphi^i - 3\bar\epsilon_i L^{ij} \chi_j + \tfrac{1}{8}\bar\epsilon_i \gamma^{ab} T_{ab}^+ \varphi_j \varepsilon^{ij} + 2\bar\eta_i \varphi^i + (3\lambda_D - i\lambda_T) G ,$$

$$\delta E_{\mu\nu} = \tfrac{1}{4}i\bar\epsilon^i \gamma_{\mu\nu}\varphi^j \varepsilon_{ij} - \tfrac{1}{4}i\bar\epsilon_i \gamma_{\mu\nu}\varphi_j \varepsilon^{ij} + \tfrac{1}{2}iL_{ij}\varepsilon^{jk}\bar\epsilon^i \gamma_{[\mu}\psi_{\nu]k} - \tfrac{1}{2}iL^{ij} \varepsilon_{jk}\bar\epsilon_i \gamma_{[\mu}\psi_{\nu]}{}^k , \tag{3.103}$$

where

$$E^\mu = e^{-1}\varepsilon^{\mu\nu\rho\sigma} \partial_\nu E_{\rho\sigma} - \tfrac{1}{2}\left(\bar\psi_\nu^i \gamma^{\mu\nu}\varphi^j \varepsilon_{ij} + \text{h.c.}\right) - \tfrac{1}{2}ie^{-1}\varepsilon^{\mu\nu\rho\sigma} L_{ij}\varepsilon^{jk}\bar\psi_\nu^i \gamma_\rho\psi_{\sigma k} . \tag{3.104}$$

A first step in building actions from this multiplet has been set in [7], but more applications can be found in [8].

---

[18]Of course the tensor multiplet for rigid supersymmetry can be obtained from (3.103) by setting to zero the fields of the Weyl multiplet (T and ψ_μ) and replacing the covariant derivatives by ordinary derivatives.

3.3 Construction of the Superconformal Actions

This section is devoted to the construction of local superconformally invariant actions for the vector and the hypermultiplet. As we have shown in the example of Sect. 3.1, later one gauge-fixes the extra symmetry such that the remaining theory has just the super-Poincaré invariance. Crucially, as we will explain in Chap. 4, for these last steps one needs to include compensating multiplets. Besides interacting matter, the resulting action from the gauge fixing will contain also the pure gravity sector.

3.3.1 Action for Vector Multiplets in $D = 4$

Let us consider the basic supergravity multiplet coupled to n vector multiplets. The physical content that one should have (from representation theory of the super-Poincaré group) can be represented in terms of particles with spin as follows:

$$
\begin{array}{ccccc}
\text{SUGRA} & & \text{vector multiplet} & & \\
2 & & & & \\
\frac{3}{2} & \frac{3}{2} & & & \\
1 & & 1 & \rightarrow n+1 & \\
& +n * & \frac{1}{2} & \frac{1}{2} & \\
& & 0 & 0 &
\end{array}
\tag{3.105}
$$

The supergravity sector contains the graviton, 2 gravitini and a so-called graviphoton, W_μ (that is a spin-1 field). When coupled to n vector multiplets, W_μ gets part of a set of $n + 1$ vectors, which will be uniformly described by the special Kähler geometry. The scalars inside these vector multiplets appear as n complex fields z^α, with $\alpha = 1, \ldots, n$.

In the framework of superconformal calculus, we consider $n + 1$ superconformal vector multiplets with scalars X^I ($I = 0, \ldots, n$) in the background of the Weyl multiplet (main formulae can be found in Sect. 3.2.1). One of these multiplets should contain the graviphoton, while we will use the missing fermions and scalars to fix superfluous gauge symmetries of the superconformal algebra.

Exploiting the fact that vector multiplets are constrained chiral multiplets (Sect. 3.2.2), we can build an action for the vector multiplet from an action for a chiral multiplet. The lowest component of the chiral multiplet should be $A = \frac{1}{2}iF(X)$,[19] being then A a new chiral superfield, given by an arbitrary holomorphic function of the scalars in vector multiplets. This function $F(X)$ will determine the

---

[19]The overall normalization is for later convenience to get a result with the normalization that is most used in the literature

action, and is called the prepotential. The further components are then defined by the transformation laws, which give, comparing with (3.21), $\Psi_i = \frac{1}{2}iF_I\Omega_i^I$, where we defined

$$F_I(X) = \frac{\partial}{\partial X^I}F(X)\,, \qquad \bar{F}_I(\bar{X}) = \frac{\partial}{\partial \bar{X}^I}\bar{F}(\bar{X})\,,$$

$$F_{IJ} = \frac{\partial}{\partial X^I}\frac{\partial}{\partial X^J}F(X) \quad \dots\,. \tag{3.106}$$

Calculating the transformation of Ψ_i one finds B_{ij}, G_{ab}^-, …

$$A = \tfrac{1}{2}iF$$

$$\Psi_i = \tfrac{1}{2}iF_I\Omega_i^I$$

$$B_{ij} = \tfrac{1}{2}iF_IY_{ij}^I - \tfrac{1}{8}iF_{IJ}\bar{\Omega}_i^I\Omega_j^J$$

$$G_{ab}^- = \tfrac{1}{2}iF_I\mathcal{F}_{ab}^{-I} - \tfrac{1}{16}iF_{IJ}\bar{\Omega}_i^I\gamma_{ab}\Omega_j^J\varepsilon^{ij}$$

$$\Lambda_i = -\tfrac{1}{2}iF_I\slashed{D}\Omega^{jI}\varepsilon_{ij} - \tfrac{1}{2}iF_If_{JK}^I\bar{X}^J\Omega_i^K - \tfrac{1}{8}iF_{IJ}\gamma^{ab}\mathcal{F}_{ab}^{-I}\Omega_i^J$$
$$- \tfrac{1}{2}iF_{IJ}\Omega_k^JY_{ij}^I\varepsilon^{jk} + \tfrac{1}{96}iF_{IJK}\gamma^{ab}\Omega_i^I\bar{\Omega}_j^J\gamma_{ab}\Omega_k^K\varepsilon^{jk}$$

$$C = -iF_ID_aD^a\bar{X}^I - \tfrac{1}{4}iF_I\mathcal{F}_{ab}^{+I}T^{+ab} - \tfrac{3}{2}iF_I\bar{\chi}_i\Omega^{iI} + \tfrac{1}{2}iF_If_{JK}^I\bar{\Omega}^{iJ}\Omega^{jK}\varepsilon_{ij}$$
$$-iF_If_{JK}^If_{LM}^J\bar{X}^K\bar{X}^LX^M - \tfrac{1}{2}iF_{IJ}Y^{ijI}Y_{ij}^J + \tfrac{1}{4}iF_{IJ}\mathcal{F}_{ab}^{-I}\mathcal{F}^{-abJ}$$
$$+\tfrac{1}{2}iF_{IJ}\bar{\Omega}_i^I\slashed{D}\Omega^{iJ} - \tfrac{1}{2}iF_{IJ}f_{KL}^I\bar{X}^K\bar{\Omega}_i^J\Omega_j^L\varepsilon^{ij} + \tfrac{1}{4}iF_{IJK}Y^{ijI}\bar{\Omega}_i^J\Omega_j^K$$
$$-\tfrac{1}{16}iF_{IJK}\varepsilon^{ij}\bar{\Omega}_i^I\gamma^{ab}\mathcal{F}_{ab}^{-J}\Omega_j^K + \tfrac{1}{48}iF_{IJKL}\bar{\Omega}_i^I\Omega_l^J\bar{\Omega}_j^K\Omega_k^L\varepsilon^{ij}\varepsilon^{kl}\,. \tag{3.107}$$

This is the composite chiral multiplet that we discussed at the end of Sect. 3.2.2, and on which we can apply the 'density formula' (3.30). As mentioned, $F(X)$ must be homogeneous of weight 2, where the X fields carry weight 1. This implies the following relations for the derivatives of F:

$$2F = F_IX^I\,, \qquad F_{IJ}X^J = F_I\,, \qquad F_{IJK}X^K = 0\,. \tag{3.108}$$

Inserting (3.107) in (3.30) leads to

$$e^{-1}\mathcal{L}_g = -iF_ID_aD^a\bar{X}^I + \tfrac{1}{4}iF_{IJ}\mathcal{F}_{ab}^{-I}\mathcal{F}^{-abJ} + \tfrac{1}{2}iF_{IJ}\bar{\Omega}_i^I\slashed{D}\Omega^{iJ}$$
$$-\tfrac{1}{2}iF_{IJ}Y^{ijI}Y_{ij}^J + \tfrac{1}{4}iF_{IJK}Y^{ijI}\bar{\Omega}_i^J\Omega_j^K$$
$$-\tfrac{1}{16}iF_{IJK}\varepsilon^{ij}\bar{\Omega}_i^I\gamma^{ab}\mathcal{F}_{ab}^{-J}\Omega_j^K + \tfrac{1}{48}iF_{IJKL}\bar{\Omega}_i^I\Omega_\ell^J\bar{\Omega}_j^K\Omega_k^L\varepsilon^{ij}\varepsilon^{k\ell}$$
$$+\tfrac{1}{2}iF_If_{JK}^I\bar{\Omega}^{iJ}\Omega^{jK}\varepsilon_{ij} - \tfrac{1}{2}iF_{IJ}f_{KL}^I\bar{X}^K\bar{\Omega}_i^J\Omega_j^L\varepsilon^{ij}$$

$$-\mathrm{i}F_I f^I_{JK} f^J_{LM} \bar{X}^K \bar{X}^L X^M$$

$$-\tfrac{1}{4}\mathrm{i}F_I \mathcal{F}^{+I}_{ab} T^{+ab} - \tfrac{3}{2}\mathrm{i}F_I \bar{\chi}_i \Omega^{iI} - \tfrac{1}{2}\mathrm{i}F_{IJ}\bar{\psi}_i \cdot \gamma \Omega^I_j Y^{ijJ}$$

$$+\tfrac{1}{2}\mathrm{i}F_I f^I_{JK} \bar{X}^J \bar{\psi}_i \cdot \gamma \Omega^K_j \varepsilon^{ij} - \tfrac{1}{2}\mathrm{i}F_I \bar{\psi}_i \cdot \gamma \slashed{D}\Omega^{iI}$$

$$+\tfrac{1}{8}\mathrm{i}F_{IJ}\mathcal{F}^{-I}_{ab} \bar{\psi}_i \cdot \gamma \gamma^{ab}\Omega^J_j \varepsilon^{ij}$$

$$+\tfrac{1}{12}\mathrm{i}F_{IJK}\bar{\Omega}^J_\ell \Omega^K_j \bar{\psi}_i \cdot \gamma \Omega^I_k \varepsilon^{ij}\varepsilon^{k\ell} + \tfrac{1}{16}\mathrm{i}F_I \bar{\psi}_{\mu i}\gamma \cdot T^+ \gamma^\mu \Omega^I_j \varepsilon^{ij}$$

$$-\tfrac{1}{8}\mathrm{i}F T^+_{ab}T^{+ab}$$

$$-\tfrac{1}{2}\mathrm{i}F_I \bar{\psi}_{\mu i}\gamma^{\mu\nu}\psi_{vj} Y^{ijI} - \tfrac{1}{2}\mathrm{i}F T^{+\mu\nu}\bar{\psi}_{\mu i}\psi_{vj}\varepsilon^{ij} + \tfrac{1}{2}\mathrm{i}F_I \mathcal{F}^{-\mu\nu I}\bar{\psi}_{\mu i}\psi_{vj}\varepsilon^{ij}$$

$$-\tfrac{1}{16}\mathrm{i}F_{IJ}\bar{\psi}_{\mu i}\psi_{vj}\bar{\Omega}^I_k \gamma^{\mu\nu}\Omega^J_\ell \varepsilon^{ij}\varepsilon^{k\ell} + \tfrac{1}{8}\mathrm{i}F_{IJ}\bar{\Omega}^I_k \Omega^J_\ell \bar{\psi}_{\mu i}\gamma^{\mu\nu}\psi_{vj}\varepsilon^{ik}\varepsilon^{j\ell}$$

$$-\tfrac{1}{4}\varepsilon^{ij}\varepsilon^{k\ell}e^{-1}\varepsilon^{\mu\nu\rho\sigma}\bar{\psi}_{\mu i}\psi_{vj}\bar{\psi}_{\rho k}\left(\gamma_\sigma F_I \Omega^I_\ell + F\psi_{\sigma\ell}\right) + \text{h.c.} \qquad (3.109)$$

The first terms of the action (3.109) are kinetic terms for the scalars X, the vectors and the fermions Ω. The following term says that Y_{ij} is an auxiliary field that can be eliminated by its field equation. The first 5 lines are the ones that we would encounter also in rigid supersymmetry, see [18, (20.15)]. For these terms, the relations (3.108) have not been used, and this part is thus the general result for rigid supersymmetry. The other lines are due to the local superconformal symmetry. For those interested in rigid symmetry, we repeat that in that case the covariant derivatives (3.19) reduce to, e.g.,

$$D_a X^I = \partial_a X^I - W^K_a X^J f_{JK}{}^I,$$

$$D_a \Omega^I_i = \partial_a \Omega^I_i - W^K_a \Omega^J_i f_{JK}{}^I,$$

$$\mathcal{F}^I_{ab} = 2\partial_{[a}W^I_{b]} + W^K_b W^J_a f_{JK}{}^I. \qquad (3.110)$$

Note that the Lagrangian is a total derivative if $F(X)$ is a quadratic function of X^I with real coefficients:

$$F(X) = C_{IJ}X^I X^J, \qquad C_{IJ} \in \mathbb{R} \qquad \rightarrow \qquad S = \int \mathrm{d}^4 x \, \mathcal{L}_g = 0. \qquad (3.111)$$

In deriving the above formulae, we assumed for simplicity that F is a gauge-invariant function such that $\mathcal{L}_g$ is invariant under gauge transformations. However, the property (3.111) suggests that there is a more general situation [37, 38] in which F transforms under the gauge transformations as

$$\delta_G(\theta)F \equiv F_I \theta^K X^J f_{JK}{}^I = -\theta^I C_{I,JK}X^J X^K, \qquad (3.112)$$

where $C_{I,JK}$ are real constants. In fact, due to (3.111), the action is invariant for rigid transformations that satisfy (3.112), transforming to

$$\delta_G \mathcal{L}_g = \tfrac{1}{3} C_{I,JK} \varepsilon^{\mu\nu\rho\sigma} \theta^I F_{\mu\nu}^J F_{\rho\sigma}^K \,.$$ (3.113)

In order to allow this extra possibility with local θ^I, one has to add to (3.109) a Chern–Simons term

$$\mathcal{L}_{CS} = \tfrac{2}{3} C_{I,JK} \varepsilon^{\mu\nu\rho\sigma} W_\mu{}^I W_\nu{}^J \left(\partial_\rho W_\sigma{}^K + \tfrac{3}{8} f_{LM}{}^K W_\rho{}^L W_\sigma{}^M \right) \,.$$ (3.114)

To prove the supersymmetry invariance of $\mathcal{L}_g + \mathcal{L}_{CS}$ one needs a few more relations that follow from (3.113). Replacing the arbitrary θ^K by X^K the variation vanishes, and thus for the consistency of (3.112) we should have

$$C_{(I,JK)} X^I X^J X^K = 0 \,,$$ (3.115)

namely the completely symmetric part of $C_{I,JK}$ must vanish.

By taking two derivatives of (3.112) we obtain

$$C_{K,IJ} = f_{K(I}{}^L F_{J)L} - \tfrac{1}{2} F_{IJL} X^M f_{MK}{}^L = f_{K(I}{}^L \bar{F}_{J)L} - \tfrac{1}{2} \bar{F}_{IJL} \bar{X}^M f_{MK}{}^L \,.$$ (3.116)

To prove the invariance of the sum of (3.109) and (3.114) one needs an identity [37]

$$f_{KL}{}^M C_{M,IJ} = 2 f_{J[K}{}^M C_{L],IM} + 2 f_{I[K}{}^M C_{L],JM} \,,$$ (3.117)

which follows from the requirement that the gauge group closes on $F(X)$. A simple example of the occurrence of a Chern–Simons term is given in [37, (3.21)].

3.3.1.1 Simplifications

In order to get a more useful form of the action, one has to make the conformal covariant derivatives explicit. The principle is explained for the bosonic case in (3.2). This leads here to

$$\Box^C \bar{X}^I = \widehat{\partial}_\mu D^\mu \bar{X}^I - \omega_\mu{}^{\mu\nu} D_\nu \bar{X}^I - i W_\mu D^\mu \bar{X}^I + 2 f_\mu{}^\mu \bar{X}^I - \tfrac{1}{2} \bar{\psi}_{\mu i} D^\mu \Omega^{iI}$$
$$+ \tfrac{1}{32} \bar{\psi}_\mu^i \gamma^\mu \gamma \cdot T^+ \Omega^{jI} \varepsilon_{ij} - \tfrac{1}{2} \bar{\Omega}^{iI} \gamma \cdot \phi_i - \tfrac{3}{4} \bar{\psi}^i \cdot \gamma \chi_i \bar{X}^I$$
$$- \tfrac{1}{2} \varepsilon^{ij} \bar{\psi}_i \cdot \gamma \Omega_j^J \bar{X}^K f_{JK}^I - \tfrac{1}{2} \varepsilon_{ij} \bar{\psi}^i \cdot \gamma \Omega^{jJ} \bar{X}^K f_{JK}^I \,.$$ (3.118)

Hence the first term of (3.109) after adding a total derivative, is

$$
-\mathrm{i}F_I\Box^C\bar{X}^I = \mathrm{i}F_{IJ}\mathcal{D}_\mu X^I\left(\mathcal{D}^\mu\bar{X}^J - \tfrac{1}{2}\bar{\psi}_i^\mu\Omega^{iJ}\right) - 2\mathrm{i}F_I f_\mu{}^\mu\bar{X}^I + \tfrac{1}{2}\mathrm{i}F_I\bar{\psi}_{\mu i}D^\mu\Omega^{iI}
$$

$$
- \tfrac{1}{32}\mathrm{i}F_I\bar{\psi}_\mu^i\gamma^\mu\gamma\cdot T^-\tfrac{1}{2}\Omega^{jI}\varepsilon_{ij} + \tfrac{1}{2}\mathrm{i}F_I\bar{\Omega}^{iI}\gamma\cdot\phi_i + \tfrac{3}{4}\mathrm{i}F_I\bar{\psi}^i\cdot\gamma\chi_i\bar{X}^I
$$

$$
+ \tfrac{1}{2}\mathrm{i}F_I\varepsilon^{ij}\bar{\psi}_i\cdot\gamma\Omega_j^J\bar{X}^K f_{JK}^I + \tfrac{1}{2}\mathrm{i}F_I\varepsilon_{ij}\bar{\psi}^i\cdot\gamma\Omega^{jJ}\bar{X}^K f_{JK}^I
$$

$$
+ \mathrm{i}F_I\bar{\psi}_{[\mu}^i\gamma^\nu\psi_{\nu]i}\left(\mathcal{D}^\mu\bar{X}^I - \tfrac{1}{2}\bar{\psi}_i^\mu\Omega^{iI}\right)
$$

$$
+ \text{total derivative.} \tag{3.119}
$$

The other term that has to be written explicitly is the covariant derivative of the fermions

$$
\slashed{\mathcal{D}}\Omega^{iJ} = \slashed{\mathcal{D}}\Omega^{iJ} - \gamma^\mu\gamma^\nu\psi_\mu^i\left(\mathcal{D}_\nu\bar{X}^J - \tfrac{1}{2}\bar{\psi}_{\nu j}\Omega^{jJ}\right)
$$

$$
- \tfrac{1}{4}\gamma_\mu\gamma\cdot\mathcal{F}^{+J}\psi_j^\mu\varepsilon^{ij} - \gamma\cdot\psi_j\left(Y^{ijJ} + \varepsilon^{ij}\bar{X}^K X^L f_{KL}^J\right) - 2\bar{X}^J\gamma\cdot\phi^i.
$$

$$
\tag{3.120}
$$

Deleting total derivatives, the action is at this point (adding also (3.114))

$$
e^{-1}\mathcal{L}_g = \mathrm{i}F_{IJ}\mathcal{D}_\mu X^I\mathcal{D}^\mu\bar{X}^J - 2\mathrm{i}F_I f_\mu{}^\mu\bar{X}^I + \tfrac{1}{4}\mathrm{i}F_{IJ}\mathcal{F}_{ab}^{-I}\mathcal{F}^{-abJ} - \tfrac{1}{8}\mathrm{i}F T_{ab}^+ T^{+ab}
$$

$$
- \tfrac{1}{4}\mathrm{i}F_I\mathcal{F}_{ab}^{+I}T^{+ab} - \tfrac{1}{2}\mathrm{i}F_{IJ}Y^{ijI}Y_{ij}^J - \mathrm{i}F_I f_{JK}^I f_{LM}^J\bar{X}^K\bar{X}^L X^M + e^{-1}\mathcal{L}_{CS}
$$

$$
+ \mathrm{i}F_I\bar{X}^I\bar{\psi}_{\mu i}\gamma^{\mu\nu}\phi_\nu^i + \tfrac{1}{2}\mathrm{i}F_{IJ}\bar{\Omega}_i^I\slashed{\mathcal{D}}\Omega^{iJ} + \tfrac{1}{2}\mathrm{i}F_I\bar{\psi}_{\mu i}\gamma^{\mu\nu}\gamma^\rho\mathcal{D}_\rho\bar{X}^I\psi_\nu^i
$$

$$
+ \tfrac{1}{2}\mathrm{i}F_I\bar{\Omega}^{iI}\gamma\cdot\phi_i
$$

$$
+ \tfrac{1}{8}\mathrm{i}F_I\bar{\psi}_{\mu i}\gamma^{\mu\nu}\gamma\cdot\mathcal{F}^{+I}\varepsilon^{ij}\psi_{\nu j} - \tfrac{1}{32}\mathrm{i}F_I\bar{\psi}_\mu^i\gamma^\mu\gamma\cdot T^-\Omega^{jI}\varepsilon_{ij}
$$

$$
+ \tfrac{3}{4}\mathrm{i}F_I\bar{\psi}^i\cdot\gamma\chi_i\bar{X}^I
$$

$$
+ \tfrac{1}{2}\mathrm{i}F_I\varepsilon^{ij}\bar{\psi}_i\cdot\gamma\Omega_j^J\bar{X}^K f_{JK}^I + \tfrac{1}{2}\mathrm{i}F_I\varepsilon_{ij}\bar{\psi}^i\cdot\gamma\Omega^{jJ}\bar{X}^K f_{JK}^I
$$

$$
+ \mathrm{i}F_I\bar{\psi}_{[\mu}^i\gamma^\nu\psi_{\nu]i}\left(\mathcal{D}^\mu\bar{X}^I - \tfrac{1}{2}\bar{\psi}_i^\mu\Omega^{iI}\right) - \tfrac{3}{2}\mathrm{i}F_I\bar{\chi}_i\Omega^{iI}
$$

$$
+ \tfrac{1}{2}\mathrm{i}F_I f_{JK}^I\bar{\Omega}^{iJ}\Omega^{jK}\varepsilon_{ij}
$$

$$
- \tfrac{1}{2}\mathrm{i}F_{IJ}\mathcal{D}_\mu X^J\bar{\psi}_i^\mu\Omega^{iI} - \tfrac{1}{2}\mathrm{i}F_I\bar{\psi}_{\mu i}\gamma^{\mu\nu}\mathcal{D}_\nu\Omega^{iI}
$$

$$
- \tfrac{1}{2}\mathrm{i}F_{IJ}\bar{\Omega}_i^I\gamma^\mu\gamma^\nu\psi_\mu^i\mathcal{D}_\nu\bar{X}^J + \tfrac{1}{2}\mathrm{i}F_I\mathcal{F}^{-\mu\nu I}\bar{\psi}_{\mu i}\psi_{\nu j}\varepsilon^{ij}
$$

$$
- \tfrac{1}{8}\mathrm{i}F_{IJ}\bar{\Omega}_i^I\gamma_\mu\gamma\cdot\mathcal{F}^{+J}\psi_j^\mu\varepsilon^{ij} - \tfrac{1}{2}\mathrm{i}F_{IJ}\bar{\Omega}_i^I\gamma\cdot\psi_j\varepsilon^{ij}\bar{X}^K X^L f_{KL}^J
$$

$$
- \mathrm{i}F_{IJ}\bar{X}^J\bar{\Omega}_i^I\gamma\cdot\phi^i - \tfrac{1}{2}\mathrm{i}F_{IJ}f_{KL}^I\bar{X}^K\bar{\Omega}_i^J\Omega_j^L\varepsilon^{ij} + \tfrac{1}{4}\mathrm{i}F_{IJK}Y^{ijI}\bar{\Omega}_i^J\Omega_j^K
$$

$$
- \tfrac{1}{16}\mathrm{i}F_{IJK}\varepsilon^{ij}\bar{\Omega}_i^I\gamma^{ab}\mathcal{F}_{ab}^{-J}\Omega_j^K
$$

$$- \tfrac{1}{2} i \varepsilon^{ij} \bar{\psi}_{\mu i} \left(F T^{+\mu\nu} \psi_{\nu j} - \tfrac{1}{8} \gamma \cdot T^{+} \gamma^{\mu} F_I \Omega_j^I \right)$$

$$+ \tfrac{1}{2} i F_I f_{JK}^I \bar{X}^J \bar{\psi}_i \cdot \gamma \Omega_j^K \varepsilon^{ij} + \tfrac{1}{8} i F_{IJ} \mathcal{F}_{ab}^{-I} \bar{\psi}_i \cdot \gamma \gamma^{ab} \Omega_j^J \varepsilon^{ij}$$

$$+ \tfrac{1}{12} i F_{IJK} \bar{\Omega}_\ell^J \Omega_j^K \bar{\psi}_i \cdot \gamma \Omega_k^L \varepsilon^{ij} \varepsilon^{k\ell} - \tfrac{1}{4} i F_I \bar{\psi}_{\mu i} \gamma^{\mu\nu} \gamma^\rho \psi_\nu^i \bar{\psi}_{\rho k} \Omega^{kI}$$

$$+ \tfrac{1}{8} i F_{IJ} \bar{\Omega}_k^I \Omega_\ell^J \bar{\psi}_{\mu i} \gamma^{\mu\nu} \psi_{\nu j} \varepsilon^{ik} \varepsilon^{j\ell} + \tfrac{1}{4} i F_{IJ} \bar{\Omega}_i^I \gamma^\mu \gamma^\nu \psi_\mu^i \bar{\psi}_{\nu j} \Omega^{jJ}$$

$$+ \tfrac{1}{48} i F_{IJKL} \bar{\Omega}_i^I \Omega_\ell^J \bar{\Omega}_j^K \Omega_k^L \varepsilon^{ij} \varepsilon^{k\ell} - \tfrac{1}{16} i F_{IJ} \bar{\psi}_{\mu i} \psi_{\nu j} \bar{\Omega}_k^I \gamma^{\mu\nu} \Omega_\ell^J \varepsilon^{ij} \varepsilon^{k\ell}$$

$$- \tfrac{1}{4} \varepsilon^{ij} \varepsilon^{k\ell} e^{-1} \varepsilon^{\mu\nu\rho\sigma} \bar{\psi}_{\mu i} \psi_{\nu j} \bar{\psi}_{\rho k} \left(\gamma_\sigma F_I \Omega_\ell^I + F \psi_{\sigma\ell} \right) + \text{h.c.} \qquad (3.121)$$

We use then (3.16) and the values of the conformal gauge fields that follow from the constraints:

$$f_\mu{}^\mu = -\tfrac{1}{12} R - \tfrac{1}{2} D$$

$$+ \left\{ \tfrac{1}{8} \bar{\psi}^i \cdot \gamma \chi_i + \tfrac{1}{24} i e^{-1} \varepsilon^{\mu\nu\rho\sigma} \bar{\psi}_\mu^i \gamma_\nu \mathcal{D}_\rho \psi_{\sigma i} + \tfrac{1}{24} \bar{\psi}_\mu^i \psi_\nu^j \varepsilon_{ij} T^{+\mu\nu} + \text{h.c.} \right\},$$

$$\phi_\mu^i = \tfrac{1}{4} \gamma_\mu \chi^i + \tfrac{1}{4} \left(\gamma^{\nu\rho} \gamma_\mu - \tfrac{1}{3} \gamma_\mu \gamma^{\nu\rho} \right) \left(\mathcal{D}_\nu \psi_\rho^i - \tfrac{1}{16} \gamma \cdot T^{-} \varepsilon^{ij} \gamma_\nu \psi_{\rho j} \right).$$

$$(3.122)$$

This leads to various simplifications, after which the vector action reduces to [18, (20.89)]:

$$e^{-1} \mathcal{L}_g = -\tfrac{1}{6} N R - N D - N_{IJ} D_\mu X^I D^\mu \bar{X}^J + N_{IJ} \mathbf{Y}^I \cdot \mathbf{Y}^J$$

$$+ N_{IJ} f_{KL}{}^I \bar{X}^K X^L f_{MN}{}^J \bar{X}^M X^N + e^{-1} \mathcal{L}_{CS}$$

$$+ \left\{ -\tfrac{1}{4} i \bar{F}_{IJ} \hat{F}_{\mu\nu}^{+I} \hat{F}^{+\mu\nu J} - \tfrac{1}{16} N_{IJ} X^I X^J T_{ab}^{+} T^{+ab} + \tfrac{1}{4} N_{IJ} X^I \hat{F}_{ab}^{+J} T^{+ab} \right.$$

$$- \tfrac{1}{4} N_{IJ} \bar{\Omega}^{iI} \slashed{D} \Omega_i^J + \tfrac{1}{6} N \bar{\psi}_{i\mu} \gamma^{\mu\nu\rho} D_\nu \psi_\rho^i$$

$$- \tfrac{1}{2} N \bar{\psi}_{ia} \gamma^a \chi^i + N_{IJ} X^I \bar{\Omega}^{iJ} \chi_i - \tfrac{1}{3} N_{IJ} X^J \bar{\Omega}^{iI} \gamma^{\mu\nu} D_\mu \psi_{\nu i}$$

$$+ \tfrac{1}{2} N_{IJ} \bar{\psi}_\mu^i \slashed{D} \bar{X}^I \gamma^\mu \Omega_i^J + \tfrac{1}{4} N_{IJ} \bar{X}^I \bar{\psi}_{ai} \gamma^{abc} \psi_b^i D_c X^J$$

$$+ \tfrac{1}{2} i \bar{F}_{IJ} \varepsilon_{ij} \left(\bar{\Omega}^{iI} \gamma_\mu - \bar{X}^I \bar{\psi}_\mu^i \right) \psi_\nu^j \tilde{\hat{F}}^{\mu\nu J} - \tfrac{1}{16} i F_{IJK} \bar{\Omega}_i^I \gamma^{\mu\nu} \Omega_j^J \varepsilon^{ij} F_{\mu\nu}^{-K}$$

$$+ \tfrac{1}{2} N_{IJ} \bar{\Omega}_i^I f_{KL}{}^J \left(\Omega_j^L + \gamma^a \psi_{aj} X^L \right) \bar{X}^K \varepsilon^{ij}$$

$$+ \left(\tfrac{1}{12} N \bar{\psi}_i^a \psi_j^b - \tfrac{1}{6} N_{IJ} \bar{X}^I \bar{\Omega}_i^J \gamma^a \psi_j^b + \tfrac{1}{32} i F_{IJK} \bar{\Omega}_i^I \gamma^{ab} \Omega_j^J \bar{X}^K \right) T_{ab}^{-} \varepsilon^{ij}$$

$$\left. - \tfrac{1}{4} i F_{IJK} D_\mu X^I \bar{\Omega}_i^J \gamma^\mu \Omega^{iK} + \tfrac{1}{4} i F_{IJK} Y^{ijI} \bar{\Omega}_i^J \Omega_j^K + \text{h.c.} \right\}$$

$$+ \text{4-fermion terms} . \qquad (3.123)$$

Here, important quantities are introduced, which will often be used below:

$$N_{IJ} = N_{IJ}(X, \bar{X}) \equiv 2 \operatorname{Im} F_{IJ} = -i F_{IJ} + i \bar{F}_{IJ}, \qquad N \equiv N_{IJ} X^I \bar{X}^J. \tag{3.124}$$

Since F_{IJ} is a function of X^K, we have the chain rule for the gauge transformations:

$$\delta F_{IJ} = F_{IJK} \delta X^K = F_{IJK} \theta^L X^M f_{ML}{}^K, \tag{3.125}$$

and therefore the gauge transformation of N_{IJ} is by (3.116)

$$\delta N_{IJ} = 2\theta^K f_{K(I}{}^L N_{J)L}. \tag{3.126}$$

Covariant derivatives are presented in (3.19) together with

$$D_\mu \psi_{vi} = \left(\partial_\mu + \tfrac{1}{4} \omega_\mu{}^{ab} \gamma_{ab} + \tfrac{1}{2} b_\mu + \tfrac{1}{2} i A_\mu \right) \psi_{vi} + V_{\mu i}{}^j \psi_{vj}. \tag{3.127}$$

3.3.2 Action for Vector Multiplets in $D = 5$

The scheme for $D = 5$ is similar to (3.105) with the only exception that vector multiplets[20] have only one real scalar, and thus at the end we will have n real scalars, which are the σ^I that we saw already in Sect. 3.2.1. For $D = 6$, which we will not treat in detail here, it is also be similar but without scalars in the vector multiplets.

The rigid superconformally invariant action $D = 5$ is determined by a prepotential $C_{IJK}\sigma^I\sigma^J\sigma^K$ cubic in the scalars σ^I [39, 40]. Since the vectors W_μ^I can gauge a group, I is also the index of the adjoint of the gauge group. As a consequence, the local superconformal action is determined by a gauge-invariant symmetric tensor C_{IJK}

$$f_{I(J}{}^M C_{KL)M} = 0. \tag{3.128}$$

---

[20]We use here and below freely the terminology 'spin 1' for vectors, spin-$\tfrac{1}{2}$ for spinors, ..., though of course only in 4 dimensions the representations of the little group of the Lorentz group can be characterized by just one number, which is called 'spin'. In higher dimensions, the representations should be characterized by more numbers, but often the same fields, like graviton as a symmetric tensor, vectors, ... occur, and we denote them freely with the terminology that is appropriate for the 4-dimensional fields.

Note that this tensor has no relation to the tensor $C_{I,JK}$ introduced for $D = 4$ in (3.112). The presence of this tensor allows Chern–Simons terms in the action. The kinetic term of the scalars defines the 'very special real geometry' (see [18, Sect. 20.3.2]). The full superconformal invariant action takes the form [5]

$$
\begin{aligned}
e^{-1}\mathcal{L}_g = C_{IJK} \Bigg[&\left(-\tfrac{1}{4}\widehat{F}_{\mu\nu}^I \widehat{F}^{\mu\nu J} - \tfrac{1}{2}\bar{\psi}^I \slashed{D}\psi^J + \tfrac{1}{3}\sigma^I \Box^c \sigma^J \right. \\
&+ \tfrac{1}{6}\mathcal{D}_a\sigma^I \mathcal{D}^a\sigma^J + 2\mathbf{Y}^I \cdot \mathbf{Y}^J \Bigg) \sigma^K \\
&- \tfrac{4}{3}\sigma^I\sigma^J\sigma^K \left(D + \tfrac{26}{3}T_{ab}T^{ab} \right) + 4\sigma^I\sigma^J \widehat{F}_{ab}^K T^{ab} \\
&- \tfrac{1}{8}\mathrm{i}\bar{\psi}^I \gamma \cdot \widehat{F}^J \psi^K - \tfrac{1}{2}\mathrm{i}\bar{\psi}^{iI} \psi^{jJ} Y_{ij}^K + \mathrm{i}\sigma^I \bar{\psi}^J \gamma \cdot T\psi^K - 8\mathrm{i}\sigma^I\sigma^J \bar{\psi}^K \chi \\
&+ \tfrac{1}{6}\sigma^I \bar{\psi}_\mu \gamma^\mu \left(\mathrm{i}\sigma^J \slashed{D}\psi^K + \tfrac{1}{2}\mathrm{i}(\slashed{D}\sigma^J)\psi^K - \tfrac{1}{4}\gamma\cdot\widehat{F}^J\psi^K + 2\sigma^J\gamma\cdot T\psi^K \right. \\
&\left. - 8\sigma^J\sigma^K\chi \right) \\
&- \tfrac{1}{6}\bar{\psi}_a\gamma_b\psi^I \left(\sigma^J\widehat{F}^{abK} - 8\sigma^J\sigma^K T^{ab} \right) - \tfrac{1}{12}\sigma^I\bar{\psi}_\lambda\gamma^{\mu\nu\lambda}\psi^J\widehat{F}_{\mu\nu}^K \\
&+ \tfrac{1}{12}\mathrm{i}\sigma^I\bar{\psi}_a\psi_b \left(\sigma^J\widehat{F}^{abK} - 8\sigma^J\sigma^K T^{ab} \right) + \tfrac{1}{48}\mathrm{i}\sigma^I\sigma^J\bar{\psi}_\lambda\gamma^{\mu\nu\lambda\rho}\psi_\rho\widehat{F}_{\mu\nu}^K \\
&- \tfrac{1}{2}\sigma^I\bar{\psi}_\mu^i\gamma^\mu\psi^{jJ}Y_{ij}^K + \tfrac{1}{6}\mathrm{i}\sigma^I\sigma^J\bar{\psi}_\mu^i\gamma^{\mu\nu}\psi_\nu^j Y_{ij}^K - \tfrac{1}{24}\mathrm{i}\bar{\psi}_\mu\gamma_\nu\psi^I\bar{\psi}^J\gamma^{\mu\nu}\psi^K \\
&+ \tfrac{1}{12}\mathrm{i}\bar{\psi}_\mu^i\gamma^\mu\psi^{jI}\bar{\psi}_i^J\psi_j^K - \tfrac{1}{48}\sigma^I\bar{\psi}_\mu\psi_\nu\bar{\psi}^J\gamma^{\mu\nu}\psi^K + \tfrac{1}{24}\sigma^I\bar{\psi}_\mu\gamma^{\mu\nu}\psi_\nu^j\bar{\psi}_i^J\psi_j^K \\
&- \tfrac{1}{12}\sigma^I\bar{\psi}_\lambda\gamma^{\mu\nu\lambda}\psi^J\bar{\psi}_\mu\gamma_\nu\psi^K + \tfrac{1}{24}\mathrm{i}\sigma^I\sigma^J\bar{\psi}_\lambda\gamma^{\mu\nu\lambda}\psi^K\bar{\psi}_\mu\psi_\nu \\
&+ \tfrac{1}{48}\mathrm{i}\sigma^I\sigma^J\bar{\psi}_\lambda\gamma^{\mu\nu\lambda\rho}\psi_\rho\bar{\psi}_\mu\gamma_\nu\psi^K + \tfrac{1}{96}\sigma^I\sigma^J\sigma^K\bar{\psi}_\lambda\gamma^{\mu\nu\lambda\rho}\psi_\rho\bar{\psi}_\mu\psi_\nu \\
&- \tfrac{1}{24}e^{-1}\varepsilon^{\mu\nu\lambda\rho\sigma}W_\mu^I \left(F_{\nu\lambda}^J F_{\rho\sigma}^K - f_{FG}{}^J W_\nu^F W_\lambda^G \left(\tfrac{1}{2}F_{\rho\sigma}^K - \tfrac{1}{10}f_{HL}{}^K W_\rho^H W_\sigma^L \right) \right) \\
&+ \tfrac{1}{4}\mathrm{i}\sigma^I\sigma^J f_{LM}{}^K \bar{\psi}^L\psi^M \Bigg],
\end{aligned}
\tag{3.129}
$$

where covariant derivatives and $\widehat{F}_{\mu\nu}^I$ are given in (3.12) and (3.13), and the superconformal d'Alembertian is defined as

$$
\begin{aligned}
\Box^c \sigma^I &= \mathcal{D}^a\mathcal{D}_a\sigma^I \\
&= \left(\partial^a - 2b^a + \omega_b{}^{ba} \right) \mathcal{D}_a\sigma^I + f_{JK}{}^I W_a^J \mathcal{D}^a\sigma^K - \tfrac{1}{2}\mathrm{i}\bar{\psi}_\mu D^\mu\psi^I - 2\sigma^I\bar{\psi}_\mu\gamma^\mu\chi \\
&\quad + \tfrac{1}{2}\bar{\psi}_\mu\gamma^\mu\gamma\cdot T\psi^I + \tfrac{1}{2}\bar{\phi}_\mu\gamma^\mu\psi^I + 2f_\mu{}^\mu\sigma^I - \tfrac{1}{2}\bar{\psi}_\mu\gamma^\mu f_{JK}{}^I\psi^J\sigma^K .
\end{aligned}
\tag{3.130}
$$

The dependent gauge fields are given in (2.100).

3.3.3   Action for Hypermultiplets

While the actions of vector multiplets were constructed using tensor calculus manipulations, for the hypermultiplets we use another procedure. The main difference is that we have already the equations of motion from the non-closure relations, e.g. (3.99) in $D = 5$, and we can therefore infer the action from the latter. To this end, we need a few ingredients that we are going to introduce in the following.

3.3.3.1   Ingredients

We first define a covariantly constant antisymmetric tensor $C_{\mathcal{AB}}(q)$ that describes the proportionality between the field equations for the fermions $\zeta^{\mathcal{A}}$ and the non-closure functions. For example, in $D = 5$,

$$\frac{\delta S_{\text{hyper}}}{\delta \bar{\zeta}^{\mathcal{A}}} = 2C_{\mathcal{AB}} i \Gamma^{\mathcal{B}} . \tag{3.131}$$

Then, once the right-hand side of (3.131) is known, one can functionally integrate the above equation in order to obtain the action. The properties of the tensor are (independent whether we consider $D = 5$ or $D = 4$):

$$\nabla_X C_{\mathcal{AB}} \equiv \partial_X C_{\mathcal{AB}} + 2\omega_{X[\mathcal{A}}{}^C C_{\mathcal{B}]C} = 0 ,$$

$$C_{\mathcal{AB}} = -C_{\mathcal{BA}} ,$$

$$C^{\bar{\mathcal{A}}\bar{\mathcal{B}}} \equiv (C_{\mathcal{AB}})^* = \rho^{\mathcal{A}C} \rho^{\mathcal{B}D} C_{CD} . \tag{3.132}$$

As will become clear below, the kinetic terms involve the Hermitian metric in tangent space

$$d^{\bar{\mathcal{A}}}{}_{\mathcal{B}} \equiv -\rho^{\mathcal{A}C} C_{C\mathcal{B}} ,$$

$$d^{\bar{\mathcal{A}}}{}_{\mathcal{B}} = (d^{\bar{\mathcal{B}}}{}_{\mathcal{A}})^* = \rho^{\bar{\mathcal{A}}C} d^{\bar{D}}{}_C \rho_{\mathcal{B}\bar{D}} , \tag{3.133}$$

such that

$$C_{\mathcal{AB}} = \rho_{\mathcal{A}\bar{C}} d^{\bar{C}}{}_{\mathcal{B}} . \tag{3.134}$$

We also define an inverse

$$C^{\mathcal{A}C} C_{\mathcal{B}C} = \delta^{\mathcal{A}}{}_{\mathcal{B}} , \tag{3.135}$$

so that we can use these matrices to raise and lower $\mathcal{A}$ indices, using the common NE–SW convention

$$V_{\mathcal{A}} = V^{\mathcal{B}} C_{\mathcal{B}\mathcal{A}}, \qquad V^{\mathcal{A}} = C^{\mathcal{A}\mathcal{B}} V_{\mathcal{B}}. \tag{3.136}$$

On the other hand, we raise or lower indices changing the holomorphicity as in (3.94). This is then consistent with changing the holomorphicity using $d^{\bar{\mathcal{A}}}{}_{\mathcal{B}}$. For example, for the gauge-transformation matrices in (3.94):

$$t_{I\mathcal{A}\mathcal{B}} = t_{I\mathcal{A}}{}^{C} C_{C\mathcal{B}} = t_{I\mathcal{A}\bar{\mathcal{B}}} d^{\bar{\mathcal{B}}}{}_{\mathcal{B}}. \tag{3.137}$$

Consistency of the transformations of the left- and right-hand side of (3.131) under the isometry group, determined by (3.82), implies that this matrix should be symmetric:

$$t_{I\mathcal{A}\mathcal{B}} = t_{I\mathcal{B}\mathcal{A}}. \tag{3.138}$$

This equation is, using (3.80), equivalent to

$$t_I{}^{\bar{\mathcal{A}}}{}_{\bar{\mathcal{B}}} d^{\bar{\mathcal{B}}}{}_{C} = t_{IC}{}^{\mathcal{B}} d^{\bar{\mathcal{A}}}{}_{\mathcal{B}}, \tag{3.139}$$

which shows more clearly that it is related to the invariance of the action with signature matrix $d^{\bar{\mathcal{A}}}{}_{\mathcal{B}}$.

With the above conditions, $d^{\bar{\mathcal{A}}}{}_{\mathcal{B}}$ respects the quaternionic structure. It has been proven in [37], using the theorems of [41], that at any point one can choose a basis such that ρ is in the form (3.35) and at the same time

$$d^{\bar{\mathcal{A}}}{}_{\mathcal{B}} = \begin{pmatrix} \eta & \\ & \eta \end{pmatrix} = \begin{pmatrix} -\mathbb{1}_p & & & \\ & \mathbb{1}_q & & \\ & & -\mathbb{1}_p & \\ & & & \mathbb{1}_q \end{pmatrix}, \qquad p+q = n_H. \tag{3.140}$$

For rigid supersymmetry, positive kinetic terms will be obtained for $p = 0$ and $q = n_H$. For supergravity we need one compensating multiplet and will use $p = 1$. These matrices should be covariantly constant. As we use a basis where they are actually constant, this implies from (3.132) that (using the lowering of indices as in (3.136))

$$2\omega_{X[\mathcal{A}}{}^{C} C_{\mathcal{B}]C} = -\omega_{X\mathcal{A}\mathcal{B}} + \omega_{X\mathcal{B}\mathcal{A}} = 0. \tag{3.141}$$

Thus the USp-connection is symmetric in such bases. From

$$\nabla_X d^{\bar{\mathcal{A}}}{}_{C} = -\bar{\omega}_X{}^{\bar{\mathcal{A}}}{}_{\bar{\mathcal{B}}} d^{\bar{\mathcal{B}}}{}_{C} - \omega_{XC}{}^{\mathcal{D}} d^{\bar{\mathcal{A}}}{}_{\mathcal{D}} = 0, \tag{3.142}$$

one finds

$$d^{\bar{\mathcal{A}}}{}_{\mathcal{C}} \omega_{X \mathcal{B}}{}^{\mathcal{C}} = -\bar{\omega}_{X}{}^{\bar{\mathcal{A}}}{}_{\bar{\mathcal{C}}} d^{\bar{\mathcal{C}}}{}_{\mathcal{B}} . \tag{3.143}$$

When $d = \mathbb{1}$ the above condition is the antihermiticity of ω. In the preferred basis with (3.35) and (3.140) we can also write

$$C_{\mathcal{A}\mathcal{B}} = \begin{pmatrix} 0 & \eta \\ -\eta & 0 \end{pmatrix} , \qquad t_{I\mathcal{A}}{}^{\mathcal{B}} = \begin{pmatrix} U_I & V_I \\ W_I & U_I{}^* \end{pmatrix} , \qquad \begin{matrix} V_I{}^T = \eta V_I \eta , \ W_I{}^T = \eta W_I \eta , \\ U^\dagger = -\eta U \eta , \ V^* = -W . \end{matrix} \tag{3.144}$$

This expresses that the transformations are in the subgroup of $G\ell(n_H, \mathbb{H})$ that preserves the antisymmetric metric $C_{\mathcal{A}\mathcal{B}}$ and the metric $d^{\bar{\mathcal{A}}}{}_{\mathcal{B}}$, which is $USp(2p, 2q)$.

We define then the metric of the manifold to be

$$g_{XY} = \left(f^{i\bar{\mathcal{A}}}{}_X \right)^* d^{\bar{\mathcal{A}}}{}_{\mathcal{B}} f^{i\mathcal{B}}{}_Y = f^{i\mathcal{A}}{}_X \varepsilon_{ij} C_{\mathcal{A}\mathcal{B}} f^{j\mathcal{B}}{}_Y , \tag{3.145}$$

such that the holonomy associated to g_{XY} is indeed $USp(2p, 2q)$.

The curvature tensor on the scalar manifold is determined in terms of a 4-index symmetric tensor in $Sp(2n_H)$, denoted by $W_{\mathcal{A}\mathcal{B}\mathcal{C}\mathcal{D}}$:

$$W_{\mathcal{A}\mathcal{B}\mathcal{C}\mathcal{D}} \equiv W_{\mathcal{A}\mathcal{B}\mathcal{C}}{}^{\mathcal{E}} C_{\mathcal{E}\mathcal{D}} = -\varepsilon^{ij} f^X{}_{i\mathcal{A}} f^Y{}_{j\mathcal{B}} R_{XYC\mathcal{D}}$$
$$= \tfrac{1}{2} f^{Xi}{}_{\mathcal{A}} f^Y{}_{i\mathcal{B}} f^{Zk}{}_{\mathcal{C}} f^W{}_{k\mathcal{D}} R_{XYZW} , \tag{3.146}$$

where we used the metric g_{XY} (3.145) to lower the indices.

3.3.3.2   Remark on the Conformal Symmetry

Due to the fact that we have now a metric available, we can invoke the homothetic Killing equation (3.68) and, similarly as in (1.45), introduce a scalar function $\tilde{k}_D$ such that

$$k_{DX} = g_{XY} k_D{}^Y = \partial_X \tilde{k}_D . \tag{3.147}$$

It is also possible to start from this scalar function, and generate the metric from

$$g_{XY} = \frac{2}{D-2} \nabla_X \partial_Y \tilde{k}_D . \tag{3.148}$$

We also define $k_D{}^2$ using the metric (3.145)

$$k_D{}^2 \equiv g_{XY} k_D^X k_D^Y . \tag{3.149}$$

It will be also useful to express $k_D{}^2$ in terms of the sections introduced in (3.70) and their complex conjugates

$$A_{i\bar{\mathcal{A}}} = \left(A^{i\mathcal{A}}\right)^* = A^{j\mathcal{B}}\rho_{\mathcal{B}\bar{\mathcal{A}}}\varepsilon_{ji}, \qquad A^{i\mathcal{A}} = -\varepsilon^{ij}\rho^{\bar{\mathcal{B}}\mathcal{A}}A_{j\bar{\mathcal{B}}}. \qquad (3.150)$$

To do so, we note that the matrix

$$M^i{}_j \equiv A^{i\mathcal{A}}d^{\bar{\mathcal{B}}}{}_{\mathcal{A}}A_{j\bar{\mathcal{B}}}, \qquad (3.151)$$

is Hermitian and equal to $\varepsilon^{ik}\varepsilon_{j\ell}M^\ell{}_k$, i.e. $\sigma_2 M^T \sigma_2$. Therefore it should be proportional to the unit matrix. Indeed, using (3.70) and (3.145)

$$M^i{}_j = \tfrac{1}{2}\delta^i_j A^{k\mathcal{A}}d^{\bar{\mathcal{B}}}{}_{\mathcal{A}}A_{k\bar{\mathcal{B}}} = \tfrac{1}{2}\delta^i_j k_D{}^2. \qquad (3.152)$$

Another way in which $k_D{}^2$ appears is in terms of an inner product of the SU(2) Killing vectors introduced in (3.69):

$$k_D{}^2 = \tfrac{1}{3}(D-2)^2 \mathbf{k}_X \cdot \mathbf{k}^X. \qquad (3.153)$$

It is useful to record the relation between these quantities for arbitrary vectors $\mathbf{A}$ and $\mathbf{B}$:

$$\mathbf{A} \cdot \mathbf{k}_X \, \mathbf{B} \cdot \mathbf{k}^X = \frac{1}{(D-2)^2} k_D{}^2 \mathbf{A} \cdot \mathbf{B}. \qquad (3.154)$$

3.3.3.3 Moment Maps

The isometries defined in (3.75) can be expressed in terms of moment maps. The definition of the latter depends on the theory. As we will discuss in Sect. 5.4.1, isometries for Kähler manifolds can be generically generated from a real moment map function using the complex structure and the metric. The hypermultiplet geometry has three complex structures, and as such have a triplet moment map for any isometry $\mathbf{P}_I$. They should satisfy

$$\partial_X \mathbf{P}_I = \mathbf{J}_X{}^Y k_{IY}. \qquad (3.155)$$

Furthermore, they satisfy an 'equivariance relation', which is necessary to build supersymmetric actions with these symmetries:

$$k_I{}^X \mathbf{J}_{XY} k_J{}^Y = f_{IJ}{}^K \mathbf{P}_K. \qquad (3.156)$$

With conformal symmetry, the solution of (3.155) is determined to be[21]

$$
\mathbf{P}_I = \mathbf{k}^X k_{IX} = \frac{1}{D-2} k_{\mathrm{D}}{}^Y \mathbf{J}_Y{}^X k_{IX} = \frac{2}{(D-2)^2} A^{i\mathcal{A}} t_{I\mathcal{A}\mathcal{B}} \tau_{ij} A^{j\mathcal{B}} .
\tag{3.157}
$$

In this context, it is also convenient to rewrite an expression that appears in the potential that occurs in these theories

$$
k_I^X k_{JX} = \frac{4}{(D-2)^2} \varepsilon_{ij} A^{i\mathcal{A}} A^{j\mathcal{B}} t_{I\mathcal{A}}{}^C C_{C\mathcal{D}} t_{J\mathcal{B}}{}^{\mathcal{D}} .
\tag{3.158}
$$

3.3.3.4   Action for Hypermultiplets in $D = 5$

The resulting action is [5]

$$
\begin{aligned}
e^{-1}\mathcal{L}_h = {}& -\tfrac{1}{2} g_{XY} D_a q^X D^a q^Y + \bar{\zeta}_A \mathcal{D}\zeta^A + \tfrac{4}{9} D k_{\mathrm{D}}^2 + \tfrac{8}{27} T^{ab} T_{ab} k_{\mathrm{D}}^2 \\
& + \tfrac{16}{3} i \bar{\zeta}_A \chi_i k_{\mathrm{D}}^X f^{iA}{}_X + 2 i \bar{\zeta}_A \gamma \cdot T \zeta^A - \tfrac{1}{4} W_{ABCD} \bar{\zeta}^A \zeta^B \bar{\zeta}^C \zeta^D \\
& - \tfrac{2}{9} \bar{\psi}_a \gamma^a \chi k_{\mathrm{D}}^2 - \tfrac{1}{3} \bar{\zeta}_A \gamma^a \gamma \cdot T \psi_{ai} k^X f^{iA}{}_X - \tfrac{1}{2} i \bar{\zeta}_A \gamma^a \gamma^b \psi_{ai} D_b q^X f^{iA}{}_X \\
& + \tfrac{2}{3} f_a{}^a k_{\mathrm{D}}^2 - \tfrac{1}{6} i \bar{\psi}_a \gamma^{ab} \phi_b k_{\mathrm{D}}^2 + \bar{\zeta}_A \gamma^a \phi_{ai} k_{\mathrm{D}}^X f^{iA}{}_X \\
& + \tfrac{1}{12} \bar{\psi}_a^i \gamma^{abc} \psi_b^j D_c q^Y J_{YX\,ij} k_{\mathrm{D}}^X - \tfrac{1}{9} i k_{\mathrm{D}}^2 \bar{\psi}^a \left(\psi^b T_{ab} - \tfrac{1}{2} \gamma^{abcd} \psi_b T_{cd} \right) \\
& + i \sigma^I t_{IB}{}^A \bar{\zeta}_A \zeta^B - 2 i k_I^X f^{iA}{}_X \bar{\zeta}_A \psi_i^I - \tfrac{1}{2} \sigma^I k_I^X f^{iA}{}_X \bar{\zeta}_A \gamma^a \psi_{ai} \\
& - \tfrac{1}{2} \bar{\psi}_a^i \gamma^a \psi^{jI} P_{Iij} + \tfrac{1}{4} i \bar{\psi}_a^i \gamma^{ab} \psi_b^j \sigma^I P_{Iij} + Y_{ij}^I P_I^{ij} - \tfrac{1}{2} \sigma^I \sigma^J k_I^X k_{JX} ,
\end{aligned}
\tag{3.159}
$$

with covariant derivatives given in (3.98).

3.3.3.5   Action for Hypermultiplets in $D = 4$

When we discuss $D = 4$, we can multiply (3.131) at both sides with a chiral projection P_R. Using the rules (A.63) we should now impose for the action S_{hyper}

$$
\frac{\delta S_{\text{hyper}}}{\delta \bar{\zeta}_{\bar{\mathcal{A}}}} = 2 d^{\bar{\mathcal{A}}}{}_{\mathcal{B}} \Gamma^{\mathcal{B}} .
\tag{3.160}
$$

---

[21]It is a nice exercise to prove that (3.155) is solved by (3.157). You may replace the ∂_X by covariant derivatives and use (3.69), (3.39), (3.68), (3.77) and (3.83).

We also want the action to generate the field equations for the scalars that we have seen in (3.95). This leads in rigid supersymmetry to

$$
\mathcal{L}_h = -\tfrac{1}{2} g_{XY} D_\mu q^X D^\mu q^Y - \left(\bar{\zeta}_{\bar{\mathcal{A}}} \slashed{\nabla} \zeta^{\mathcal{B}} d^{\bar{\mathcal{A}}}{}_{\mathcal{B}} + \text{h.c.} \right)
$$

$$
+ \tfrac{1}{2} W_{\mathcal{A}\mathcal{B}}{}^{\mathcal{E}\mathcal{F}} d^{\bar{\mathcal{C}}}{}_{\mathcal{E}} d^{\bar{\mathcal{D}}}{}_{\mathcal{F}} \bar{\zeta}_{\bar{\mathcal{C}}} \bar{\zeta}_{\bar{\mathcal{D}}} \bar{\zeta}^{\mathcal{A}} \zeta^{\mathcal{B}}
$$

$$
+ \left(2 X^I t_{I\mathcal{A}\mathcal{B}} \bar{\zeta}^{\mathcal{A}} \zeta^{\mathcal{B}} + 2\mathrm{i} f_X^{i\mathcal{A}} k_I^X \bar{\zeta}_{\bar{\mathcal{B}}} \Omega^{jI} \varepsilon_{ij} d^{\bar{\mathcal{B}}}{}_{\mathcal{A}} + \text{h.c.} \right)
$$

$$
+ 2\mathbf{P}_I \cdot \mathbf{Y}^I - 2\bar{X}^I X^J k_I{}^X k_{JX} , \tag{3.161}
$$

with the covariant derivatives in (3.87), which satisfies (3.160), and also

$$
\frac{\delta S_{\text{hyper}}}{\delta q^X} = g_{XY} \Delta^Y - \left(2\bar{\zeta}_{\bar{\mathcal{A}}} \Gamma^{\mathcal{B}} \omega_{X\mathcal{B}}{}^{\mathcal{C}} d^{\bar{\mathcal{A}}}{}_{\mathcal{C}} + \text{h.c.} \right) . \tag{3.162}
$$

See [18, Exercises 20.8 and 20.9] for a concrete example.

After gauge covariantization and using the values of the conformal gauge fields as in (3.122) and the covariant derivatives (3.100), the superconformal hypermultiplet action with gauged isometries in $D = 4$ is [18, (20.93)]

$$
e^{-1} \mathcal{L}_h = -\tfrac{1}{12} k_{\mathrm{D}}{}^2 R + \tfrac{1}{4} k_{\mathrm{D}}{}^2 D - \tfrac{1}{2} g_{XY} D_\mu q^X D^\mu q^Y - 2\bar{X}^I X^J k_I{}^X k_{JX} + 2\mathbf{P}_I \cdot \mathbf{Y}^I
$$

$$
+ \Big\{ -\bar{\zeta}_{\bar{\mathcal{A}}} \widehat{\slashed{D}} \zeta^{\mathcal{B}} d^{\bar{\mathcal{A}}}{}_{\mathcal{B}} + \tfrac{1}{12} k_{\mathrm{D}}{}^2 \bar{\psi}_{i\mu} \gamma^{\mu\nu\rho} D_\nu \psi_\rho^i
$$

$$
+ \tfrac{1}{8} k_{\mathrm{D}}{}^2 \bar{\psi}_{ia} \gamma^a \chi^i - 2\mathrm{i} d^{\bar{\mathcal{A}}}{}_{\mathcal{B}} A^{i\mathcal{B}} \bar{\zeta}_{\bar{\mathcal{A}}} \chi_i
$$

$$
+ \tfrac{1}{2} \mathrm{i} \bar{\zeta}_{\bar{\mathcal{A}}} \gamma^a \slashed{D} q^X \psi_{ai} f^{i\mathcal{B}}{}_X d^{\bar{\mathcal{A}}}{}_{\mathcal{B}} - \tfrac{1}{3} \mathrm{i} d^{\bar{\mathcal{A}}}{}_{\mathcal{B}} A^{i\mathcal{B}} \bar{\zeta}_{\bar{\mathcal{A}}} \gamma^{\mu\nu} D_\mu \psi_{vi}
$$

$$
+ \left(\tfrac{1}{12} \mathrm{i} d^{\bar{\mathcal{A}}}{}_{\mathcal{B}} A^{i\mathcal{B}} \bar{\zeta}_{\bar{\mathcal{A}}} \gamma_a \psi_b^j - \tfrac{1}{48} k_{\mathrm{D}}{}^2 \bar{\psi}_a^i \psi_b^j \right) T^{+ab} \varepsilon_{ij}
$$

$$
- \tfrac{1}{8} \bar{\zeta}_{\bar{\mathcal{A}}} \gamma^{ab} T_{ab}^+ \zeta_{\bar{\mathcal{B}}} C^{\bar{\mathcal{A}}\bar{\mathcal{B}}} + 2\mathrm{i} \bar{X}^I k_I^X \bar{\zeta}_{\bar{\mathcal{A}}} \gamma^a \psi_a^j \varepsilon_{ij} d^{\bar{\mathcal{A}}}{}_{\mathcal{B}} f^{i\mathcal{B}}{}_X
$$

$$
+ 2 X^I \bar{\zeta}^{\mathcal{A}} \zeta^{\mathcal{B}} t_{I\mathcal{A}\mathcal{B}} + 2\mathrm{i} k_I^X f^{i\mathcal{A}}{}_X \bar{\zeta}_{\bar{\mathcal{B}}} \Omega^{jI} \varepsilon_{ij} d^{\bar{\mathcal{B}}}{}_{\mathcal{A}}
$$

$$
+ \tfrac{1}{2} \bar{\psi}_{aj} \gamma^a \Omega_i^I P_I{}^{ij} + \tfrac{1}{2} \bar{X}^I \bar{\psi}_a^i \gamma^{ab} \psi_b^j P_{Iij} + \text{h.c.} \Big\}
$$

$$
+ \tfrac{1}{2} \bar{\psi}_a^i \gamma^{abc} \psi_{bj} D_c q^X \mathbf{k}_X \cdot \boldsymbol{\tau}_i{}^j + 4\text{-fermion terms} . \tag{3.163}
$$

We can rewrite the kinetic terms for the scalars q^X in terms of the sections (3.70) using the bosonic part of (3.102)

$$
g_{XY} D_\mu q^X D^\mu q^Y = \varepsilon_{ij} C_{\mathcal{A}\mathcal{B}} \left(\widehat{D}_\mu A^{i\mathcal{A}} \right) \left(\widehat{D}^\mu A^{j\mathcal{B}} \right) . \tag{3.164}
$$

3.3.4 Splitting the Hypermultiplets and Example

In general we did not use the basis (3.35). Sometimes, however, it will be convenient to use such basis in examples. To do this, we split the index $\mathcal{A} = 1, \ldots, 2n_H$ into $\mathcal{A} = (\alpha a)$, with $\alpha = 1, 2$ and $a = 1, \ldots, n_H$. The index $\bar{\mathcal{A}}$ will then have the same form, but with a in the opposite (up-down) position. We can then write the canonical basis with (3.35) and (3.140) as

$$\rho_{\mathcal{A}\bar{\mathcal{B}}} = \varepsilon_{\alpha\beta}\delta_a^b, \qquad d^{\bar{\mathcal{A}}}{}_{\mathcal{B}} = \eta_{ab}\delta_\beta^\alpha, \qquad C_{\mathcal{A}\mathcal{B}} = \eta_{ab}\varepsilon_{\alpha\beta}. \tag{3.165}$$

The components of $A^{i\mathcal{A}}$ can then be written as $A^{i\alpha a}$ and $(A^{i\alpha a})^* = A^{j\beta b}\varepsilon_{ji}\varepsilon_{\beta\alpha}$. Upon this splitting the action (3.163) starts with

$$e^{-1}\mathcal{L}_h = -\tfrac{1}{12}k_D^2 R + \tfrac{1}{4}k_D^2 D - \tfrac{1}{2}\widehat{D}_\mu A^{i\alpha a}\widehat{D}^\mu A^{j\beta b}\varepsilon_{ij}\varepsilon_{\alpha\beta}\eta_{ab} + \cdots,$$

$$k_D^2 = A^{i\alpha a}A^{j\beta b}\varepsilon_{ij}\varepsilon_{\alpha\beta}\eta_{ab} = A^{i\alpha a}\left(A^{i\beta b}\right)^*\eta_{ab}. \tag{3.166}$$

The conditions on the symmetry matrices $t_{I\mathcal{A}}{}^{\mathcal{B}}$ (see (3.144)) are such that they can be decomposed as

$$t_{I\alpha a}{}^{\beta b} = t_{I0a}{}^b\delta_\alpha{}^\beta + \mathbf{t}_{Ia}{}^b\tau_\alpha{}^\beta, \qquad t_{I0a}{}^b, \mathbf{t}_{Ia}{}^b \in \mathbb{R},$$

$$t_{I0a}{}^b = -\eta_{ac}t_{I0d}{}^c\eta^{db}, \qquad \mathbf{t}_{Ia}{}^b = \eta_{ac}\mathbf{t}_{Id}{}^c\eta^{db}. \tag{3.167}$$

As an example, we may consider

$$t_{I\alpha a}{}^{\beta b} = iQ_{Ia}{}^b(\sigma_3)_\alpha{}^\beta, \qquad Q_{Ia}{}^b \in \mathbb{R}, \qquad Q_{Ia}{}^b = \eta^{bc}Q_{Ic}{}^d\eta_{da}. \tag{3.168}$$

Then from (3.84) and (3.157) we have

$$k_I{}^X = if^X{}_{I(\alpha a)}A^{i\beta b}Q_{Ib}{}^a(\sigma_3)_\beta{}^\alpha,$$

$$\mathbf{P}_I = \tfrac{1}{2}iA^{i\alpha a}Q_{Iab}(A^{j\beta b})^*\tau_i{}^j(\sigma_3)_\alpha{}^\beta, \tag{3.169}$$

with $Q_{Iab} = Q_{Ia}{}^c\eta_{cb} = Q_{Iba}$.

References

1. M. Günaydin, M. Zagermann, The gauging of five-dimensional, $N = 2$ Maxwell–Einstein supergravity theories coupled to tensor multiplets. Nucl. Phys. **B572**, 131–150 (2000). https://doi.org/10.1016/S0550-3213(99)00801-9, arXiv:hep-th/9912027 [hep-th]

2. M. Günaydin, M. Zagermann, The vacua of $5d$, $N = 2$ gauged Yang–Mills/Einstein/tensor supergravity: Abelian case. Phys. Rev. **D62**, 044028 (2000) . https://doi.org/10.1103/PhysRevD.62.044028, arXiv:hep-th/0002228 [hep-th]

3. M. Günaydin, M. Zagermann, Gauging the full R-symmetry group in five-dimensional, $N = 2$ Yang–Mills/Einstein/tensor supergravity. Phys. Rev. **D63**, 064023 (2001). https://doi.org/10.1103/PhysRevD.63.064023, arXiv:hep-th/0004117 [hep-th]

4. A. Ceresole, G. Dall'Agata, General matter coupled $N = 2$, $D = 5$ gauged supergravity. Nucl. Phys. **B585**, 143–170 (2000). https://doi.org/10.1016/S0550-3213(00)00339-4, arXiv:hep-th/0004111 [hep-th]

5. E. Bergshoeff, S. Cucu, T. de Wit, J. Gheerardyn, S. Vandoren, A. Van Proeyen, $N = 2$ supergravity in five dimensions revisited. Class. Quant. Grav. **21**, 3015–3041 (2004). https://doi.org/10.1088/0264-9381/23/23/C01,10.1088/0264-9381/21/12/013, arXiv:hep-th/0403045[hep-th], erratum **23** (2006) 7149

6. B. de Wit, J.W. van Holten, A. Van Proeyen, Structure of $N = 2$ supergravity. Nucl. Phys. **B184**, 77–108 (1981). https://doi.org/10.1016/0550-3213(83)90548-5,10.1016/0550-3213(81)90211-X, [Erratum: Nucl. Phys.B222,516(1983)]

7. B. de Wit, R. Philippe, A. Van Proeyen, The improved tensor multiplet in $N = 2$ supergravity. Nucl. Phys. **B219**, 143–166 (1983). https://doi.org/10.1016/0550-3213(83)90432-7

8. B. de Wit, F. Saueressig, Off-shell $N = 2$ tensor supermultiplets. J. High Energy Phys. **09**, 062 (2006). https://doi.org/10.1088/1126-6708/2006/09/062, arXiv:hep-th/0606148 [hep-th]

9. G. Dall'Agata, R. D'Auria, L. Sommovigo, S. Vaulá, $D = 4$, $N = 2$ gauged supergravity in the presence of tensor multiplets. Nucl. Phys. **B682**, 243–264 (2004). https://doi.org/10.1016/j.nuclphysb.2004.01.014, arXiv:hep-th/0312210 [hep-th]

10. L. Sommovigo, S. Vaulà, $D = 4$, $N = 2$ supergravity with abelian electric and magnetic charge. Phys. Lett. **B602**, 130–136 (2004). https://doi.org/10.1016/j.physletb.2004.09.058, arXiv:hep-th/0407205 [hep-th]

11. R. D'Auria, L. Sommovigo, S. Vaulà, $N = 2$ supergravity Lagrangian coupled to tensor multiplets with electric and magnetic fluxes. J. High Energy Phys. **0411**, 028 (2004). https://doi.org/10.1088/1126-6708/2004/11/028, arXiv:hep-th/0409097 [hep-th]

12. N. Cribiori, G. Dall'Agata, On the off-shell formulation of $N = 2$ supergravity with tensor multiplets. J. High Energy Phys. **08**, 132 (2018). https://doi.org/10.1007/JHEP08(2018)132, arXiv:1803.08059 [hep-th]

13. J. De Rydt, B. Vercnocke, *De Lagrangiaan van vector- en hypermultipletten in $N = 2$ supergravitatie*, Thesis Licenciaat, Katholieke Universiteit Leuven, Leuven, 2006

14. E. Bergshoeff, S. Cucu, T. de Wit, J. Gheerardyn, R. Halbersma, S. Vandoren, A. Van Proeyen, Superconformal $N = 2$, $D = 5$ matter with and without actions. J. High Energy Phys. **10**, 045 (2002). https://doi.org/10.1088/1126-6708/2002/10/045, arXiv:hep-th/0205230 [hep-th]

15. T. Kugo, K. Ohashi, Supergravity tensor calculus in $5D$ from $6D$. Prog. Theor. Phys. **104**, 835–865 (2000). https://doi.org/10.1143/PTP.104.835, arXiv:hep-ph/0006231 [hep-ph]

16. T. Kugo, K. Ohashi, Off-shell $d = 5$ supergravity coupled to matter–Yang–Mills system. Prog. Theor. Phys. **105**, 323–353 (2001). https://doi.org/10.1143/PTP.105.323, arXiv:hep-ph/0010288 [hep-ph]

17. C. Pope, *Lectures on Kaluza-Klein*. http://people.physics.tamu.edu/pope/

18. D.Z. Freedman, A. Van Proeyen, *Supergravity* (Cambridge University, Cambridge, 2012). http://www.cambridge.org/mw/academic/subjects/physics/theoretical-physics-and-mathematical-physics/supergravity?format=AR

19. A. Van Proeyen, Vector multiplets in $N = 2$ supersymmetry and its associated moduli spaces, in *1995 Summer school in High Energy Physics and Cosmology*, eds. by E. Gava et al. The ICTP series in theoretical physics, vol.12 (World Scientific, Singapore, 1997), p.256. hep-th/9512139

20. I. Antoniadis, H. Partouche, T.R. Taylor, Spontaneous breaking of $N = 2$ global supersymmetry. Phys. Lett. **B372**, 83–87 (1996). https://doi.org/10.1016/0370-2693(96)00028-7, arXiv:hep-th/9512006 [hep-th]

21. N. Cribiori, S. Lanza, On the dynamical origin of parameters in $\mathcal{N} = 2$ supersymmetry. Eur. Phys. J. **C79**(1), 32 (2019). https://doi.org/10.1140/epjc/s10052-019-6545-6, arXiv:1810.11425 [hep-th]

22. I. Antoniadis, H. Jiang, O. Lacombe, $\mathcal{N} = 2$ supersymmetry deformations, electromagnetic duality and Dirac-Born-Infeld actions. J. High Energy Phys. **07**, 147 (2019). https://doi.org/10.1007/JHEP07(2019)147, arXiv:1904.06339 [hep-th]

23. M. de Roo, J.W. van Holten, B. de Wit, A. Van Proeyen, Chiral superfields in $\mathcal{N} = 2$ supergravity. Nucl. Phys. **B173**, 175–188 (1980). https://doi.org/10.1016/0550-3213(80)90449-6

24. A. Galperin, E. Ivanov, S. Kalitsyn, V. Ogievetsky, E. Sokatchev, Unconstrained $\mathcal{N} = 2$ matter, Yang–Mills and supergravity theories in harmonic superspace. Class. Quant. Grav. **1**, 469–498 (1984). https://doi.org/10.1088/0264-9381/1/5/004, [Erratum: Class. Quant. Grav.2,127(1985)]

25. A.S. Galperin, E.A. Ivanov, V.I. Ogievetsky, E.S. Sokatchev, Harmonic superspace, in *Cambridge Monographs on Mathematical Physics* (Cambridge University, Cambridge, 2007). https://doi.org/10.1017/CBO9780511535109, http://www.cambridge.org/mw/academic/subjects/physics/theoretical-physics-and-mathematical-physics/harmonic-superspace?format=PB

26. A. Karlhede, U. Lindström, M. Rocek, Selfinteracting tensor multiplets in $\mathcal{N} = 2$ superspace. Phys. Lett. **147B**, 297–300 (1984). https://doi.org/10.1016/0370-2693(84)90120-5

27. U. Lindström, M. Roček, New hyperkähler metrics and new supermultiplets. Commun. Math. Phys. **115**, 21 (1988). https://doi.org/10.1007/BF01238851

28. U. Lindström, M. Roček, $\mathcal{N} = 2$ super Yang–Mills theory in projective superspace. Commun. Math. Phys. **128**, 191 (1990). https://doi.org/10.1007/BF02097052

29. U. Lindström, M. Roček, Properties of hyperkähler manifolds and their twistor spaces. Commun. Math. Phys. **293**, 257–278 (2010). https://doi.org/10.1007/s00220-009-0923-0, arXiv:0807.1366 [hep-th]

30. S.M. Kuzenko, Lectures on nonlinear sigma-models in projective superspace. J. Phys. **A43**, 443001 (2010). https://doi.org/10.1088/1751-8113/43/44/443001, arXiv:1004.0880 [hep-th]

31. B. de Wit, B. Kleijn, S. Vandoren, Rigid $\mathcal{N} = 2$ superconformal hypermultiplets, in *Supersymmetries and Quantum Symmetries*. Proceeding of International Seminars, Dubna (1997), eds. by J. Wess, E.A. Ivanov. Lecture Notes in Physics, vol. 524 (Springer, Berlin, 1999), p. 37. hep-th/9808160

32. B. de Wit, B. Kleijn, S. Vandoren, Superconformal hypermultiplets. Nucl. Phys. **B568**, 475–502 (2000). https://doi.org/10.1016/S0550-3213(99)00726-9, arXiv:hep-th/9909228 [hep-th]

33. F. Cordaro, P. Frè, L. Gualtieri, P. Termonia, M. Trigiante, $\mathcal{N} = 8$ gaugings revisited: an exhaustive classification. Nucl. Phys. **B532**, 245–279 (1998). https://doi.org/10.1016/S0550-3213(98)00449-0, arXiv:hep-th/9804056

34. H. Nicolai, H. Samtleben, Compact and noncompact gauged maximal supergravities in three dimensions. J. High Energy Phys. **04**, 022 (2001). https://doi.org/10.1088/1126-6708/2001/04/022, arXiv:hep-th/0103032 [hep-th]

35. B. de Wit, H. Samtleben, M. Trigiante, Magnetic charges in local field theory. J. High Energy Phys. **09**, 016 (2005). https://doi.org/10.1088/1126-6708/2005/09/016, arXiv:hep-th/0507289 [hep-th]

36. H. Samtleben, Lectures on gauged supergravity and flux compactifications. Class. Quant. Grav. **25**, 214002 (2008). https://doi.org/10.1088/0264-9381/25/21/214002, arXiv:0808.4076 [hep-th]

37. B. de Wit, P.G. Lauwers, A. Van Proeyen, Lagrangians of $\mathcal{N} = 2$ supergravity–matter systems. Nucl. Phys. **B255**, 569–608 (1985). https://doi.org/10.1016/0550-3213(85)90154-3

38. B. de Wit, C.M. Hull, M. Roček, New topological terms in gauge invariant actions. Phys. Lett. **B184**, 233–238 (1987). https://doi.org/10.1016/0370-2693(87)90573-9

39. M. Günaydin, G. Sierra, P.K. Townsend, The geometry of $\mathcal{N} = 2$ Maxwell–Einstein supergravity and Jordan algebras. Nucl. Phys. **B242**, 244–268 (1984). https://doi.org/10.1016/0550-3213(84)90142-1

40. N. Seiberg, Five dimensional SUSY field theories, non-trivial fixed points and string dynamics. Phys. Lett. **B388**, 753–760 (1996). https://doi.org/10.1016/S0370-2693(96)01215-4, arXiv:hep-th/9608111 [hep-th]
41. B. Zumino, Normal forms of complex matrices. J. Math. Phys. **3**, 1055–1057 (1962)

Chapter 4
Gauge Fixing of Superconformal Symmetries

Abstract In this chapter we combine the actions for vector multiplets and hyper-multiplets, containing also compensating multiplets. We gauge fix the superconformal symmetries that are not necessary for the super-Poincaré theory, and in such way obtain matter-coupled Poincaré supergravity theories. We extensively discuss pure supergravity (using also other compensating multiplets leading to the off-shell theory) and discuss its reduction to $\mathcal{N} = 1$. Then we discuss appropriate variables for the gauge fixing of general matter couplings.

4.1 General Considerations

In the previous chapters, we constructed the local superconformal invariant actions involving vector multiplets and hypermultiplets. The total action is then given by the sum of the vector multiplet and hypermultiplet actions, given respectively by (3.123) and (3.163) [1][1]

$$\mathcal{L} = \mathcal{L}_g + \mathcal{L}_h . \tag{4.1}$$

It remains to break the superfluous symmetries in order to obtain a super-Poincaré invariant action. This can be done by taking a 'gauge choice' for the parts of the superconformal symmetry that are not in the super-Poincaré algebra: dilatations, K-symmetries, S-supersymmetry and the R-symmetry. As we illustrated at the end of Sect. 3.1, classical gauge invariance implies that some degree of freedom in the set of fields $\{\phi^i\}$ in fact disappears from the action $S[\phi^i]$. More precisely, from the gauge symmetry requirement one has

$$0 = \delta S[\phi] = \frac{\delta S[\phi]}{\delta \phi^i} \delta \phi^i , \tag{4.2}$$

---

[1] We concentrate here on 4 dimensions. The principles are the same for $D = 5$ and for a large part also for $D = 6$.

© Springer Nature Switzerland AG 2020

E. Lauria, A. Van Proeyen, $\mathcal{N} = 2$ *Supergravity in D = 4, 5, 6 Dimensions*,
Lecture Notes in Physics 966, https://doi.org/10.1007/978-3-030-33757-5_4

and if one can redefine the basis of fields such that all $\delta\phi^i = 0$ $\forall i$ except, for example, $i = 0$, then (4.2) clearly implies that S does not depend on ϕ^0. This argument immediately shows that, after writing out all covariant derivatives and dependent gauge fields, b_μ should disappear from the action. Indeed, as clear from Eq, (2.73), from our fundamental fields only b_μ transforms under K and therefore

$$\delta(\lambda_K)S = \frac{\delta S}{\delta b_\mu} 2\lambda_{K\mu} = 0 \,. \tag{4.3}$$

Since the action does not depend on b_μ in this basis, we can arbitrarily fix a gauge for K-invariance. Conventionally

$$K\text{-gauge:} \quad b_\mu = 0. \tag{4.4}$$

We could repeat a similar procedure for the other gauged symmetries that we want to fix.[2] However, this procedure is often cumbersome (field redefinitions are often non-local) while the result is just the same as taking a 'gauge choice' for this symmetry. The example presented in Sect. 1.2.2 can illustrate these remarks.

Note that if one imposes a gauge condition, it does not imply that one should forget about the transformations that are gauge fixed, e.g. the K transformations after the choice (4.4). The correct conclusion is that now the K transformations are dependent on the other ones, in such a way that the gauge condition is respected. In other words, the K transformation must act in such a way on b_μ so that it eliminates all the other transformations that act on it. Only then will (4.4) be a consistent gauge condition that does maintain the invariance of the action, where the field truly vanishes. In $D = 4$, using (2.25) and (2.90) while demanding that $\delta_I(\epsilon^I)b_\mu = 0$, we obtain a *decomposition law*:

$$\lambda_K^a(\lambda_D, \epsilon, \eta) = -\tfrac{1}{2}e^{\mu a}\left[\partial_\mu \lambda_D + \tfrac{1}{2}\left(\bar{\epsilon}^i \phi_{\mu i} - \tfrac{3}{4}\bar{\epsilon}^i \gamma_\mu \chi_i - \bar{\eta}^i \psi_{\mu i} + \text{h.c.}\right)\right]. \tag{4.5}$$

With the above expression for the parameter of K transformations, (4.4) will be a consistent condition. The action of a K transformation is now dependent on the other transformations, via the decomposition law (4.5), e.g. the first term in (4.5) is a contribution to local dilatations of the action where b_μ is omitted.

The present chapter is structured as follows: In Sect. 4.2 we explain the gauge fixing procedure for *pure $\mathcal{N} = 2$ supergravity in $D = 4$*, i.e. where the physical fields are just the graviton, the doublet gravitino and the so-called graviphoton. We will discuss 3 different ways of obtaining an off-shell representation using various sets of compensating multiplets. In Sect. 4.3 we consider the reduction of

---

[2]If there is one field that has the same number of degrees of freedom than the gauge symmetry itself, one can often make redefinitions proportional to that field such that only that one field transforms.

the $\mathcal{N} = 2$, $D = 4$ pure supergravity theory to $\mathcal{N} = 1$. In Sect. 4.4 we will re-introduce vector multiplets and hypermultiplets (both as matter and compensators). We derive the equations of motion for the auxiliary fields of these actions in $D = 4, 5$ and choose appropriate gauge fixing conditions, focussing mostly on $D = 4$ supergravity. Then we will introduce appropriate (projective) coordinates, which will parameterize the Poincaré theory. In Sect. 4.5 this will be done for the scalars of the vector multiplets, which will allow us to discuss already the main properties of special Kähler geometry. In Sect. 4.6 we will in the same way introduce appropriate coordinates to discuss the quaternionic-Kähler geometry for the scalars of the hypermultiplets. We end this chapter in Sect. 4.7 with some remarks on the gauge fixing for $D = 5$ and $D = 6$ supergravity, which follows similar patterns as what we discussed for $D = 4$.

All these ingredients will be intensively exploited in following chapters on special geometry and to define the final actions for $\mathcal{N} = 2$ matter-coupled Poincaré supergravity in Chap. 6.

4.2   Pure $\mathcal{N} = 2$ Supergravity

This section is devoted to the construction of the pure $\mathcal{N} = 2$ supergravity, i.e. there are no other physical fields than the graviton, the gravitino and the graviphoton. We will also pay attention to the off-shell structure of the theory, in particular discussing the auxiliary field structure. Though the principles that we discuss here are applicable in general, *we will concentrate in this section on $D = 4$.*

We start by introducing the so-called minimal field representations. As we will explain in Sect. 4.2.1, the latter is sufficient to fix the dilatation symmetry, but it leads to an inconsistent action. In order to overcome this problem we need a second compensating multiplet, for which there are several known choices: the hypermultiplet, the so-called nonlinear multiplet [2–4] and the tensor multiplet. In each case one gets a different set of auxiliary fields for the $\mathcal{N} = 2$ super-Poincaré theory, with the same number of field components ($40 + 40$ off-shell degrees of freedom) [5]. We will discuss these in following subsections.

4.2.1   The Minimal Field Representation

It is natural to perform the gauge fixing of dilatations and of the U(1) part of the R-symmetry group using the scalars of a vector multiplet. They are rather similar to the scalars of chiral multiplets in $\mathcal{N} = 1$, which are used in that case to define the so-called old-minimal sets of auxiliary fields, see [6, 7] and reviewed in [8, Sect. 16.2].

The starting point is (3.123) for only one vector multiplet with complex scalar X:

$$F(X) = -\tfrac{1}{4}\mathrm{i}X^2 \qquad \rightarrow \qquad N_{00} = -1, \qquad N = -X\bar{X}. \tag{4.6}$$

The minus sign reflects that X defines a compensating multiplet and not a physical multiplet, similar to the fact that the action (1.10) has negative kinetic terms for its scalar. We fix the $U(1)$ gauge by restricting X to be real, and the dilatational gauge by taking $X = \kappa^{-1}$, then we are left with

$$e^{-1}\mathcal{L}_g = \tfrac{1}{6}\kappa^{-2}R + \kappa^{-2}D + \cdots. \tag{4.7}$$

The Weyl multiplet with the $24 + 24$ field components in Table 2.3 and a compensating off-shell vector multiplet with $8 + 8$ components as in Table 3.1 forms a 'minimal field representation' of $\mathcal{N} = 2$ supergravity. However, when considering this minimal field representation for building an action, we are faced with two issues. First, the action (4.7) contains a term linear in the D field, which would impose $X\bar{X} = \kappa^{-2} = 0$ as a consequence of the equation of motion for D, therefore removing the Einstein–Hilbert term. Second, there is still a remaining gauged $SU(2)$ R-symmetry, because the scalars of the vector multiplets are invariant under these transformations. The super-Poincaré group does not contain this group, and thus a matter-coupled $\mathcal{N} = 2$ supergravity does not necessarily have such an $SU(2)$ gauge symmetry.

Both problems can be solved by introducing a second 'compensating' multiplet in the background of the vector multiplet, involving scalars that do transform under $SU(2)$ and that can be therefore used to fix the $SU(2)$. In this book we concentrated a lot on the hypermultiplets, whose scalars indeed transform under $SU(2)$ and whose action (3.163) contains another term for the auxiliary field and for an Einstein–Hilbert term. This is indeed one of the possibilities, but in this section we will present also the other known solutions.

Off-shell fields form a massive representation of supersymmetry, which become propagating in a super-Weyl gravity [9]. Massive representations contain representations of $USp(4)$ [10]. Let us recall the relevant massive representations of $\mathcal{N} = 2$ supersymmetry[3]:

$$\text{spin 2 multiplet} : 24 + 24 \ : \{2^1, (\tfrac{3}{2})^4, 1^5, 1^1, (\tfrac{1}{2})^4, 0^1\},$$

$$\text{spin 1 multiplet} : \ \ 8 + 8 \ : \ \{1^1, (\tfrac{1}{2})^4, 0^5\},$$

$$\text{spin } \tfrac{1}{2} \text{ multiplet} : \ \ 4 + 4 \ : \ \{(\tfrac{1}{2})^2, 0^4\}. \tag{4.8}$$

---

[3]We first write the number of bosonic + fermionic components ($2j + 1$ components for spin j), and then the spin content of the fields with notation $(j)^{\#}$, separating $USp(4)$ representations.

The fields of the Weyl multiplet represent the components of a massive spin 2 multiplet:

field	dof	spin 2	spin $\frac{3}{2}$	spin 1	spin $\frac{1}{2}$	spin 0
$e_\mu{}^a$	5	1				
b_μ	0					
$V_{\mu i}{}^j$	9			3		
A_μ	3			1		
T_{ab}^-	6			2		
D	1					1
$\psi_\mu{}^i$	16		4			
χ^i	8				4	

$$(4.9)$$

Note that in this and the following tables, the off-shell number of degrees of freedom (dof) of fields are given, subtracting all the superconformal gauge dof. The following columns order these according to the massive spin representations. We thus find indeed the multiplet as written in the first line of (4.8). Similarly, the vector multiplet of the minimal field representation of $\mathcal{N}=2$ supergravity represents a spin 1 massive multiplet:

field	dof	spin 1	spin $\frac{1}{2}$	spin 0
X	2			2
W_μ	3	1		
Y_{ij}	3			3
Ω^i	8		4	

$$(4.10)$$

We can use some of these fields to fix superconformal symmetries. The first of these was already mentioned in (4.4). The complex scalar X can be put to a constant to gauge fix the dilatation and the U(1) R-symmetry. Similar as in the example in Sect. 1.2.2 we can choose a convenient value to obtain the standard normalization of the Einstein–Hilbert action. But we will choose this only after we have considered the full action. Similarly, the fermion Ω^i of the compensating multiplet can be fixed to a value to remove S-supersymmetry as an independent symmetry (similar to K in (4.5)). The combination used for pure Poincaré supergravity is thus of the form (a spacetime index μ indicates that the field is a gauge field, while those with a are not gauge, and thus have 4 components)

Weyl multiplet: $e_\mu{}^a$, b_μ, ψ_μ^i, $V_{\mu i}{}^j$, A_μ, T_{ab}, χ^i, D

$+$

vector multiplet: X, Ω^i, W_μ, Y_{ij}

$$\downarrow \quad \begin{array}{l} \text{gauge fixing symmetries} \ \ K_a, \ D, \ \text{U(1)}, \ S^i \\ \text{by choosing values for} \ \ b_\mu, \ X, \ \ \Omega^i \end{array}$$

$\mathcal{N} = 2$ super-Poincaré multiplet $(32 + 32)$

$$e_\mu{}^a \ (6), \ W_\mu \ (3), \ V_{\mu i}{}^j \ (9), \ A_a \ (4), \ T_{ab} \ (6), \ D \ (1), \ Y_{ij} \ (3)$$
$$\psi_\mu^i \ (24), \ \chi^i \ (8).$$

$$(4.11)$$

The graviton, gravitino and graviphoton (the latter originating in the vector multiplet) are the physical fields. On the other hand, $V_{\mu i}{}^j$, T_{ab}, χ^i and D are auxiliary fields in the Weyl multiplet, while Y_{ij} is the SU(2) triplet auxiliary field originating in the vector multiplet.

We now consider the three possibilities for the second compensating multiplet.

4.2.2 Version with Hypermultiplet Compensator

We can use a hypermultiplet as a second compensating multiplet [11]. We use for this multiplet constant frame fields $f^{iA}{}_X$, such that (3.68) is solved by $k_D{}^X = q^X$. Then (3.70) just defines another parameterization of the 4 physical scalars in the form A^{iA}. They are thus written as a doublet of the R-symmetry group SU(2) and $\mathcal{A} = 1, 2$ are indices for a priori another USp(2) = SU(2) group. To define the metric as in Sect. 3.3.3 we take $d^{\bar{A}}{}_B = -\delta^A{}_B$, where again the negative sign reflects that this is a compensating multiplet. With this choice, removing a bar from the index induces a minus sign and this is in accordance with Sect. 3.3.4, where a now runs only over one value and $\eta_{11} = -1$. Normalizing $C_{AB} = -\varepsilon_{AB}$, we thus have e.g. $\rho_{A\bar{B}} = \varepsilon_{AB}$, such that $(A^{iA})^* = A^{jB} \varepsilon_{ji} \varepsilon_{BA}$. In this way the action (3.163) starts with

$$e^{-1}\mathcal{L}_h = -\tfrac{1}{12}k_D{}^2 R + \tfrac{1}{4}k_D{}^2 D + \tfrac{1}{2}\partial_\mu A^{iA}\partial^\mu A^{jB} \varepsilon_{ij}\varepsilon_{AB} + \cdots,$$
$$k_D{}^2 = -A^{iA}A^{jB}\varepsilon_{ij}\varepsilon_{AB} = -A^{iA}(A^{iA})^*. \qquad (4.12)$$

In the case of a single hypermultiplet it is known how to close the supersymmetry algebra off-shell by including auxiliary fields F^{iA}, appearing in the transformations of the fermions ζ^A [1, 12–15]. In particular, one adds to the symmetry group transformations (3.75) a non-compact generator Z, such that $k_Z{}^X f^{iA}{}_X = F^{iA}$. This generator is then gauged by the vector multiplet in the minimal field representation, leading to central charge transformations. We do not discuss here the full procedure.[4]

The off-shell multiplet is then of the form

field	dof	spin $\frac{1}{2}$	spin 0
$A^{i\mathcal{A}}$	4		4
$F^{i\mathcal{A}}$	4		4
$\zeta^{\mathcal{A}}$	8	4	

(4.13)

To gauge fix the SU(2), we restrict the quaternion to be a real number.[5] We write

$$\text{SU(2) gauge:} \qquad A^{i\mathcal{A}} = \sqrt{2}\kappa^{-1}\varepsilon^{i\mathcal{A}}e^u, \qquad \rightarrow \qquad k_{\text{D}}{}^2 = -4\kappa^{-2}e^{2u},$$

(4.14)

being $\varepsilon^{i\mathcal{A}}$ the Levi-Civita antisymmetric symbol, which breaks the local SU(2) on the i indices while leaving a global SU(2) group acting now on both indices. Here u is an auxiliary field, which in the sum of (4.7) and (4.12) appears together with the auxiliary field D:

$$e^{-1}\mathcal{L} = \tfrac{1}{6}\kappa^{-2} R \left(1 + 2e^{2u}\right) R + \kappa^{-2}\left(1 - e^{2u}\right) D + 2\kappa^{-2}e^{2u}\partial_\mu u\, \partial^\mu u + \cdots.$$

(4.15)

On-shell $u \approx 0$ and the Einstein–Hilbert term has the canonical normalization.

The scheme that extends (4.11) is thus:

Minimal field representation

+

second compensating multiplet: $A^{i\mathcal{A}}$, $\zeta^{\mathcal{A}}$, $F^{i\mathcal{A}}$

$\downarrow$ gauge fixing SU(2) symmetry
by choosing $A^{i\mathcal{A}}$ in terms of real u

$\mathcal{N} = 2$ super-Poincaré gravity $(40 + 40)$

$$e_\mu{}^a\ (6),\ W_\mu\ (3),\ V_{ai}{}^j\ (12),\ A_a\ (4),\ T_{ab}\ (6),\ D\ (1),\ Y_{ij}\ (3),\ u\ (1),\ F^{i\mathcal{A}}\ (4)$$
$$\psi_\mu^i\ (24),\ \chi^i\ (8),\ \zeta^{\mathcal{A}}\ (8).$$

(4.16)

---

[4]It is not clear whether such a procedure can also be done when the quaternionic geometry is not flat.

[5]This is of course similar to the gauge fixing of U(1), which fixes the phase of the complex X of the compensating multiplet.

In the remaining set of fields of the $\mathcal{N} = 2$ super-Poincaré theory, $e_\mu{}^a$ and W_μ are the physical graviton and the graviphoton, and ψ_μ^i are the gravitini. All the others are auxiliary fields, and the full set includes $40 + 40$ off-shell degrees of freedom.

One denotes by gauged $\mathcal{N} = 2$ supergravity the situation where the gravitinos transform under a local SU(2) and where this induces then also a cosmological constant (anti-de Sitter supergravity). As an example, in the present formalism we can gauge a $U(1) \subset SU(2)$ if the scalar of the compensating multiplet transforms under a U(1) subgroup of the USp(2) group acting on the indices $\mathcal{A}$. We see from the first line in (3.163) that such a gauging produces a potential. For example, since g_{XY} for the compensating multiplet is negative definite, the term $2\bar{X}^I X^J k_I{}^X g_{XY} k_J{}^Y$ produces a negative term in the potential. There will be another negative term after elimination $\mathbf{Y}^0$. Since the involved fields are now constants, this is a negative cosmological term, whose size is determined by the choice of the Killing vector. Thus we have anti-de Sitter supergravity. Due to the gauge condition (4.14), the gauge U(1) is then identified with a subgroup of the SU(2) R-symmetry group.

If more vector multiplets are included, the full automorphism group SU(2) can be gauged by having the second compensating multiplet transforming under an SU(2) group. Also in this case, the gauge fixing (4.14) mixes the SU(2) factor of the superconformal group with the group gauged by vector multiplets. The first compensating multiplet and two physical vector multiplets can even gauge a non-compact SO(2, 1) gauge group [16] leading to a massive vector multiplet.

4.2.3   Version with Tensor Multiplet Compensator

The second choice is to use a tensor multiplet as second compensating multiplet [16]. The tensor multiplet, shortly discussed in Sect. 3.2.5, contains a triplet of scalars L_{ij}, a gauge antisymmetric tensor $E_{\mu\nu}$, a doublet of Majorana spinors φ^i and a complex scalar G. As massive fields they represent a spin 1 multiplet:

field	dof	spin 1	spin $\frac{1}{2}$	spin 0
L_{ij}	3			3
$E_{\mu\nu}$	3	1		
G	2			2
φ^i	8		4	

(4.17)

The scalars L_{ij} form a triplet of the SU(2) part of the R-symmetry group. Therefore, a gauge choice on this field, e.g.

$$\text{SU(2)}/\,\text{U(1) gauge:} \qquad L_{ij} = \delta_{ij} e^u \qquad\qquad (4.18)$$

breaks this group to a U(1) subgroup. This remaining subgroup is gauged by one of the components of $V_{\mu i}{}^j$: with the choice (4.18) this is proportional to $V_\mu' =$

$V_{\mu 1}{}^2 - V_{\mu 2}{}^1$. On-shell, the latter is identified with the graviphoton W_μ. The other 2 components of $V_{\mu i}{}^j$ are then no more gauge fields, and indicated below as $V_{ai}{}^j$ (hence with 2×4 dof). The scheme that extends (3.5) is thus:

Minimal field representation

$+$

second compensating multiplet: L_{ij}, φ^i, G, $E_{\mu\nu}$

$\downarrow$ gauge fixing $SU(2)/U(1)$ symmetry
by choosing L_{ij} in terms of real u

$\mathcal{N} = 2$ super-Poincaré gravity $(40 + 40)$

$$e_\mu{}^a\ (6),\ W_\mu\ (3),\ V'_\mu\ (3),\ V_{ai}{}^j\ (8),\ A_a\ (4),\ T_{ab}\ (6),\ D\ (1),\ Y_{ij}\ (3),$$
$$E_{\mu\nu}\ (3),\ u\ (1),\ G\ (2) \tag{4.19}$$
$$\psi_\mu^i\ (24),\ \chi^i\ (8),\ \varphi^i\ (8).$$

Again the physical fields are $e_\mu{}^a$, W_μ, ψ_μ^i and the others are auxiliary. With the gauge antisymmetric tensor and the remaining $U(1)$ gauge group, this version has similar properties as the 'new minimal' set of auxiliary fields of $\mathcal{N} = 1$ supergravity [17].

Similarly to the previous case, the gauged $\mathcal{N} = 2$ supergravity can be obtained by coupling the first compensating multiplet (vector multiplet) to the compensating tensor multiplet, which will include in the Lagrangian a term $\varepsilon^{\mu\nu\rho\sigma} E_{\mu\nu} F_{\rho\sigma}(W)$.

4.2.4 Version with Nonlinear Multiplet Compensator

The third choice (historically, it is the first version [2, 11]) is to use as a second compensating multiplet a *nonlinear tensor multiplet* [18], consisting of the fields $\Phi_i^{\mathcal{A}}$, λ_i, $M^{[ij]}$, V_a. The scalar field $\Phi_i^{\mathcal{A}}$, $\mathcal{A} = 1, 2$ is a 2×2 special unitary matrix, where the indices $\mathcal{A}$ are unrelated to the R-symmetry $SU(2)$ indices i. This multiplet is called nonlinear because its transformation rules are nonlinear in the fields. The fields Φ_i^α represent 3 degrees of freedom due to the nonlinear constraint. The linearized version of this multiplet is again the tensor multiplet. The vector field V_a is constrained in the superconformal background due to the following (linearized) relation:

$$\partial_a V^a = D + \tfrac{1}{3} R, \tag{4.20}$$

where R is the Ricci scalar. One may view this equation as eliminating one of the 4 degrees of freedom of the vector in the nonlinear multiplet, which with the two real components in the complex $M^{[ij]}$ and $\Phi_i^{\mathcal{A}}$ complete the 8 bosonic degrees of freedom of this multiplet. However, in the superconformal framework we will use it to eliminate the auxiliary field D, which was the cause of the first problem of the minimal field representation mentioned at the start of Sect. 4.2. The multiplet is also a spin 1 massive multiplet:

field	dof	spin 1	spin $\frac{1}{2}$	spin 0
Φ_i^A	3			3
$M^{[ij]}$	2			2
V_a	3	1		
λ_i	8		4	

$$(4.21)$$

To break the local SU(2) invariance one imposes on the 3 components of Φ_i^A

$$\text{SU(2) gauge:} \qquad \Phi_i^A = \delta_i^A, \qquad (4.22)$$

such that the distinction between indices $\mathcal{A}$ and $i, j, \ldots$ is lost. The scheme that extends (3.5) is thus:

Minimal field representation

$+$

second compensating multiplet: Φ_i^α, λ_i, M_{ij}, V_a

$\downarrow$ gauge fixing SU(2) symmetry
by choosing Φ_i^A and solve for D

$\mathcal{N} = 2$ super-Poincaré gravity $(40 + 40)$

$e_\mu{}^a$ (6), W_μ (3), $V_{ai}{}^j$ (12), A_a (4), T_{ab} (6), Y_{ij} (3), V_a (4), $M_{[ij]}$ (2)
ψ_μ^i (24), χ^i (8), λ_i (8) .

$$(4.23)$$

The theory remains invariant under rigid SU(2), corresponding to the residual group of (4.22). Finally, a U(1) subgroup of this residual SU(2) (acting on the index $\mathcal{A}$) can be gauged via the first compensating multiplet. This produces gauged $\mathcal{N} = 2$ supergravity.

4.3 Reduction from $\mathcal{N} = 2$ to $\mathcal{N} = 1$

We now discuss the reduction of the 3 sets of auxiliary fields to $\mathcal{N} = 1$. This has been first considered in the discussion section of [5]. We will be more explicit here. For any version of auxiliary fields, we have to start from the Weyl multiplet with the vector compensating multiplet. We treat here just the pure supergravity theory. More general reductions in the Poincaré formulation have been discussed in [19, 20]. A conformal treatment has been given in [21].

4.3.1   Reduction of the $\mathcal{N} = 2$ Weyl Multiplet

To reduce the $D = 4$, $\mathcal{N} = 2$ Weyl multiplet, we consider (2.90) and the Majorana spinors $\epsilon = \epsilon^1 + \epsilon_1$, $\eta = \eta^1 + \eta_1$, and put ϵ^2 and η_2 to zero. We then compare with the transformations in [8, Chap. 17] for the reduction to $\mathcal{N} = 1$. Obviously this reduction breaks the SU(2) part of the R-symmetry group of $\mathcal{N} = 2$ to a U(1) part, i.e. the part where

$$\lambda_1{}^1 = -\lambda_2{}^2 \in i\mathbb{R}, \qquad \lambda_1{}^2 = \lambda_2{}^1 = 0. \tag{4.24}$$

The frame field transforms to $\psi_{\mu 1} + \psi_\mu^1$, which is thus the $\mathcal{N} = 1$ gravitino. Its transformation is

$$\delta\psi_\mu^1 = \left(\partial_\mu + \tfrac{1}{2}b_\mu + \tfrac{1}{4}\gamma^{ab}\omega_{\mu ab} - \tfrac{1}{2}iA_\mu - V_{\mu 1}^1\right)\epsilon^1 - \gamma_\mu\eta^1. \tag{4.25}$$

This identifies the real combination

$$A_\mu^{(1)} = \tfrac{1}{3}A_\mu - \tfrac{2}{3}iV_{\mu 1}{}^1, \tag{4.26}$$

as the A_μ field of $\mathcal{N} = 1$, which transforms into

$$\delta A_\mu^{(1)} = -\tfrac{1}{2}i\epsilon^1\phi_{\mu 1} + \tfrac{1}{8}i\epsilon^1\gamma_\mu\chi_1 + \tfrac{1}{2}i\eta_1\psi_\mu^1 + \text{h.c.} \tag{4.27}$$

The composite field ϕ_μ for $\mathcal{N} = 1$ is as in (2.92), but without the χ-term, i.e.

$$\phi_{\mu 1} = P_L\phi_\mu^{(1)} + \tfrac{1}{4}\gamma_\mu\chi_1, \qquad \phi_\mu^{(1)} = \phi_\mu - \tfrac{1}{4}\gamma_\mu\chi, \tag{4.28}$$

so that the (Q-susy part of the) right-hand side of (4.27) is indeed proportional to $\phi_\mu^{(1)}$.

The transformation of b_μ in (2.90), rewritten using (4.28), is

$$\delta b_\mu = \tfrac{1}{2}\bar{\epsilon}^1\phi_\mu^{(1)} - \tfrac{1}{4}\bar{\epsilon}^1\gamma_\mu\chi_1 - \tfrac{1}{2}\bar{\eta}^1\psi_{\mu 1} + \text{h.c.} \tag{4.29}$$

But b_μ transforms also under special conformal transformations, see (2.73). Therefore we identify the $\mathcal{N} = 1$ supersymmetry as a linear combination of the ϵ^1 supersymmetry with a special conformal transformation:

$$\delta^{(1)}(\epsilon) = \delta(\epsilon^1) + \delta_K(\lambda_{K\mu}), \qquad \lambda_{K\mu} = \tfrac{1}{8}\bar{\epsilon}^1 \gamma_\mu \chi_1 + \text{h.c.}, \qquad (4.30)$$

which eliminates the terms with χ_1 in (4.29). The fields of the $\mathcal{N} = 1$ Weyl multiplet are thus $\left\{ e_\mu{}^a,\, b_\mu,\, \psi_\mu^1,,\, A_\mu^{(1)} \right\}$. The combination (4.27) implies that the superconformal U(1) of $\mathcal{N} = 1$ is a diagonal subgroup of the U(1) × SU(2) of $\mathcal{N} = 2$:

$$\lambda_T^{(1)} = \tfrac{1}{3}\left(\lambda_T - 2\mathrm{i}\lambda_1{}^1\right). \qquad (4.31)$$

Similarly, one can check that there is an $\mathcal{N} = 1$ vector multiplet containing[6]

$$\left\{ B_\mu,\, \mathrm{i}(\chi^1 - \chi_1),\, D \right\}, \qquad B_\mu = -\tfrac{2}{3}\left(A_\mu + \mathrm{i}V_{\mu 1}{}^1\right). \qquad (4.32)$$

The gauge field gauges a combination of λ_T and $\mathrm{i}\lambda_1{}^1$, complementary to (4.31)

$$\theta = -\tfrac{2}{3}\left(\lambda_T + \mathrm{i}\lambda_1{}^1\right). \qquad (4.33)$$

The remaining fields of the $\mathcal{N} = 2$ Weyl multiplet form a double spin-3/2 multiplet:

$\mathcal{N} = 2$ Weyl	Weyl multiplet	spin 1 multiplet	2 spin $\frac{3}{2}$ multiplets
$e_\mu{}^a$	$e_\mu{}^a$		
b_μ	b_μ		
$A_\mu,\, V_{\mu 1}{}^1$	$A_\mu^{(1)}$	B_μ	
$V_{\mu 1}{}^2$			$V_{\mu 1}{}^2 \in \mathbb{C}$
T_{ab}^-			T_{ab}
D		D	
$\psi_\mu{}^i$	$\psi_\mu{}^1$		$\psi_\mu{}^2$
χ^i		χ^1	χ^2
$24 + 24$	$8 + 8$	$4 + 4$	$12 + 12$

$$(4.34)$$

In conclusion, the $24 + 24$ fields, reduce to $8 + 8$ Weyl and a $4 + 4$ vector multiplet under the truncation. The remainder are two $6 + 6$ spin 3/2 multiplets. However,

---

[6]The factors, e.g. i in the second component, are introduced such that the transformations agree with [8, (17.1)].

note that the spin 3/2 multiplets cannot be treated separately beyond the linear level, since the nonlinear transformations involve the frame field and the $\mathcal{N} = 1$ gravitino.

4.3.2 Reduction of the Compensating Vector Multiplet

The superconformal transformations of the vector multiplet are in (3.15). We now need the abelian version, i.e. we can omit the index I, and $f_{JK}{}^I = 0$. The complex scalar X transforms chiral and is thus the basic field of a chiral multiplet:

$$\left\{ X, \frac{1}{\sqrt{2}} \Omega_1, Y_{11} \right\}. \tag{4.35}$$

The scalar X transforms under the θ transformation (4.33). Indeed, considering the U(1) transformations:

$$\delta X = i\lambda_T X = i\left(\lambda_T^{(1)} + \theta\right) X. \tag{4.36}$$

Thus the chiral multiplet transforms under the gauge transformation of the vector multiplet (4.32). This is also important to understand the following contribution to the Q-transformation of the auxiliary field Y_{11}. According to (3.19), $\mathcal{D}_\mu \Omega_1$ contains a term $-2X\phi_{\mu 1}$, which, after use of (4.28) implies that the transformation of Y_{11} contains a term $\bar{\epsilon}_1 \chi_1 X$. This term thus depends on the fermion of the gauge multiplet (4.32). The latter is the transformation term that the auxiliary fields of chiral multiplets obtain in Wess–Zumino gauge in $\mathcal{N} = 1$ supersymmetry due to the θ symmetry (4.36). For the normalization, see e.g.[8, (17.3)].

The remaining fields of the $\mathcal{N} = 2$ vector multiplet form the $\mathcal{N} = 1$ vector multiplet:

$$\left\{ W_\mu, -\Omega_2 - \Omega^2, 2iY_{12} \right\}. \tag{4.37}$$

Thus the reduction goes as follows:

$\mathcal{N} = 2$ vector	spin $\frac{1}{2}$ multiplet	spin 1 multiplet
X	$X \in \mathbb{C}$	
W_μ		W_μ
Y_{ij}	$Y_{11} \in \mathbb{C}$	$iY_{12} \in \mathbb{R}$
Ω_i	Ω_1	Ω_2
$8 + 8$	$4 + 4$	$4 + 4$

$$(4.38)$$

For clarity, we indicated the reality properties of the Y_{11} and Y_{12}, see (A.21). The component Y_{22} is the complex conjugate of Y_{11}.

One consistent reduction to $\mathcal{N} = 1$ consists in keeping only the Weyl multiplet of (4.34) and the chiral multiplet (4.35), which provides then the compensating fields for dilatation, U(1) and S-supersymmetry. The conformal action of the chiral multiplet in the background of the Weyl multiplet gives the $\mathcal{N} = 1$ pure supergravity with the 'old-minimal' set of auxiliary fields.

If we keep also the spin-1 multiplet of (4.34) in the background, since the chiral multiplet transforms under θ, see (4.36), this will induce a Fayet–Iliopoulos term $e\, X\, \bar{X}\, D$. However, to describe pure supergravity, there are no kinetic terms for the spin-1 multiplet. Therefore the field equation for D is inconsistent with a gauge choice that gives a non-zero value to X. That is why we need the second compensating multiplet as viewed from the $\mathcal{N} = 1$ perspective.

4.3.3 Reduction of the Second Compensating Multiplet

We comment here on the reduction of the two versions that have linear transformation laws. When we use the hypermultiplet, there are two $\mathcal{N} = 1$ chiral multiplets:

$$\left\{ A^{1\mathcal{A}},\ -\sqrt{2}\mathrm{i}\zeta^{\mathcal{A}},\ -\frac{1}{\sqrt{2}}\bar{X}F^{2\mathcal{A}} \right\}. \tag{4.39}$$

These are chiral multiplets for each value of $\mathcal{A}$. Remember that $A^{2\mathcal{A}}$ is dependent on the complex conjugate of $A^{1\mathcal{A}}$. In the $\mathcal{N} = 1$ reduction we can include one of these two chiral multiplets, e.g. $\mathcal{A} = 1$. Under the U(1) transformations:

$$\delta A^{1\mathcal{A}} = A^{1\mathcal{A}}\lambda_1{}^1 = \mathrm{i}\left(\lambda_T^{(1)} + \tfrac{1}{2}\theta\right)A^{1\mathcal{A}}. \tag{4.40}$$

Hence, these multiplets transform also under θ and there is another Fayet–Iliopoulos term, such that the field equation of D relates X to $A^{1\mathcal{A}}$ and a consistent dilatational gauge (and gauge for $\lambda_T^{(1)}$ and θ) can be chosen.

When we take for the second compensating multiplet of $\mathcal{N} = 2$ the linear multiplet, we find that the $\mathcal{N} = 1$ reduction leads to a chiral multiplet

$$\left\{ L^{11},\ \sqrt{2}\varphi^1,\ -G \right\}, \tag{4.41}$$

and an $\mathcal{N} = 1$ linear multiplet

$$\left\{ \mathrm{i}L^{12},\ \phi^2 + \phi_2,\ E_{\mu\nu} \right\}. \tag{4.42}$$

In Sect. 4.2.3 we chose the gauge fixing (4.18). The proportionality with δ_{ij} was just an arbitrary choice in SU(2). Here we could choose the same (which means L^{11} real, but as well iL^{12} to be the non-zero component. L^{11} transforms under the remaining part of the SU(2), (4.24), and hence under the θ of (4.33). On the other hand L^{12} is invariant under that remaining U(1). Therefore, if we keep the chiral multiplet in the $\mathcal{N} = 1$ reduction, it performs the same role as one of the multiplets (4.39) to allow a consistent field equation for the auxiliary D. If we keep the linear multiplet in the reduction, then there is also a Fayet–Iliopoulos-like coupling of that field with D, but moreover a local U(1) remains. Hence we find the situation as in new minimal supergravity, having a preserved local U(1) (and gauge auxiliary tensor $E_{\mu\nu}$) coupled to the other matter multiplets of the reduction of the Weyl and first compensating multiplet.

4.4 Matter-Coupled Supergravity

In this section we consider again the full action (4.1), which includes vector multiplets and hypermultiplets, as well as the compensating multiplets. In light of the discussion in Sect. 4.2, the latter is thus the extension of the action with an hypermultiplet as second compensating multiplet, considered in Sect. 4.2.2. We choose the hypermultiplet as compensating multiplet such that the couplings of the physical hypermultiplets can be obtained as a projection from a conformal hyper-Kähler manifold, similar to the way in which the couplings of the vector multiplets are obtained as a projection from a conformal special Kähler manifold. The disadvantage of this choice is that it is not known how to keep the structure with auxiliary fields $F^{i,A}$ mentioned in Sect. 4.2.2 for a non-trivial hyper-Kähler geometry without introducing an infinite number of fields as in harmonic or projective superspace. In total we therefore have $n + 1$ vector multiplets and $n_H + 1$ hypermultiplets, described by the action (4.1), being $\mathcal{L}_g$, $\mathcal{L}_h$ the action for the vector multiplets and hypermultiplets, respectively. We thus redefine the range of the indices X and $\mathcal{A}$ for the hypermultiplets to, respectively, $4(n + 1)$ and $2(n_H + 1)$ values.

Though in this section, as in the previous one, we will mostly focus on $D = 4$ dimensions, some results for $D = 5$ will also be presented along the way, exploiting the fact that the treatment is very similar. Full results for $D = 5$ in the conformal setting are given in [22–25] and with other methods in [26–28].

In Sect. 4.2 we took gauge choices, keeping the auxiliary fields. In the present section we first eliminate the auxiliary fields (Sect. 4.4.1) and then we consider the gauge fixing (Sect. 4.4.2). We conclude with the final formulae for the resulting action (Sect. 4.4.3) and corresponding transformations (Sect. 4.4.4).

4.4.1 Elimination of Auxiliary Fields

We now present the equations of motion for the auxiliary fields of the matter-coupled supergravity in $D = 4, 5$. It has been expected for a long time, but proven only recently [29], that equations of motion can always be represented in the form of covariant equations. This means that gauge fields (in our case mostly the gravitino) appear only within covariant derivatives.[7]

4.4.1.1 Case $D = 5$

The vector multiplet and hypermultiplet actions are given by (3.129) and (3.159). The theory then contains several auxiliary fields. In particular, the matter D, T_{ab} and χ^i, and gauge field V_μ^{ij} are inherited from the Weyl multiplet, while the gauge triplet Y_{ij}^I belongs to the vector multiplet. Both D and χ^i appear as Lagrange multipliers in the action, leading to the following covariant equations of motion, where the symbol $\approx$ denotes the 'on-shell' relation[8]:

$$D: \qquad C_{IJK}\sigma^I\sigma^J\sigma^K - \tfrac{1}{3}k_{\mathrm D}^2 \approx 0\,,$$

$$\chi^i: \qquad -8\mathrm{i}\left(C_{IJK}\sigma^I\sigma^J\psi^{iK} - \tfrac{2}{3}\mathrm{i}A^{iA}\zeta_A\right) \approx 0\,, \qquad (4.43)$$

where we used A^{iA} as defined by (3.70). The covariant equation of motion for the SU(2) gauge field V_μ^{ij} is

$$0 \approx k_X^{ij}\mathcal{D}_\mu q^X - \tfrac{1}{2}C_{IJK}\sigma^K\,\bar\psi^{iI}\gamma_\mu\psi^{jJ}\,. \qquad (4.44)$$

The first term, using (3.98) and (3.154) contains, according to (3.98), $-\tfrac{2}{9}k_{\mathrm D}^2\,V_\mu^{ij}$, so that this equation can be read as

$$\tfrac{2}{9}k_{\mathrm D}^2\,V_\mu^{ij} \approx k_X^{ij}\partial_\mu q^X - W_\mu^I P_I^{ij} + \tfrac{2}{3}\mathrm{i}\bar\psi_\mu^{(i}A^{j)A}\zeta_A - \tfrac{1}{2}C_{IJK}\sigma^K\,\bar\psi^{iI}\gamma_\mu\psi^{jJ}\,, \qquad (4.45)$$

using (3.71) and the moment maps (3.157).

---

[7]The proper field equation of a field ϕ^i is in general not covariant if the derivative of that field $\partial_\mu\phi^i$ appears in the transformation of another field $\delta\phi^j$. However, using the field equation of ϕ^j removes the non-covariant terms in the field equation of ϕ^i.

[8]Since the supersymmetry transformation of D contains the spacetime derivative of χ, see (2.98), the proper field equation of χ contains a non-covariant term, which is however proportional to the field equation of D and we have thus omitted here. Similar covariantizations have been done below and will not be indicated anymore.

The covariant equations of motion of the other auxiliary fields Y_{ij}^I and T_{ab} are given by

$$Y^{ij\,J}C_{IJK}\sigma^K \approx -\tfrac{1}{2}P_I^{ij} + \tfrac{1}{4}iC_{IJK}\bar{\psi}^{ij}\psi^{jK} \,,$$

$$\frac{64}{9}k_{\mathrm{D}}{}^2 T_{ab} \approx 4\sigma^I\sigma^J \hat{F}_{ab}^K C_{IJK} + i\sigma^I\,\bar{\psi}^J\gamma_{ab}\psi^K C_{IJK} + 2i\bar{\zeta}_A\gamma_{ab}\zeta^A \,.$$

$$(4.46)$$

4.4.1.2 Case $D = 4$

The vector multiplet and hypermultiplet actions are given, respectively, by (3.123) and (3.163). The fields D and χ^i appear in the total action as Lagrange multipliers. Therefore, their (covariant) field equations imply the following conditions:

$$D: \ -N + \tfrac{1}{4}k_{\mathrm{D}}{}^2 \approx 0,$$

$$\chi^i: \ N_{IJ}\bar{X}^I\Omega_i^J + id^{\bar{B}}{}_A A_{i\bar{B}}\zeta^A \approx 0.$$

$$(4.47)$$

The covariant equations of motion for the U(1) gauge field A_μ and SU(2) gauge field $V_{\mu i}{}^j$ are

$$0 \approx \tfrac{1}{2}iN_{IJ}(X^I\mathcal{D}_\mu\bar{X}^J - \bar{X}^J\mathcal{D}_\mu X^I) + \tfrac{1}{8}iN_{IJ}\bar{\Omega}^{iI}\gamma_\mu\Omega_i^J - \tfrac{1}{2}i\bar{\zeta}_A\gamma_\mu\zeta^B d^A{}_B \,,$$

$$0 \approx g_{XY}\mathcal{D}_\mu q^X \mathbf{k}^Y - \tfrac{1}{4}N_{IJ}\bar{\Omega}_i^I\gamma_\mu\Omega^{Jj}\tau_j{}^i \,,$$

$$(4.48)$$

which can be solved in the forms

$$N\,A_\mu \approx \tfrac{1}{2}iN_{IJ}(X^I\partial_\mu\bar{X}^J - \bar{X}^J\partial_\mu X^I) + W_\mu^I P_I^0 + \tfrac{1}{4}iN_{IJ}$$
$$\times \left(\bar{X}^I\bar{\psi}_\mu^i\Omega_i^J - X^I\bar{\psi}_{\mu i}\Omega^{iJ}\right) + \tfrac{1}{8}iN_{IJ}\bar{\Omega}^{iI}\gamma_\mu\Omega_i^J - \tfrac{1}{2}i\bar{\zeta}_A\gamma_\mu\zeta^B d^A{}_B \,,$$

$$-\tfrac{1}{2}k_{\mathrm{D}}{}^2\mathbf{V}_\mu \approx g_{XY}\partial_\mu q^X\mathbf{k}^Y - W_\mu^I\mathbf{P}_I + \tfrac{1}{2}i\tau_i{}^j d^{\bar{A}}{}_B\left(A_{j\bar{A}}\bar{\psi}_\mu^i\zeta^A - A^{iB}\bar{\psi}_{\mu j}\zeta_{\bar{A}}\right)$$
$$- \tfrac{1}{4}N_{IJ}\bar{\Omega}_i^I\gamma_\mu\Omega^{Jj}\tau_j{}^i \,,$$

$$(4.49)$$

where for the SU(2) part again we used (3.71) and the moment maps (3.157). For the U(1) part, the moment maps for the Kähler isometries P_I^0 appear. They are real functions, defined[9] such that

$$N_{IJ}\delta(\theta)X^J = -i\theta^I\partial_{\bar{j}}P_I^0 \,.$$

$$(4.50)$$

---

[9]For a general discussion of moment maps in Kähler manifolds, see e.g. [8, Sect. 13.4.1], and also Sect. 5.4.1 below.

The condition above is satisfied by

$$P_I^0 = \mathrm{i} N_{JL} f_{KI}{}^J X^K \bar{X}^L \,, \tag{4.51}$$

due to the definition (3.124) and the homogeneity relations (3.108). Inserting again (3.124) and (3.116) it can be rewritten in various ways:

$$\begin{aligned}
P_I^0 &= f_{IJ}{}^K (X^J \bar{F}_K + \bar{X}^J F_K) - 2C_{I,JK} X^J \bar{X}^K \\
&= -\mathrm{i} N_{IJ} f_{KL}{}^J X^K \bar{X}^L = \mathrm{i} N_{JL} f_{KI}{}^J X^K \bar{X}^L = -\mathrm{i} N_{JK} f_{LI}{}^J X^K \bar{X}^L \quad (4.52)
\end{aligned}$$

and

$$X^I P_I^0 = \bar{X}^I P_I^0 = 0 \,. \tag{4.53}$$

The covariant field equations of the other auxiliary fields, Y^{ij} and T_{ab}, are

$$N_{IJ} Y_{ij}^J \approx -P_{Iij} - \tfrac{1}{4} \mathrm{i} F_{IJK} \bar{\Omega}_i^J \Omega_j^K + \tfrac{1}{4} \mathrm{i} \bar{F}_{IJK} \varepsilon_{ik} \varepsilon_{j\ell} \bar{\Omega}^{kJ} \Omega^{\ell K} \,,$$

$$\tfrac{1}{8} N_{IJ} X^I X^J T_{ab}^+ \approx \tfrac{1}{4} N_{IJ} X^I \widehat{F}_{ab}^{+J} - \tfrac{1}{32} \mathrm{i} F_{IJK} \bar{\Omega}^{iI} \gamma_{ab} \Omega^{jJ} X^K \varepsilon_{ij} - \tfrac{1}{8} \bar{\zeta}_{\bar{A}} \gamma_{ab} \zeta_{\bar{B}} C^{\bar{A}\bar{B}} \,. \tag{4.54}$$

4.4.2 Gauge Fixing for Matter-Coupled Supergravity

We now move to the gauge fixing of the total action (4.1). Though the principles are the same for $D = 5$ (and for a large part also for $D = 6$) for simplicity here we concentrate on 4 dimensions, and comment on $D = 5$ and $D = 6$ in Sect. 4.7.

As mentioned before, the conformal coupling of vector multiplets and hypermultiplets, has been chosen such that there is in each sector one multiplet with negative kinetic energy, used for gauge fixing, and all the others have positive kinetic energies.[10] We recall that (4.1) contains the fields of the Weyl multiplet as a background. Dependent fields in this Weyl multiplet have already been solved for in the final actions that we presented. As mentioned in (4.4), the dilatation gauge field b_μ can be eliminated by a gauge choice for the special conformal transformations K. We will combine the field equation of the auxiliary field D, (4.47), with a dilatation gauge choice to eliminate the modulus of the scalars of the compensating multiplets. The U(1) gauge eliminates the phase of the scalar

---

[10]The positivity of the kinetic energy is a further constraint on the choices of $N(X, \bar{X})$ and the quaternionic metric, which we discuss further in Sect. 4.5.4.

of the compensating vector multiplet.[11] Similarly the quaternionic phases of the scalars of the compensating hypermultiplet are fixed by gauge choices of the SU(2).

On the fermionic side the χ^i field equation will be combined with a gauge for S-supersymmetry to eliminate the fermions in the compensating multiplets.

The only remaining field of the compensating multiplets is the vector field W_μ, which becomes the graviphoton in the final super-Poincaré action. The result will be the coupling of n_H hypermultiplets and n vector multiplets to $\mathcal{N} = 2$, $D = 4$ Poincaré supergravity, see Sect. 4.4.3.

It is important to remark that the fixing of gauge symmetries leads to a change in the definitions of covariant derivatives. Indeed, the fields that we use in the Poincaré supergravity are chosen such that they do not transform any more under the broken symmetries. Moreover, the parameters of the broken symmetries will be expressed in terms of the independent parameters by the decomposition laws, similar to (4.5).

We now give the explicit expressions for the gauge fixing of dilatations and S-supersymmetry. For the dilatations, we have already mentioned in the simple example (1.11) that our aim is to have a standard kinetic term for gravity. Hence, we just collect the terms with the scalar curvature and obtain[12]

$$D\text{-gauge:} \qquad -\tfrac{1}{6}N - \tfrac{1}{12}k_D{}^2 = \tfrac{1}{2}\kappa^{-2}. \qquad (4.55)$$

Combining this with the field equation for D, (4.47), we obtain

$$N = -\kappa^{-2}, \qquad k_D{}^2 = -4\kappa^{-2}. \qquad (4.56)$$

The condition that fixes the dilatation gauge is physically the requirement that the kinetic terms of the scalars and the spin-2 field are not mixed.

The total action does contain terms of the form $\gamma^{\mu\nu}\partial_\mu\psi_\nu^i$, multiplied with fermion fields of the vector and hypermultiplet. These would imply a mixing of the kinetic terms of spin 3/2 and 1/2. We choose, in analogy to the bosonic sector, the S-gauge condition such that such a mixing does not occur. Hence, we put to zero the coefficient of $\gamma^{\mu\nu}\partial_\mu\psi_\nu^i$:

$$S\text{-gauge:} \qquad N_{IJ}\bar{X}^I\Omega_i^J - 2\mathrm{i}A_{i\bar{B}}d^{\bar{B}}{}_A\zeta^A = 0. \qquad (4.57)$$

---

[11] Here is the only difference with 5 dimensions. In that case the superconformal algebra does not contain a U(1), but the scalar of the vector multiplet is also real, such that this step is not necessary, see Sect. 4.7.

[12] The dimensionful coupling constant κ in fact appears here for the first time, thus breaking the dilatation invariance.

Combining (4.57) with the field equation of the auxiliary field χ_i, (4.47), leads to

$$N_{IJ}\bar{X}^I \Omega_i^J = 0, \qquad A_{i\bar{B}}d^{\bar{B}}{}_A \zeta^A = 0. \tag{4.58}$$

Importantly, at this point the R-symmetry is still gauged. The gauge fixing of the latter will have some surprising consequences on the Nature of our $\mathcal{N} = 2$ theories, leading to the so-called *Special geometries*. This important passage will be therefore discussed separately, in Sect. 4.5.

4.4.3 Full Action for $D = 4$

We now have all the ingredients to write down the full action. Using the gauge conditions and the definition (3.16), we find

$$
\begin{aligned}
e^{-1}\mathcal{L} = {} & \kappa^{-2}\left(\tfrac{1}{2}R - \bar{\psi}_{i\mu}\gamma^{\mu\nu\rho}D_\nu\psi_\rho^i\right) - N_{IJ}D_\mu X^I D^\mu \bar{X}^J - \tfrac{1}{2}g_{XY}D_\mu q^X D^\mu q^Y \\
& + \left\{-\tfrac{1}{4}i\bar{F}_{IJ}F_{\mu\nu}^{+I}F^{+\mu\nu J} + \tfrac{1}{16}N_{IJ}X^I X^J T_{ab}^+ T^{+ab} + \text{h.c.}\right\} \\
& - N_{IJ}\mathbf{Y}^I \cdot \mathbf{Y}^J - N^{-1|IJ}P_I^0 P_J^0 - 2\bar{X}^I X^J k_I{}^X k_{JX} \\
& + \tfrac{2}{3}C_{I,JK}e^{-1}\varepsilon^{\mu\nu\rho\sigma}W_\mu^I W_\nu^J \left(\partial_\rho W_\sigma^K + \tfrac{3}{8}f_{LM}{}^K W_\rho^L W_\sigma^M\right) \\
& + \tfrac{1}{2}\bar{\psi}_{ai}\gamma^{abc}\psi_b^j \left(\delta_j^i N_{IJ}\bar{X}^I D_c X^J + D_c q^X k_X \cdot \tau_j{}^i\right) \\
& + \left\{-\tfrac{1}{4}N_{IJ}\bar{\Omega}^{iI}\hat{D}\!\!\!/\,\Omega_i^J - \bar{\zeta}_{\bar{A}}\hat{D}\!\!\!/\,\zeta^B d^{\bar{A}}{}_B\right. \\
& + \tfrac{1}{2}N_{IJ}\bar{\psi}_{ia}D\!\!\!/\,X^I\gamma^a \Omega^{iJ} + i\bar{\psi}_{ia}D\!\!\!/\,q^X\gamma^a\zeta_{\bar{A}}f^{iB}{}_X d^{\bar{A}}{}_B \\
& - \tfrac{1}{2}N_{IJ}\varepsilon_{ij}\left(\bar{\Omega}^{iI}\gamma_\mu - \bar{X}^I\bar{\psi}_\mu^i\right)\psi_\nu^j F^{-\mu\nu J} - \tfrac{1}{16}i F_{IJK}\bar{\Omega}_i^I\gamma^{\mu\nu}\Omega_j^J\varepsilon^{ij}F_{\mu\nu}^{-K} \\
& + \tfrac{1}{2}\bar{\psi}_{ai}\gamma^a\left[\Omega_j^I P_I{}^{ij} + \Omega_j^I N_{IJ}f_{KL}{}^J X^L\bar{X}^K\varepsilon^{ij} - 4iX^I k_I{}^X f^{iB}{}_X C_{BA}\zeta^A\right] \\
& + \tfrac{1}{2}N_{IJ}\bar{\Omega}_i^I f_{KL}{}^J \Omega_j^L \bar{X}^K\varepsilon^{ij} + \tfrac{1}{2}\bar{X}^I\bar{\psi}_a^i\gamma^{ab}\psi_b^j P_{Iij} \\
& \left. + 2X^I\bar{\zeta}^A\zeta^B t_{IAB} + 2ik_I^X f^{iB}{}_X\varepsilon_{ij}d^{\bar{A}}{}_B\bar{\zeta}_{\bar{A}}\Omega^{jI} + \text{h.c.}\right\} + \text{4-fermion terms.}
\end{aligned}
\tag{4.59}
$$

We have used the field equations to determine the auxiliary fields, which are thus no longer to be considered as independent fields, but stand for their values in (4.49)–(4.54). For example, the term that was $+N_{IJ}\mathbf{Y}^I \cdot \mathbf{Y}^J+$ other $\mathbf{Y}$ terms in (3.123) and (3.163) are now written as $-N_{IJ}\mathbf{Y}^I \cdot \mathbf{Y}^J$. The covariant derivatives are (omitting contributions that give rise to 4-fermion terms)[13]

$$D_\mu X^I = \left(\partial_\mu - \mathrm{i}\mathcal{A}_\mu\right) X^I + W_\mu{}^J X^K f_{JK}{}^I,$$

$$D_\mu q^X = \partial_\mu q^X + 2\mathcal{V}_\mu \cdot \mathbf{k}^X - W_\mu{}^I k_I{}^X,$$

$$\hat{D}_\mu \Omega_i^I = D_\mu \Omega_i^I + \Gamma^I{}_{JK} \Omega^K D_\mu X^I,$$

$$D_\mu \Omega_i^I = \left(\partial_\mu + \tfrac{1}{4}\omega_\mu{}^{ab}(e)\gamma_{ab} - \tfrac{1}{2}\mathrm{i}\mathcal{A}_\mu\right) \Omega_i^I + \mathcal{V}_{\mu i}{}^j \Omega_j{}^I + W_\mu{}^J \Omega_i^K f_{JK}{}^I,$$

$$\hat{D}_\mu \zeta^A = \left(\partial_\mu + \tfrac{1}{4}\omega_\mu{}^{ab}(e)\gamma_{ab} + \tfrac{1}{2}\mathrm{i}\mathcal{A}_\mu\right) \zeta^A - W_\mu^I t_{IB}{}^A \zeta^B + \partial_\mu q^X \omega_{XB}{}^A \zeta^B,$$

$$D_\mu \psi_{vi} = \left(\partial_\mu + \tfrac{1}{4}\omega_\mu{}^{ab}(e)\gamma_{ab} + \tfrac{1}{2}\mathrm{i}\mathcal{A}_\mu\right) \psi_{vi} + \mathcal{V}_{\mu i}{}^j \psi_{vj}. \tag{4.60}$$

The connection Γ^I_{JK} refers to the completely holomorphic connection in a Kähler manifold with metric N_{IJ}. See below, (4.73), for more detail. $\mathcal{A}_\mu$ and $\mathcal{V}_{\mu i}{}^j$ are the bosonic parts of their expressions in (4.49):

$$A_\mu = \mathcal{A}_\mu + A_\mu^{\mathrm{F}},$$

$$\mathcal{A}_\mu = -\tfrac{1}{2}\mathrm{i}\kappa^2 N_{IJ}(X^I \hat{\partial}_\mu \bar{X}^J - \bar{X}^J \hat{\partial}_\mu X^I)$$

$$= \mathcal{A}_\mu^0 - \kappa^2 W_\mu^I P_I^0,$$

$$\mathcal{A}_\mu^0 = -\tfrac{1}{2}\mathrm{i}\kappa^2 N_{IJ}(X^I \partial_\mu \bar{X}^J - \bar{X}^J \partial_\mu X^I),$$

$$A_\mu^{\mathrm{F}} = -\tfrac{1}{8}\mathrm{i}\kappa^2 N_{IJ}\bar\Omega^{iI}\gamma_\mu\Omega_i^J + \tfrac{1}{2}\mathrm{i}\kappa^2\bar\zeta_{\bar A}\gamma_\mu\zeta^B d^{\bar A}{}_B,$$

$$\mathbf{V}_\mu = \mathcal{V}_\mu + \mathbf{V}_\mu^{\mathrm{F}},$$

$$\mathcal{V}_\mu = \tfrac{1}{2}\kappa^2 \mathbf{k}_X \hat{\partial}_\mu q^X = \tfrac{1}{2}\kappa^2 \left(\mathbf{k}_X \partial_\mu q^X - W_\mu^I \mathbf{P}_I\right),$$

$$\mathbf{V}_\mu^{\mathrm{F}} = \tfrac{1}{8}\kappa^2 N_{IJ}\bar\Omega^{iI}\gamma_\mu\Omega_j^J \boldsymbol{\tau}_i{}^j, \tag{4.61}$$

where we used

$$\hat{\partial}_\mu X^I = \partial_\mu X^I + W_\mu{}^J X^K f_{JK}{}^I,$$

$$\hat{\partial}_\mu q^X = \partial_\mu q^X - W_\mu{}^I k_I{}^X. \tag{4.62}$$

Another way to split the covariant derivative $D_\mu X^I$ in (4.60) splitting the geometric and the gauge parts is

---

[13]In the previous chapter, the auxiliary fields were considered independent, and appeared as such in the 'linear' part of the covariant derivatives D_μ, see e.g. (3.19) or (3.100). Here we consider e.g. A_μ as its solution in (4.61), and the linear part written as D_μ contains only $\mathcal{A}_\mu$.

$$D_\mu X^I = \nabla_\mu X^I + W_\mu^K \left(i\kappa^2 P_K^0 X^I + f_{KJ}{}^I X^J \right),$$

$$\nabla_\mu X^I \equiv \partial_\mu X^I - iA_\mu^0 X^I. \tag{4.63}$$

4.4.4   Supersymmetry Transformations

The transformation laws for physical fields can be obtained by plugging into the superconformal transformations the values for the gauge fixing as well as the parameters as fixed by the decomposition laws. The superconformal transformations from which we have to start are those of the Weyl multiplet, (see Sect. 2.6.2), the vector multiplets, (3.15), and the hypermultiplets (3.100). Restricting ourselves to the physical fields, the Q- and S-supersymmetries are (using also (3.101))

$$\delta e_\mu{}^a = \tfrac{1}{2}\bar{\epsilon}^i \gamma^a \psi_{\mu i} + \text{h.c.},$$

$$\delta\psi_\mu^i = \left(\partial_\mu + \tfrac{1}{4}\gamma^{ab}\omega_{\mu ab} - \tfrac{1}{2}iA_\mu \right)\epsilon^i + V_\mu{}^i{}_j\epsilon^j - \tfrac{1}{16}\gamma^{ab}T_{ab}^-\varepsilon^{ij}\gamma_\mu\epsilon_j - \gamma_\mu\eta^i,$$

$$\delta X^I = \tfrac{1}{2}\bar{\epsilon}^i\Omega_i^I,$$

$$\delta\Omega_i^I = \slashed{D}X^I\epsilon_i + \tfrac{1}{4}\gamma^{ab}\mathcal{F}_{ab}{}^I\varepsilon_{ij}\epsilon^j + Y_{ij}{}^I\epsilon^j + X^J\bar{X}^K f_{JK}{}^I\varepsilon_{ij}\epsilon^j + 2X^I\eta_i,$$

$$\delta W_\mu^I = \tfrac{1}{2}\varepsilon^{ij}\bar{\epsilon}_i\gamma_\mu\Omega_j{}^I + \varepsilon^{ij}\bar{\epsilon}_i\psi_{\mu j}X^I + \text{h.c.},$$

$$\delta q^X = -i\bar{\epsilon}^i\zeta^A f^X{}_{iA} + i\varepsilon^{ij}\rho^{\bar{A}B}\bar{\epsilon}_i\zeta_{\bar{A}}f^X{}_{jB},$$

$$\widehat{\delta}A^{iA} = -i\bar{\epsilon}^i\zeta^A + i\bar{\epsilon}_j\zeta_{\bar{B}}\varepsilon^{ji}\rho^{\bar{B}A},$$

$$\widehat{\delta}\zeta^A = \tfrac{1}{2}i\slashed{\widehat{D}}A^{iA}\epsilon_i + i\bar{X}^I k_I{}^X f^{iA}{}_X\varepsilon_{ij}\epsilon^j + iA^{iA}\eta_i, \tag{4.64}$$

where we use the covariant transformations as defined in (3.52). Covariant derivatives and covariant gauge field strengths are given in (3.16), (3.19), (3.102). The relevant bosonic symmetry transformations are the U(1) and SU(2) R-symmetries and the Yang–Mills gauge symmetries:

$$\delta\psi_\mu^i = \tfrac{1}{2}i\lambda_T\psi_\mu^i + \psi_\mu^j\lambda_j{}^i,$$

$$\delta X^I = i\lambda_T X^I + \theta^J X^K f_{KJ}{}^I,$$

$$\delta\Omega_i^I = \tfrac{1}{2}i\lambda_T\Omega_i^I - \lambda_i{}^j\Omega_j^I + \theta^J\Omega_i^K f_{KJ}{}^I,$$

$$\delta W_\mu^I = \partial_\mu\theta^I + \theta^J W_\mu^K f_{KJ}{}^I,$$

$$\delta q^X = -2\lambda \cdot \mathbf{k}^X + \theta^I k_I{}^X,$$

$$\hat{\delta} A^{i\mathcal{A}} = A^{j\mathcal{A}} \lambda_j{}^i + A^{i\mathcal{B}} \theta^I t_{I\mathcal{B}}{}^{\mathcal{A}},$$

$$\hat{\delta}\zeta^{\mathcal{A}} = -\tfrac{1}{2} i \lambda_T \zeta^{\mathcal{A}} + \theta^I \zeta^{\mathcal{B}} t_{I\mathcal{B}}{}^{\mathcal{A}}. \tag{4.65}$$

4.4.4.1 Decomposition Laws

The gauge choices from Sect. 4.4.2 imply that the parameters of the gauge fixed symmetries are functions of the remaining gauge symmetries. The decomposition law for the K symmetry is given in (4.5). But since none of the fields in (4.64) transforms under special conformal transformations, we do not need this decomposition law of λ_K^μ.

The S gauge (4.57) is the Q-supersymmetry transformed of the D-gauge (4.55). Therefore, with this choice, the latter is invariant under ordinary supersymmetry, and the invariance then implies that the dilatation parameter can be put to zero:

$$\lambda_D = 0. \tag{4.66}$$

We also need the decomposition law for the S-supersymmetry. We can calculate the variation of any of the two equations in (4.58), which should lead to the same result. These equations are gauge invariant. We use the field equations of Sect. 4.4.1. When calculating the Q-supersymmetry of the first one, a useful relation due to the field equation (4.54) is

$$N_{IJ} \bar{X}^J \mathcal{F}_{ab}^{-I} = -\tfrac{1}{8} i F_{IJK} \bar{\Omega}_i^I \gamma_{ab} \Omega_j^J \bar{X}^K \varepsilon^{ij} + \tfrac{1}{2} C_{AB} \bar{\zeta}^A \gamma_{ab} \zeta^B. \tag{4.67}$$

In both ways we obtain

$$\kappa^{-2} \eta^i(\epsilon) = -\tfrac{1}{2} P_I^{ij} X^I \epsilon_j + \tfrac{1}{2} \kappa^{-2} \gamma^a \epsilon^j V_{aj}^{\mathrm{F}\, i}$$
$$+ \tfrac{1}{4} \gamma_a \epsilon^i \bar{\zeta}_{\bar{A}} \gamma^a \zeta^B d^{\bar{A}}{}_B + \tfrac{1}{16} \gamma_{ab} \varepsilon^{ij} \epsilon_j C^{\bar{A}\bar{B}} \bar{\zeta}_{\bar{A}} \gamma^{ab} \zeta_{\bar{B}}, \tag{4.68}$$

where $V_{ai}^{\mathrm{F}\, j}$ is the traceless expression in (4.61):

$$V_{ai}^{\mathrm{F}\, j} = -\tfrac{1}{4} \kappa^2 \left(\bar{\Omega}_i^I \gamma_a \Omega^{jJ} - \tfrac{1}{2} \delta_i^j \bar{\Omega}_k^I \gamma_a \Omega^{kJ} \right) N_{IJ}. \tag{4.69}$$

Finally, the gauge fixing of the R-symmetries will be discussed in Sect. 4.5. The U(1) and SU(2) decomposition laws are presented (6.43) and (4.168), respectively.

4.5 Vector Multiplet Scalars: Special Kähler Geometry

The constraints (4.56), obtained by combining the dilatational gauge fixing with the D-field equation, determine a real condition on the manifold spanned by the scalars $\{X^I\}$ and $\{q^X\}$. As we will see, the gauge fixings of the U(1) and SU(2) R-symmetries remove another real degree of freedom of the $\{X^I\}$ and 3 real degrees of freedom of the $\{q^X\}$. This procedure defines a complex n-dimensional submanifold of the manifold spanned by the $(n + 1)$ variables X^I, as well as a quaternionic n_H-dimensional submanifold of the real $4(n_H + 1)$-dimensional manifold spanned by the q^X. These submanifolds can be identified by choosing an appropriate system of (projective) coordinates. This identification, for what regards the scalars of the vector multiplets which defines special Kähler geometry, will be the subject of this section. Similarly, as we will present in Sect. 4.6, the scalars of the hypermultiplets define quaternionic-Kähler manifolds.

As a further simplification, in this section we will focus on the geometric part for the scalars of the vector multiplet, hence reducing $D_\mu X^I$ to $\nabla_\mu X^I$ as in (4.63), and leaving the connection with the gauge vectors W_μ^I to the following chapters. Upon this reduction, the relevant term of (4.59) are

$$e^{-1}\mathcal{L}_X = -g^{\mu\nu} N_{IJ} \nabla_\mu X^I \nabla_\nu \bar{X}^J \,, \qquad \nabla_\mu X^I \equiv \partial_\mu X^I - i\mathcal{A}_\mu^0 X^I \,, \qquad (4.70)$$

where N_{IJ} is defined in (3.124). Due to the homogeneity of $F(X)$—see (3.108) and (3.124)—one also has

$$N_{IJ} = \frac{\partial}{\partial X^I} \frac{\partial}{\partial \bar{X}^J} N \equiv \partial_I \bar{\partial}_J N \,. \qquad (4.71)$$

4.5.1 Rigid Special Kähler Manifold

We first look back to the vector multiplets in rigid supersymmetry with conformal symmetry. Hence, we just consider $\partial_\mu X^I$ in (4.70). We also do not consider the constraints (4.56) and the fields X^I are thus independent. Then (4.70) defines X^I as complex coordinates of a Kähler manifold with Kähler metric[14]:

$$G_{I\bar{J}} = N_{IJ} \,, \qquad G_{IJ} = G_{\bar{I}\bar{J}} = 0 \,, \qquad (4.72)$$

---

[14]To apply the general relations of Kähler manifolds we have to distinguish indices I and $\bar{I}$, which cannot be done consistently for the relations that follow from the prepotential F, see e.g. (3.124) or the first of (4.72) where at the right-hand side N_{IJ} is symmetric and at the left-hand side there is a holomorphic and an anti-holomorphic index. In (4.73) we use the metric connection with all holomorphic indices, considering just $G_{I\bar{J}}$ as metric, ignoring these relations to a prepotential F.

and Kähler potential N. Because the manifold is Kähler, the affine connection is given by [8, (13.19)]

$$\Gamma_{JK}^{I} = G^{I\bar{L}}\partial_{J}G_{K\bar{L}} = N^{-1|IL}\partial_{J}N_{KL} = -\mathrm{i}N^{-1|IL}F_{JKL}\,. \tag{4.73}$$

The homogeneity equations also imply

$$X^{K}\partial_{K}N_{IJ} = -\mathrm{i}X^{K}F_{IJK} = 0\,, \qquad X^{J}\Gamma_{JK}^{I} = 0\,, \tag{4.74}$$

and their complex conjugates. Therefore $k_{\mathrm{D}}{}^{I} = X^{I}$ is a closed homothetic Killing vector of this metric (look at (1.30) with $w = 1$).

This Kähler geometry encountered in rigid supersymmetry is in the mathematics literature also indicated as affine special Kähler geometry [30–33].

4.5.2   Coordinates in the Projective Special Kähler Manifold

We now turn to the supergravity case, where a key ingredient is the gauged U(1) T-symmetry. First, there is the contribution of the T-gauge field $\mathcal{A}^{0}$ in (4.70). Furthermore, we will impose the first relation of (4.56), which was related to the gauge choice for dilatations, and a gauge choice for the T-symmetry.[15] We will see how this projection from the $n + 1$-dimensional (complex) Kähler manifold of Sect. 4.5.1 leads to n-dimensional special Kähler geometry [34–36], which is sometimes also denoted as 'projective' special Kähler, and has a non-trivial U(1) curvature.[16]

This mechanism for the emergence of projective special Kähler manifolds is similar to what appears in $\mathcal{N} = 1$ and has as such been reviewed in [8, Sect. 17.3].

To exhibit the projective nature of the manifold, we first consider the on-shell value of the auxiliary fields as it follows from (4.70):

$$\begin{aligned}
\mathcal{A}_{\mu}^{0} &= \frac{\mathrm{i}}{2N}\left[X^{I}N_{IJ}(\partial_{\mu}\bar{X}^{J}) - (\partial_{\mu}X^{I})N_{IJ}\bar{X}^{J}\right] \\
&= \frac{\mathrm{i}}{2N}\left[(\partial_{\mu}\bar{X}^{J})\bar{\partial}_{J}N - (\partial_{\mu}X^{I})\partial_{I}N\right].
\end{aligned} \tag{4.75}$$

---

[15]Note that such a U(1) symmetry, here included as part of the superconformal algebra, in fact follows from the presence of a complex structure and the dilatational symmetry, see e.g. [8, Sect. 17.3.2]. Similarly the SU(2) symmetry in the hypermultiplet sector follows from the hypercomplex structure and the dilatation.

[16]The reader will find in Chap. 5 a more detailed discussion on special Kähler geometry independent of this construction procedure.

One can check that this agrees with (4.61) upon using the gauge condition (4.56). The value of (4.70) using (4.75) is

$$
\begin{aligned}
e^{-1}\mathcal{L}_X &= -N_{IJ}\partial_\mu X^I \partial^\mu \bar{X}^J + \frac{1}{4N}\left[\partial_\mu \bar{X}^J \bar{\partial}_J N - \partial_\mu X^I \partial_I N\right]^2 \\
&= -\partial_\mu X^I \partial^\mu \bar{X}^J \left[\partial_I \bar{\partial}_J N - \frac{1}{N}(\partial_I N)(\bar{\partial}_J N)\right] - \frac{1}{4N}\partial_\mu N\,\partial^\mu N \\
&= -N\partial_\mu X^I \partial^\mu \bar{X}^J \partial_I \bar{\partial}_J \log|N| - \frac{1}{4N}\partial_\mu N\,\partial^\mu N\,.
\end{aligned}
\tag{4.76}
$$

On the surface defined by (4.56), the last term vanishes. But the first one contains derivatives N_I and $N_{\bar{I}}$ that are non-vanishing. This will lead to another Kähler manifold as compared to the rigid Kähler geometry.

4.5.2.1 Projective Coordinates

As we anticipated, the proper way to manifest this new structure is to select appropriate (projective) coordinates, in which we split between the direction orthogonal to the surface and along the surface. For this to be done, it is useful to split $(n + 1)$ complex variables $\{X^I\}$ as $\{y, z^\alpha\}$, $\alpha = 1, \ldots, n$ (and their complex conjugates $\{\bar{y}, \bar{z}^{\bar{\alpha}}\}$), where the variables y will serve for defining the surface and z^α for coordinates on the n-dimensional surface. We will therefore refer to the space spanned by the $\{X^I\}$ as the embedding space, while that spanned by the $\{z^\alpha\}$ as the projective space. The latter will describe the physical scalars.

The embedding in the $(n + 1)$ dimensional manifold is given by functions $Z^I(z)$, $\bar{Z}^I(\bar{z})$ defined as follows:

$$
X^I = y\,Z^I(z)\,,\qquad \bar{X}^I = \bar{y}\,\bar{Z}^I(\bar{z})\,.
\tag{4.77}
$$

As we require invariance under reparameterizations $z^\alpha \to z'^\alpha(z)$, the $Z^I(z)$ must be $n + 1$ non-degenerate[17] arbitrary holomorphic functions of the z^α.

We can now assign to y the dilatation and U(1) charges carried by X^I (according to Table 3.1):

$$
\delta_{\mathrm{D},T} X^I = (\lambda_\mathrm{D} + \mathrm{i}\lambda_T)X^I \;\to\; \delta_{\mathrm{D},T}\,y = (\lambda_\mathrm{D} + \mathrm{i}\lambda_T)y\,,\qquad \delta_{\mathrm{D},T}z^\alpha = 0\,.
\tag{4.78}
$$

The non-uniqueness of the splitting (4.77) will be related to the Kähler transformations, which will be discussed in Sect. 4.5.6. A practical choice for the functions

---

[17]The matrix $\partial_\alpha Z^I$ has to be of rank n and the matrix $(Z^I, \partial_\alpha Z^I)$ has to be of rank $n + 1$.

Z^I, which is often used, is called '*special coordinates*' and corresponds to

$$Z^0(z) = 1, \qquad Z^\alpha(z) = z^\alpha \text{ for } \alpha = 1, \ldots, n. \tag{4.79}$$

For these '*special coordinates*' one can obviously write

$$z^\alpha = \frac{X^\alpha}{X^0} = \frac{Z^\alpha}{Z^0}. \tag{4.80}$$

4.5.2.2   Gauge Fixing of Dilatations and U(1) R-Symmetry

The dilatation gauge fixing amounts in the vector multiplet to the first of the conditions (4.56). We will rewrite it using (3.124), and use the homogeneity properties of N_{IJ} (of degree 0 in X^I and $\bar{X}^I$) to define $N_{IJ}(z, \bar{z})$:

$$N_{IJ}(X^I, \bar{X}^I) = N_{IJ}(Z^I(z), \bar{Z}^I(\bar{z})) = N_{IJ}(z, \bar{z}). \tag{4.81}$$

Therefore we can write the effective dilatational gauge condition as

$$N(X, \bar{X}) = |y|^2 Z^I(z) N_{IJ}(z, \bar{z}) \bar{Z}^I(\bar{z}) = -a, \qquad a = \begin{cases} 3\kappa^{-2} & \text{for } \mathcal{N} = 1, \\ \kappa^{-2} & \text{for } \mathcal{N} = 2. \end{cases} \tag{4.82}$$

We introduce here the arbitrary constant a, such that the treatment is the same for $\mathcal{N} = 1$ and $\mathcal{N} = 2$ supergravity, but for the purpose of this book $a = \kappa^{-2}$. The condition (4.82) fixes the modulus of y in terms of z^α and their complex conjugates.

With these definitions and transformations (4.78) a consistent U(1) gauge choice is

$$T\text{-gauge:} \quad y = \bar{y}, \tag{4.83}$$

which removes the phase of y as an independent variable and fixes completely y (and $\bar{y}$)

$$y = \left[-a^{-1} Z^I(z) N_{IJ}(z, \bar{z}) \bar{Z}^J(\bar{z}) \right]^{-1/2}. \tag{4.84}$$

As a result of the D, U(1) gauge fixing, we are left with only the z^α (and their complex conjugates) as unconstrained variables, which in turn will be related to the physical vector multiplet scalars in the Poincaré supergravity.

Before proceeding, we still want to make two remarks on the previous results:

• At least as long as we consider only the bosonic part of the vector multiplet (4.70), in practice we will not need to specify a T-gauge choice once in the projective basis (4.77). Indeed, the phase of y is the only bosonic quantity that transforms under T and therefore—as shown at the beginning of Chap. 4—gauge invariance implies that the latter cannot appear in (4.70). As a consequence, we will not need a T-decomposition law for the bosonic sector. On the other hand, fermions of the vector multiplet do transform under the T-symmetry.[18]

• The value (4.84) for y is a function of z and $\bar{z}$. Therefore the transition from X^I to z^α does not respect the holomorphicity. In other words, the complex structure that is relevant in the submanifold is not the same as the one in the embedding manifold.

4.5.3 The Kähler Potential

It follows from our derivation that, after the projection, the resulting action for the scalars z^α is the first term of (4.76). We will now explicitly show that this defines again a Kähler manifold with a Kähler potential:

$$\mathcal{K}(z, \bar{z}) = -a \ln \left[-a^{-1} Z^I(z) N_{IJ}(z, \bar{z}) \bar{Z}^J(\bar{z}) \right] . \tag{4.85}$$

To verify that (4.85) is indeed correct, we can use the homogeneity properties of N_{IJ} to obtain its derivatives. See e.g. (4.74), which implies that $\partial_\alpha (Z^I N_{IJ} \bar{Z}^J) = (\partial_\alpha Z^I)(N_{IJ} \bar{Z}^J)$, and hence we find the useful relations

$$\partial_\alpha \mathcal{K} = -a \frac{N_{IJ} \bar{Z}^J \partial_\alpha Z^I}{N_{KL} Z^K \bar{Z}^L} ,$$

$$\partial_\alpha \partial_{\bar{\beta}} \mathcal{K} = a \partial_\alpha Z^I \partial_{\bar{\beta}} \bar{Z}^J y \bar{y} \left[-\frac{N_{IJ}}{N} + \frac{N_{IK} \bar{X}^K X^L N_{LJ}}{N^2} \right]$$

$$= -a \partial_\alpha Z^I \partial_{\bar{\beta}} \bar{Z}^J y \bar{y} \, \partial_I \partial_J \ln N . \tag{4.86}$$

Hence we can write

$$e^{-1} \mathcal{L}_X = -\partial_\mu z^\alpha \partial^\mu \bar{z}^{\bar{\beta}} \partial_\alpha \partial_{\bar{\beta}} \mathcal{K} , \tag{4.87}$$

confirming that $\mathcal{K}$ is the Kähler potential of the projective manifold.

---

[18] We will relate these also to Kähler transformations in Sect. 4.5.6, and then this has a consequence on the global structure of the Kähler manifold, see footnote 24.

It will also be useful to write the on-shell value of the T-gauge field in terms of the Kähler potential

$$\mathcal{A}^0_\mu = \tfrac{1}{2}ia^{-1}\left(\partial_\alpha\mathcal{K}\,\partial_\mu z^\alpha - \partial_{\bar\alpha}\mathcal{K}\,\partial_\mu \bar z^{\bar\alpha}\right) - \tfrac{1}{2}i\partial_\mu \ln \tfrac{y}{\bar y}\,. \tag{4.88}$$

Note that we have not implemented yet the gauge condition (4.83), which would eliminate the last term, which is anyway a pure gauge term. Note also that the $\partial_\mu y$, $\partial^\mu \bar y$, $\partial_\mu y \partial^\mu \bar y$ terms cancel as a consequence of the relations (4.74). Furthermore, we can now rewrite (4.84) in terms of $\mathcal{K}$

$$y = \bar y = e^{\mathcal{K}/(2a)}\,, \tag{4.89}$$

and thus, with (4.77), we have

$$X^I = e^{\mathcal{K}/(2a)} Z^I(z)\,, \qquad \bar X^I = e^{\mathcal{K}/(2a)} \bar Z^I(\bar z)\,. \tag{4.90}$$

In view of the importance of the Kähler potential, we still give an alternative way to present (4.85). First we remark that due to the homogeneity, $F_I(X)$ is first order in X, and we can thus write

$$F_I(X) = y\mathcal{F}_I(Z(z))\,, \qquad \bar F_I(\bar X) = y\bar{\mathcal{F}}_I(\bar Z(\bar z))\,, \tag{4.91}$$

where $\mathcal{F}_I(Z)$ is the same functional dependence as $F_I(X)$ and can thus also be written as[19]

$$\mathcal{F}_I(z) = \frac{\partial}{\partial Z^I}\mathcal{F}(Z(z))\,, \tag{4.92}$$

where $\mathcal{F}(Z)$ is the same function as $F(X)$. Finally

$$\mathcal{K}(z,\bar z) = -\kappa^{-2}\ln\left[i\kappa^2 \bar Z^I(\bar z)\,\mathcal{F}_I(z) - i\kappa^2 Z^I(z)\,\bar{\mathcal{F}}_I(\bar z)\right]\,, \tag{4.93}$$

and thus the Kähler potential is written in terms of derivatives of the holomorphic function $\mathcal{F}(Z(z))$. Note that the last equations, starting from (4.91), are only valid for $\mathcal{N} = 2$, and thus we used $a = \kappa^{-2}$ in (4.93).

4.5.3.1 Interpretation as Sasakian Cone

There is an alternative interpretation of the metric (4.87), which gives another geometrical view on the projective manifold.[20] Before any gauge fixing we can

---

[19]Note that $\overline{\mathcal{F}}_I$ is obtained as $\bar F_I(\bar Z(\bar z))$ using for $\bar F_I(\bar Z)$ the same functional dependence as $\bar F_I(\bar X)$.

[20]This will not be important for what follows, and the reader may thus skip this part.

write the action (4.70) as follows:

$$e^{-1}\mathcal{L}_X = -\tfrac{1}{4}N^{-1}\left(\partial_\mu N\right)^2 - N\left(A_\mu - \mathcal{A}^0_\mu\right)^2 + \tfrac{N}{a}\left(\partial_{\bar\beta}\partial_\alpha\mathcal{K}\right)\partial_\mu z^\alpha\,\partial^\mu z^{\bar\beta}\,, \qquad (4.94)$$

where $\mathcal{A}^0_\mu$ is the geometric part of A_μ that we wrote in (4.75).

The obtained metric in (4.94) is a cone [37, 38]. Indeed, starting from the $n+1$ complex variables $\{X^I\}$ we have defined $\{y, z^\alpha\}$, where we can choose $y = |y|e^{i\theta}$. As follows from (4.82), the modulus $|y|$ appears in N (in order to obtain a canonical parameterization of a cone, it is useful to define $r^2 = -N$). The resulting coordinates are thus:

- a radius r (gauge degree of freedom for dilatations)
- an angle θ (U(1) R- symmetry degree of freedom);
- n complex variables z^α.

In terms of $(r, \theta, z^\alpha, \bar z^{\bar\alpha})$ the metric (4.94) takes the form

$$ds^2 = dr^2 + r^2\left[A + d\theta + \tfrac{1}{2}i\left(\partial_\alpha\mathcal{K}(z,\bar z)\,dz^\alpha - \partial_{\bar\alpha}\mathcal{K}(z,\bar z)\,d\bar z^{\bar\alpha}\right)\right]^2$$
$$- r^2\partial_\alpha\partial_{\bar\alpha}\mathcal{K}(z,\bar z)\,dz^\alpha d\bar z^{\bar\alpha}\,, \qquad (4.95)$$

where A is the one-form gauging of the U(1) group, and $\mathcal{K}(z,\bar z)$ is a function of the holomorphic prepotential $F(X)$. When U(1) is not gauged ($A_\mu = 0$), the base of the cone (the manifold with fixed N or r) is a Sasakian manifold with a U(1) invariance.[21] When the U(1) is gauged, the auxiliary field A_μ can be redefined such that the whole expression in square brackets is the field equation of A_μ itself, and it drops out on-shell. In that case, with fixed r (gauge fixing the superfluous dilatations), the remaining manifold is Kähler, with the Kähler potential $\mathcal{K}$ determined by $F(X)$. Therefore in the situation where U(1) is gauged, the geometry is further constrained to special Kähler, living on a Kähler submanifold of the $(n+1)$-complex-dimensional manifold defined by a constant value of r. This is a real condition, but the U(1) invariance implies that the other real variable θ has disappeared.

4.5.4 Positivity Requirements

A standard requirement for a suitable Lagrangian of a physical system is that its kinetic terms should define positive kinetic energy. The kinetic terms of the spin 2 particle, the graviton, are included in the Einstein–Hilbert term, the first term

---

[21]This has been remarked first in a similar situation with hypermultiplets in $\mathcal{N} = 2$ in [38], and has been looked at systematically in [37].

in (4.59). This is positive due to the gauge choices in (4.55), and for the vector multiplet part this is the choice of the sign in (4.82). In fact, according to (4.82), this choice requires that the prepotential $F(X)$ and the range of the scalars allow a domain where

$$Z^I N_{IJ} \bar{Z}^J < 0. \tag{4.96}$$

Thus $N_{IJ}(z, \bar{z})$ should have at least one negative eigenvalue for all values of the scalars in the domain. The positivity of the kinetic energy of the scalars in $\mathcal{L}_0$ requires $g_{\alpha\bar{\beta}}$ to be a positive definite matrix. Using (4.86) with N negative, this implies that N_{IJ} should be a matrix with n positive eigenvalues and 1 negative one. The negative one has the significance of the direction of the 'compensating multiplet' in the $(n + 1)$-dimensional embedding space $\{X^I\}$. The separation of the positive and negative definite parts will become explicit in the next chapter, see (5.65). These conditions define a so-called positivity domain for the scalars $\{z^\alpha\}$. The requirement that the positivity domain is non-empty, restricts the space of prepotentials F that can be used. Finally, these two conditions also imply the positivity of the kinetic terms of the spin-1 part, namely that $\mathrm{Im}\,\mathcal{N}_{IJ}$ is a negative-definite matrix [39]. The mentioned matrix equation (5.65) will make this property explicit. Since a symmetric matrix with a negative-definite imaginary part is invertible, these remarks prove that the inverse of $\mathcal{N}_{IJ}$ is well defined. Several theorems on these matrices are collected in [31].

4.5.5 Examples

We give here some examples of functions $F(X)$ and their corresponding target spaces:

$$F = -\mathrm{i}\, X^0 X^1 \qquad \frac{\mathrm{SU}(1, 1)}{\mathrm{U}(1)} \tag{4.97}$$

$$F = (X^1)^3 / X^0 \qquad \frac{\mathrm{SU}(1, 1)}{\mathrm{U}(1)} \tag{4.98}$$

$$F = -2\sqrt{X^0 (X^1)^3} \qquad \frac{\mathrm{SU}(1, 1)}{\mathrm{U}(1)} \tag{4.99}$$

$$F = \frac{1}{4}\mathrm{i} X^I \eta_{IJ} X^J \qquad \frac{\mathrm{SU}(1, n)}{\mathrm{SU}(n) \otimes \mathrm{U}(1)} \tag{4.100}$$

$$F = \frac{d_{ABC} X^A X^B X^C}{X^0} \qquad \text{'very special'} \tag{4.101}$$

The first three functions give rise to the manifold $\mathrm{SU}(1, 1)/\mathrm{U}(1)$. However, the first one is not equivalent to the other two as the manifolds have a different value of the curvature [40]. The latter two are, however, equivalent by means of a symplectic transformation, as we will show in Sect. 5.2.2. In the fourth example η is a constant non-degenerate real symmetric matrix. In order that the manifold has a non-empty positivity domain, the signature of this matrix should be $(- + \cdots +)$.

The last example, defined by a real symmetric tensor d_{ABC}, with $A, B, C = 1, \ldots n$, defines a class of special Kähler manifolds, which are denoted as 'very special' Kähler manifolds. These can be obtained by a dimensional reduction of $D = 5$ supergravity–vector multiplet couplings [41] where the tensor d_{ABC} is identified with the tensor C_{IJK} introduced in Sect. 3.3.2. We will say more on these in Sect. 5.7.

Exercise 4.1 We go through the example (4.100) in some detail. We get easily the second derivative

$$F_{IJ} = \tfrac{1}{2} i \eta_{IJ} , \qquad N_{IJ} = \eta_{IJ} . \tag{4.102}$$

We will now specify to the case

$$\eta_{IJ} = \begin{pmatrix} -1 & 0 \\ 0 & 1 \end{pmatrix} . \tag{4.103}$$

We thus write the formulae for $n = 1$, but they can be easily generalized for arbitrary n by taking the lower-right entry to be a unit $n \times n$ matrix. From (4.90) we get

$$e^{-\kappa^2 \mathcal{K}(z,\bar{z})} = Z^0(z)\bar{Z}^0(\bar{z}) - Z^1(z)\bar{Z}^1(\bar{z}) . \tag{4.104}$$

The right-hand side should be positive. A convenient parameterization consists in using the special coordinates (4.79), i.e. $Z^0 = 1$, $Z^1 = z$. Indeed, then $\partial_z Z^I$ is of rank 1, while the 2×2 matrix (Z^I, ∂_z^I) is of rank 2. The domain for z is then $|z|^2 < 1$.

This leads to

$$\kappa^2 \partial_z \mathcal{K} = \frac{\bar{z}}{1 - z\bar{z}} , \qquad \kappa^2 g_{z\bar{z}} = \kappa^2 \partial_z \partial_{\bar{z}} \mathcal{K} = \frac{1}{(1 - z\bar{z})^2} . \tag{4.105}$$

You may check that the indefinite signature of (4.103) was necessary to have a positive metric. $\square$

4.5.6 Kähler Reparameterizations

The careful reader may have noticed that the splitting (4.77) is not unique. The projected coordinates are defined up to Kähler reparameterizations that leave invariant X and $\bar{X}$:

$$y' = y \, e^{f(z)/a} , \qquad Z'^I = Z^I \, e^{-f(z)/a} , \tag{4.106}$$

where $f(z)$ is an arbitrary holomorphic function.[22] The Kähler potential is not invariant under these reparameterizations. From (4.85) we see that

$$\mathcal{K}'(z, \bar{z}) = \mathcal{K}(z, \bar{z}) + f(z) + \bar{f}(\bar{z}), \qquad (4.107)$$

where we consider $f(z)$ and $\bar{f}(\bar{z})$ as independent transformations. If we choose the U(1) gauge (4.83), then this leaves a combination of the U(1) and the Kähler transformation. The decomposition law for Kähler transformations is thus[23]

$$\delta_K^{\square}[f(z), \bar{f}(\bar{z})] = \delta_K[f(z), \bar{f}(\bar{z})] + \delta_T[\lambda_T = \tfrac{1}{2}a^{-1}(f(z) - \bar{f}(\bar{z}))], \qquad (4.108)$$

where $\delta_K[f(z), \bar{f}(\bar{z})]$ are the transformations in the conformal setting, i.e. induced by (4.106) for small f. The remaining Kähler transformation can, e.g., be used to choose one of the Z^I, say Z^0, equal to 1. In any case, one can choose the parameterization of the n physical scalars z^α (with $\alpha = 1, \ldots, n$) at random, as stressed in [42–44].[24]

4.5.7   The Kähler Covariant Derivatives

Let us consider the U(1)-covariant derivative $\nabla_\mu X^I$ as defined in (4.63). This object is of course invariant under Kähler transformations since it was defined before splitting the variables $\{X^I\}$ as in (4.77). However, after defining y and Z^I, we have to split (4.63) into derivatives of these new variables, which are not invariant under Kähler transformations. To do so in a Kähler-covariant way, it is convenient to introduce a connection for the Kähler transformations [8, (17.71)] (see also more details in [47]) as follows:

$$\nabla_\mu y \equiv \partial_\mu y - i\mathcal{A}_\mu^0 y - a^{-1}\omega_\mu y, \qquad \nabla_\mu Z^I \equiv \partial_\mu Z^I + a^{-1}\omega_\mu Z^I, \qquad (4.109)$$

where

$$\omega_\mu = \omega_\alpha \partial_\mu z^\alpha, \qquad (4.110)$$

and ω_α is a Kähler connection, which means that under small holomorphic Kähler transformations it should transform as

$$\delta\omega_\alpha = \partial_\alpha f. \qquad (4.111)$$

---

[22]We stress the fact that Kähler reparameterizations are transformations of the target space functions, like $Z^I(z)$. They do not act on the coordinates $\{z, \bar{z}\}$.

[23]The symbol $\square$ for Point-carré is used to indicate transformations in the Poincaré theory.

[24]Note that, as it is clear from Sect. 4.4.4, the fermions transform under the superconformal U(1) factor, and hence, by (4.108), under the (finite) Kähler transformations. This implies that the Kähler form should be of even integer cohomology (*Kähler–Hodge manifold*) [45, 46]. This is similar to what is needed in $\mathcal{N} = 1$ and has been explained in several steps in Appendix 17.A of [8].

In this way $\nabla_\mu y$ and $\partial_\mu Z^I$ transform as y and Z^I. Since the Kähler potential transforms as in (4.107), we can identify

$$\omega_\alpha = \partial_\alpha \mathcal{K}. \tag{4.112}$$

We can then also write

$$\nabla_\mu Z^I = \nabla_\alpha Z^I \partial_\mu z^\alpha, \qquad \nabla_\alpha Z^I \equiv \left(\partial_\alpha + a^{-1}\omega_\alpha\right) Z^I. \tag{4.113}$$

Using the expression of $\mathcal{A}^0_\mu$ in (4.88), the covariant derivative $\nabla_\mu y$ is given by

$$\nabla_\mu y = \frac{1}{2} y \partial_\mu (\ln y \bar{y}) - \tfrac{1}{2} a^{-1} y \partial_\mu \mathcal{K}. \tag{4.114}$$

So far we have only redefined the fields X^I. If we now perform the gauge fixing for dilatations and T-symmetry: (4.89) we find

$$\nabla_\mu y = 0. \tag{4.115}$$

We can then also write the value of (4.88) as a pullback of derivatives on the projective manifold:

$$\mathcal{A}^0_\mu = \mathcal{A}_\alpha \partial_\mu z^\alpha + \mathcal{A}_{\bar{\alpha}} \partial_\mu \bar{z}^{\bar{\alpha}},$$

$$\mathcal{A}_\alpha = \mathrm{i}\partial_\alpha \ln y = \frac{1}{2a}\mathrm{i}\partial_\alpha \mathcal{K} = \frac{1}{2a}\mathrm{i}\omega_\alpha,$$

$$\mathcal{A}_{\bar{\alpha}} = -\mathrm{i}\partial_{\bar{\alpha}} \ln y = -\frac{1}{2a}\mathrm{i}\partial_{\bar{\alpha}} \mathcal{K} = -\frac{1}{2a}\mathrm{i}\bar{\omega}_{\bar{\alpha}}. \tag{4.116}$$

The Kähler covariant derivatives on a scalar quantity $G(z, \bar{z})$ on the projective space can be defined in general as

$$\nabla_\mu G(z, \bar{z}) = \nabla_\alpha G \, \partial_\mu z^\alpha + \nabla_{\bar{\alpha}} G \partial_\mu \bar{z}^{\bar{\alpha}}, \tag{4.117}$$

with[25]

$$\nabla_\alpha G = \partial_\alpha G + a^{-1}\hat{w}_+ G \left(\partial_\alpha \mathcal{K}\right), \qquad \nabla_{\bar{\alpha}} G = \partial_{\bar{\alpha}} G + a^{-1}\hat{w}_- G \left(\partial_{\bar{\alpha}} \mathcal{K}\right), \tag{4.118}$$

---

[25]From (4.116), $\mathcal{A}_\alpha$ and ω_α are both related on $\partial_\alpha \mathcal{K}$, hence (4.117) should contain only combinations of weight $w_\pm$ and c.

and

$$\hat{w}_+ = w_+ + \tfrac{1}{2}c, \qquad \hat{w}_- = w_- - \tfrac{1}{2}c, \tag{4.119}$$

where c is the chiral weight, and w_+ and w_- are Kähler weights for a transformation with infinitesimal $f(z)$ and $\bar{f}(\bar{z})$, according to

$$\left(\delta_T[\lambda_T] + \delta_K[f, \bar{f}]\right) G(z, \bar{z}) = \left(ic\lambda_T - w_+ a^{-1} f(z) - w_- a^{-1} \bar{f}(\bar{z})\right) G(z, \bar{z}). \tag{4.120}$$

In the Poincaré frame, (4.108) leads thus to

$$\delta_K^{\square}[f, \bar{f}] G(z, \bar{z}) = -a^{-1} \left(\hat{w}_+ f(z) + \hat{w}_- \bar{f}(\bar{z})\right) G(z, \bar{z}). \tag{4.121}$$

For example,

$$
\begin{array}{c|ccc|cc}
G & c & w_+ & w_- & \hat{w}_+ & \hat{w}_- \\
\hline
X^I, F_I & 1 & 0 & 0 & 1/2 & -1/2 \\
\bar{X}^I, \bar{F}_I & -1 & 0 & 0 & -1/2 & 1/2 \\
y & 1 & -1 & 0 & -1/2 & -1/2 \\
\bar{y} & -1 & 0 & -1 & -1/2 & -1/2 \\
Z^I, \mathcal{F}_I & 0 & 1 & 0 & 1 & 0 \\
\bar{Z}^I, \overline{\mathcal{F}}_I & 0 & 0 & 1 & 0 & 1.
\end{array}
\tag{4.122}
$$

The weights of $\nabla_\alpha G$ are equal to the weights of G. Furthermore,

$$\left[\nabla_\alpha, \nabla_{\bar{\beta}}\right] G = a^{-1}(\hat{w}_- - \hat{w}_+) g_{\alpha\bar{\beta}} G. \tag{4.123}$$

Note that (4.115) together with (4.117) imply that

$$\nabla_\alpha y(z, \bar{z}) = \nabla_{\bar{\alpha}} y(z, \bar{z}) = 0, \tag{4.124}$$

and hence

$$\nabla_\alpha X^I = y \nabla_\alpha Z^I,$$
$$\nabla_{\bar{\alpha}} \bar{X}^I = y \nabla_{\bar{\alpha}} \bar{Z}^I,$$
$$\nabla_\alpha \bar{X}^I = \nabla_{\bar{\alpha}} X^I = 0. \tag{4.125}$$

The covariant derivative of $N = -a$, using (4.74), implies

$$(\nabla_\alpha X^I) N_{IJ} \bar{X}^J = 0. \tag{4.126}$$

These relations will be very useful when rewriting the conformal theory in Poincaré language. In particular we will use 'covariant transformations' [47] for all ordinary transformations that are defined on functions of z and $\bar{z}$ by

$$\widehat{\delta} G(z, \bar{z}) = \left(\delta z^\alpha \nabla_\alpha + \delta \bar{z}^{\bar{\alpha}} \nabla_{\bar{\alpha}} \right) G(z, \bar{z}). \qquad (4.127)$$

See the similarity with (3.52). Note that this differs from the ordinary transformation in Poincaré frame:

$$\delta^\square G(z, \bar{z}) = \left(\delta z^\alpha \partial_\alpha + \delta \bar{z}^{\bar{\alpha}} \partial_{\bar{\alpha}} \right) G(z, \bar{z}),$$

$$\widehat{\delta} G(z, \bar{z}) = \delta^\square G(z, \bar{z}) + a^{-1} G(z, \bar{z}) \left(\hat{w}_+ \delta z^\alpha \partial_\alpha + \hat{w}_- \delta \bar{z}^{\bar{\alpha}} \partial_{\bar{\alpha}} \right) \mathcal{K}$$

$$= \delta^\square G(z, \bar{z}) - \delta^\square_K \left[f = \delta z^\alpha \partial_\alpha \mathcal{K}, \ \bar{f} = \delta \bar{z}^{\bar{\alpha}} \partial_{\bar{\alpha}} \mathcal{K} \right]. \qquad (4.128)$$

Since both terms in the right-hand side are symmetries, also this covariant transformation is a symmetry. Note that (4.124) implies that for whatever symmetry transformation

$$\widehat{\delta} y = 0. \qquad (4.129)$$

4.6 Coordinates in the Quaternionic-Kähler Manifold

We will now introduce also convenient coordinates for the hypermultiplet side. This concerns the scalars q^X, where the index X runs over $4(n_H + 1)$ values. In this case, we want to take coordinates that take into account the second parts of (4.56) and (4.58). We want to project out the four directions defined by the homothetic and SU(2) Killing vectors in the hyper-Kähler manifold (they are, respectively, $k_D{}^X$ and $\mathbf{k}^X$, see (3.67) and (3.69)).

4.6.1 Projective Coordinates

We denote the direction of the dilatation generator with the coordinate q^0 and the directions of the SU(2) R-symmetry generators with q^r ($r = 1, 2, 3$). On these coordinates we will apply gauge fixing conditions for, respectively, dilatations and the SU(2) transformations (3.69). The remaining $4n_H$ real variables are indicated as q^u. Our new basis for the scalars of the hypermultiplet is therefore

$$\{q^X\} = \{q^0, q^r, q^u\},\tag{4.130}$$

and

$$k_D{}^X = \{2q^0, 0, 0\}, \qquad \mathbf{k}^X = \{0, \mathbf{k}^r, 0\}.\tag{4.131}$$

In analogy with the special Kähler manifolds, we will refer to the $\{q^X\}$ as the coordinates of the embedding space, and to the $\{q^u\}$ as those of the projective space. Note that we use the index r for the choice of coordinates and the vector sign for the 3 directions of the SU(2) vectors. The vector $\mathbf{k}^r$ connects these as a 3-bein and can be a function of q^r and q^u.

This choice of coordinates leads to equations for the complex structures in this basis. Using (4.131), from (3.69) it follows that $\mathbf{J}_0{}^X$ has only components in the $X = r$ directions. Similarly, after some algebraic manipulation one can obtain the following expressions for coordinates of the complex structures [48]:

$$
\begin{aligned}
&\mathbf{J}_0{}^0 = 0, \qquad \mathbf{J}_0{}^r = (q^0)^{-1}\mathbf{k}^r, \qquad \mathbf{J}_0{}^u = 0, \\
&\mathbf{J}_r{}^0 = \mathbf{k}_r, \qquad \mathbf{J}_r{}^s = (q^0)^{-1}\mathbf{k}_r \times \mathbf{k}^s, \qquad \mathbf{J}_r{}^u = 0, \\
&\mathbf{J}_u{}^0 = \mathbf{k}_u, \qquad \mathbf{J}_u{}^s = (q^0)^{-1}\left(\mathbf{k}_u \times \mathbf{k}^s + J_u{}^v \mathbf{k}_v \cdot \mathbf{k}^s\right),
\end{aligned}
\tag{4.132}
$$

where

$$\mathbf{k}^r \cdot \mathbf{k}_s = -q^0 \delta_s^r,\tag{4.133}$$

and $J_u{}^v$ separately satisfy the quaternionic algebra (3.38). Using the constraints of dilatation invariance and general properties of the quaternionic frame fields, one can derive relations between the connection coefficients [48]. The quaternionic metric (3.148) takes the form[26]

$$
\begin{aligned}
g_{XY}\mathrm{d}q^X\mathrm{d}q^Y &= -\frac{(\mathrm{d}q^0)^2}{q^0} + h_{uv}(q)\mathrm{d}q^u\mathrm{d}q^v - \frac{1}{q^0}\mathbf{k}_X \cdot \mathbf{k}_Y \mathrm{d}q^X\mathrm{d}q^Y, \\
\mathbf{k}_X\mathrm{d}q^X &= \mathbf{k}_r\mathrm{d}q^r + \mathbf{k}_u\mathrm{d}q^u.
\end{aligned}
\tag{4.134}
$$

Note that although $\mathbf{k}^X$ has only 3 non-zero components according to (4.131), the vector $\mathbf{k}_X$ has components in the r- and u-directions. One can now first check

---

[26]We normalize here h_{XY} with a factor q^0 different from [48].

from (3.153) that

$$k_D{}^2 = -4q^0,$$ (4.135)

and (3.154) then implies for arbitrary vectors **A** and **B**:

$$\mathbf{A} \cdot \mathbf{k}^X \mathbf{k}_X \cdot \mathbf{B} = \mathbf{A} \cdot \mathbf{k}^r \mathbf{k}_r \cdot \mathbf{B} = -q^0 \mathbf{A} \cdot \mathbf{B}.$$ (4.136)

With the present notation we have that $\mathbf{k}_u = g_{ur} \mathbf{k}^r$ and $\mathbf{k}_r = g_{rs} \mathbf{k}^s$.

For the fermionic side of the hypermultiplet, we denote by $\mathcal{A}$ the coordinates on the target space enumerating the $2(n_H + 1)$ fermions, and these can be split in $2 + 2n_H$. This is consistent, since the distinction of q^0 and q^r splits the structure group $G\ell(n_H + 1, \mathbb{H})$ to $SU(2) \times G\ell(n_H, \mathbb{H})$. We thus write

$$\{\mathcal{A}\} = \{i, A\},$$ (4.137)

where $i = 1, 2$ is an SU(2)-index. We will use this split also for distinguishing compensating and physical fermions in Sect. 6.1.2

$$\{\zeta^{\mathcal{A}}\} = \{\zeta^i, \zeta^A\}.$$ (4.138)

It has been shown in detail in [48] that the coordinates can be chosen such that some components of $f^{i\mathcal{A}}{}_X$ and $f^X{}_{i\mathcal{A}}$ vanish and one obtains, e.g.

$$f^{i A}{}_0 = f^{i A}{}_r = f^u{}_{ij} = f^0{}_{iA} = 0, \qquad f^r{}_{iA} = (q^0)^{-1} f^u{}_{iA} \mathbf{k}^r \cdot \mathbf{k}_u,$$

$$f^{ij}{}_0 = \mathrm{i}\varepsilon^{ij} \sqrt{\frac{1}{2q^0}}, \qquad f^0{}_{ij} = -\mathrm{i}\varepsilon_{ij} \sqrt{\frac{q^0}{2}}.$$ (4.139)

In these coordinates, the $\rho_{A\bar{B}}$ introduced in (3.33) as well as the C_{AB} and $d^{\bar{A}}{}_B$ from Sect. 3.3.3 are block-diagonal, e.g.

$$C_{ij} = \varepsilon_{ij}, \qquad C_{iA} = 0, \qquad d^{\bar{\imath}}{}_j = -\delta^i{}_j, \qquad d^{\bar{\imath}}{}_A = 0,$$

$$\rho_{i\bar{\jmath}} = -\varepsilon_{ij}, \qquad \rho_{i\bar{A}} = 0.$$ (4.140)

The $A^{i\mathcal{A}}$ introduced in (3.70) has only components for $\mathcal{A}$ in the doublet range:

$$A^{ij} = \mathrm{i}\sqrt{2q^0}\varepsilon^{ij}, \qquad A^{iA} = 0, \qquad A_{ij} = -\mathrm{i}\sqrt{2q^0}\varepsilon_{ij}, \qquad A_{iA} = 0.$$ (4.141)

We can now reduce the fundamental relations of the quaternionic structures in the embedding space to relations in the projective space, making use of (4.139):

$$f^{iA}{}_v f^u{}_{iA} = \delta^u_v, \qquad\qquad f^{iA}{}_u f^u{}_{jB} = \delta^i_j \delta^A_B,$$
$$\left(f^{iA}{}_u\right)^* = f^{jB}{}_u \varepsilon_{ji} \rho_{B\bar{A}}, \qquad (f^u{}_{iA})^* = \varepsilon^{ij} \rho^{\bar{A}B} f^u{}_{jB},$$
$$\rho_{A\bar{B}} \rho^{\bar{B}C} = -\delta^C_A, \qquad\qquad \rho^{\bar{A}B} = \left(\rho_{A\bar{B}}\right)^*,$$
$$2 f^{iA}{}_u f^v{}_{jA} = \delta^v_u \delta^i_j + \tau_j{}^i \cdot \mathbf{J}_u{}^v, \qquad \mathbf{J}_u{}^v = (\mathbf{J}_u{}^v)^* = -f^{iA}{}_u f^v{}_{jA} \tau_i{}^j,$$
$$d^{\bar{A}}{}_B = (d^{\bar{B}}{}_A)^* = \rho^{\bar{A}C} d^{\bar{D}}{}_C \rho_{B\bar{D}},$$
$$C_{AB} = -C_{BA} = \rho_{A\bar{C}} d^{\bar{C}}{}_B, \qquad\qquad C^{\bar{A}\bar{B}} = (C_{AB})^* = \rho^{\bar{A}C} d^{\bar{B}}{}_C,$$
$$h_{uv} = f^{iA}{}_u \varepsilon_{ij} C_{AB} f^{jB}{}_v = \left(f^{iA}{}_u\right)^* d^{\bar{A}}{}_B f^{iB}{}_v.$$

$$(4.142)$$

Using (3.38) and (4.132) the authors of [48] found that the condition (3.39) is projected to[27]

$$\widetilde{\nabla}_w \mathbf{J}_u{}^v \equiv \nabla_w \mathbf{J}_u{}^v + 2\boldsymbol{\omega}_w \times \mathbf{J}_u{}^v = \partial_w \mathbf{J}_u{}^v - \Gamma^z_{wu} \mathbf{J}_z{}^v + \Gamma^v_{wz} \mathbf{J}_u{}^z + 2\boldsymbol{\omega}_w \times \mathbf{J}_u{}^v = 0,$$

$$(4.143)$$

where $\Gamma^w_{uv} = \Gamma^w_{vu}$ is the torsionless Levi-Civita connection of the metric h_{uv}. There is a new term, originating in the embedding space from $\Gamma_{wr}{}^v \mathbf{J}_v{}^r$, containing

$$\boldsymbol{\omega}_u \equiv -\frac{1}{2q^0} \mathbf{k}_u. \qquad (4.144)$$

The triplet $\boldsymbol{\omega}_u$ is an SU(2) connection on the projected manifold, characteristic of the quaternionic-Kähler geometry. Another important projection from (3.39), namely from the antisymmetric part of $\nabla_{[u} J_{v]}{}^0 = 0$ is[28]

$$\mathbf{R}_{uv} \equiv 2\partial_{[u}\boldsymbol{\omega}_{v]} + 2\boldsymbol{\omega}_u \times \boldsymbol{\omega}_v = -\tfrac{1}{2}(q^0)^{-1} \mathbf{J}_u{}^w h_{wv}. \qquad (4.145)$$

Thus, the SU(2) curvature becomes proportional to the complex structure. This is a main property of a quaternionic-Kähler manifold, as will be further discussed in Sect. 5.6.

---

[27]For these projections one needs the Levi-Civita connection of the embedding metric g_{XY} expressed in the quantities of the projective manifold. They are all given in [48]. We repeat here a few components useful for the projections:

$$\Gamma_{00}{}^X = -\tfrac{1}{2}(q^0)^{-1}\delta^X_0, \qquad \Gamma_{0r}{}^0 = \Gamma_{0u}{}^0 = 0, \qquad \Gamma_{0r}{}^s = \tfrac{1}{2}(q^0)^{-1}\delta^s_r, \qquad \Gamma_{0u}{}^v = \tfrac{1}{2}(q^0)^{-1}\delta^v_u,$$

$$\Gamma_{uv}{}^0 = \tfrac{1}{2}g_{uv}, \qquad \Gamma_{ur}{}^0 = -\tfrac{1}{2}(q^0)^{-1}\mathbf{k}_u \cdot \mathbf{k}_r.$$

[28]We write in the right-hand side explicitly $\mathbf{J}_u{}^w h_{wv}$ since so far all raising and lowering of indices was done with g_{XY}, e.g. $\mathbf{J}_{uv} = \mathbf{J}_u{}^r g_{rv} + \mathbf{J}_u{}^v g_{wv} = \mathbf{J}_u{}^w h_{wv} + (q^0)^{-1}\mathbf{k}_u \times \mathbf{k}_v.$

Finally, from (4.142) the condition (3.32) is projected to

$$\widetilde{\nabla}_v f^{iA}{}_u \equiv \partial_v f^{iA}{}_u - \Gamma^w_{vu} f^{iA}{}_w + f^{jA}{}_u \omega_{vj}{}^i + f^{iB}{}_u \omega_{vB}{}^A = 0. \qquad (4.146)$$

4.6.2   S-Supersymmetry, Dilatations and SU(2) Gauge Fixing

Having chosen a convenient set of coordinates, it is now easy to proceed with the gauge fixings to the Poincaré theory. We start from the dilatations: using (4.131) and (4.134), the second condition in (4.56) becomes

$$q^0 = \kappa^{-2} \equiv -\nu^{-1}. \qquad (4.147)$$

The parameter ν will play a role in the characterisation of Quaternionic-Kähler manifolds, see Sect. 5.6. Similarly, we gauge fix the SU(2) by choosing the phases of the compensating quaternion in the hypermultiplet to be some constants q_0^r:

$$\text{SU(2)-gauge:} \qquad q^r = q_0^r. \qquad (4.148)$$

With the above conditions, the bosonic part of $\mathbf{V}_\mu$ in (4.61) becomes

$$\mathcal{V}_\mu = \mathbf{V}_\mu \big|_{\text{bos}} = \tfrac{1}{2}\kappa^2 \left(\mathbf{k}_u \partial_\mu q^u - W^I_\mu \mathbf{P}_I \right) = -\boldsymbol{\omega}_u \partial_\mu q^u - \tfrac{1}{2}\kappa^2 W^I_\mu \mathbf{P}_I. \qquad (4.149)$$

Inserting the above expression into the covariant derivatives (4.60) and plugging in the SU(2)-gauge (4.148) we obtain

$$D_\mu q^X = \nabla_\mu q^X - W^I_\mu \left(k^X_I + \kappa^2 \mathbf{P}_I \cdot \mathbf{k}^X \right), \qquad \nabla_\mu q^X \equiv \partial_\mu q^X - 2(\boldsymbol{\omega}_u \partial_\mu q^u) \cdot \mathbf{k}^X. \qquad (4.150)$$

Furthermore, due to (4.131), the second term in $\nabla_\mu q^X$ only contributes for $X = r$. The various cases give

$$\nabla_\mu q^0 = 0, \qquad \nabla_\mu q^r = -2(\partial_\mu q^u)\boldsymbol{\omega}_u \cdot \mathbf{k}^r, \qquad \nabla_\mu q^u = \partial_\mu q^u. \qquad (4.151)$$

If we now restrict to the non-gauge related part in $D_\mu q^X$, upon using the above relations together with (4.136), the kinetic terms of the q^X in (4.59) reduce to

$$\mathcal{L}_q = -\tfrac{1}{2} g_{XY} \nabla_\mu q^X \nabla^\mu q^Y = -\tfrac{1}{2} \partial_\mu q^u \partial_\mu q^v \left(g_{uv} + \kappa^2 \mathbf{k}_u \cdot \mathbf{k}_v \right). \qquad (4.152)$$

A quick comparison with (4.134) shows therefore that the final metric on the $4n_H$-dimensional space is h_{uv}, used in (4.134):

$$h_{uv} = g_{uv} + 4\kappa^{-2}\boldsymbol{\omega}_u \cdot \boldsymbol{\omega}_v. \tag{4.153}$$

The S-supersymmetry gauge fixing together with the χ^i field equation (4.47) lead to (4.58). Upon using (4.141), the second of these implies the S-gauge

$$S\text{-gauge:} \quad \zeta^i = 0. \tag{4.154}$$

4.6.3 Isometries in the Projective Space

Let us now explain how the isometries of the target space (defined in (3.75)) get projected onto the spaces of the $\{q^u\}$. Upon requiring the corresponding Killing vectors in the target manifold $k_I{}^X$ to commute with the closed homothetic Killing vector (this is the condition (3.83)), one can show that there exists a frame in which the components of these $k_I{}^X$ in the projection are [48]

$$k_I{}^X(q) = \{k_I{}^0 = 0, \ k_I{}^r = \mathbf{k}^r \cdot \mathbf{r}_I, \ k_I{}^u\}, \tag{4.155}$$

where $\mathbf{r}_I(q)$ are arbitrary holomorphic vectors.[29] The moment maps in the embedding space satisfy the condition (3.155), which can be rewritten in the projective space using (4.155) and (4.132) as

$$\widetilde{\nabla}_u \mathbf{P}_I \equiv \partial_u \mathbf{P}_I + 2\boldsymbol{\omega}_u \times \mathbf{P}_I = \mathbf{J}_u{}^v h_{vw} k_I{}^w. \tag{4.156}$$

The value of the moment map $\mathbf{P}_I$ in (3.157) can be written using (4.136) and (4.144) as

$$\mathbf{P}_I = k_I{}^X \mathbf{k}_X = k_I{}^r \mathbf{k}_r + k_I{}^u \mathbf{k}_u = -\kappa^{-2}\left(\mathbf{r}_I + 2k_I{}^u \boldsymbol{\omega}_u\right). \tag{4.157}$$

A similar relation for the special Kähler geometry will appear in Sect. 5.4, where also a r_Λ (or r_I for the gauged symmetries) will describe the transformation of the compensating fields, see (5.89). Using the metric in (4.134) and (4.136), the relation (4.157) implies

$$k_{Ir} = -\kappa^2 \mathbf{k}_r \cdot \mathbf{P}_I. \tag{4.158}$$

---

[29]We can make this choice since the $\mathbf{k}^r$ introduced in (4.133) can be considered as invertible 3×3 matrices. The first 0 in (4.155) is then the statement that the isometries commute with dilatations, which according to (4.131) act only on the 0-component.

The equivariance relation (3.156) gets also an extra part when we split the index X and use the metric h_{uv}:

$$k_I^{\ u} J_u^{\ w} h_{wv} k_J^{\ v} + \kappa^2 \mathbf{P}_I \times \mathbf{P}_J = f_{IJ}^{\ K} \mathbf{P}_K. \qquad (4.159)$$

This equation admits unique solutions for $\mathbf{P}_I$ if the quaternionic-Kähler manifold is non-trivial ($n_H \neq 0$):

$$2 n_H \kappa^2 \mathbf{P}_I = -J_u^{\ v} \nabla_v k_I^{\ u}. \qquad (4.160)$$

On the other hand, when $n_H = 0$ the first term in (4.159) drops and there are two possible solutions for the moment maps, which are then called Fayet–Iliopoulos (FI) terms. First, in the case where the gauge group contains an SU(2) factor, we can have

$$\mathbf{P}_I = \mathbf{e}_I \xi, \qquad (4.161)$$

for any arbitrary constant ξ, and $\mathbf{e}_I$ being non-zero constants only for I in the range of the SU(2) factor and satisfying

$$\kappa^2 \xi \mathbf{e}_I \times \mathbf{e}_J = f_{IJ}^{\ K} \mathbf{e}_K, \qquad (4.162)$$

in order that (4.159) is verified.

The second case are the U(1) FI terms. In that case the only remaining term in (4.159) tells that $\mathbf{P}_I$ and $\mathbf{P}_J$ should be in the same direction in SU(2) space. Hence

$$\mathbf{P}_I = \mathbf{e}\,\xi_I, \qquad (4.163)$$

where $\mathbf{e}$ is an arbitrary vector in SU(2) space and ξ_I are constants for the I corresponding to U(1) factors in the gauge group.

4.6.4 Decomposition Rules

To find the decomposition laws, we start from the complete transformations of the hypermultiplet scalars in the conformal setting (from (3.91), (3.67), (3.69), and (3.75)):

$$\delta q^X = \left(\delta_Q[\epsilon] + \delta_D[\lambda_D] + \delta_{SU(2)}[\boldsymbol{\lambda}] + \delta_G[\theta] \right) q^X$$

$$= \left[-\mathrm{i}\bar{\epsilon}^i \zeta^A f^X_{\ iA} + \mathrm{h.c.} \right] + \lambda_D k_D^{\ X} - 2\boldsymbol{\lambda} \cdot \mathbf{k}^X + \theta^I k_I^X. \qquad (4.164)$$

We now consider what remains from (4.164) when we restrict to Poincaré transformations (denoted as $\delta^{\square}$), namely those combinations of (4.164) that preserve the gauge conditions. As we explained previously, under projection and gauge fixing, among the q^X, the only remaining physical fields are q^u. Thus $\delta^{\square}$ are the transformations induced by the dependence on q^u, times the transformations of the latter.

First, for $X = 0$, since q^0 is a constant in the Poincaré frame and $\mathbf{k}^0 = \zeta^A f^X{}_{i\mathcal{A}} = k_I{}^0 = 0$ (see (4.131), (4.139), (4.154), (4.155)), only the dilatation term remains, and we thus find $0 = \delta_D(\lambda_D)q^0$, i.e. implying that we should put $\lambda_D = 0$ in order to obtain Poincaré transformations. Also q^r is a constant by (4.148), and thus we get for $X = r$ the decomposition law for the SU(2) symmetry:

$$0 = \delta^{\square} q^r = \left[-i\bar{\epsilon}^i \zeta^A f^r{}_{iA} + \text{h.c.} \right] - 2\lambda \cdot \mathbf{k}^r + \theta^I k_I^r . \tag{4.165}$$

Using (4.139) and (4.155) all terms are proportional to $\mathbf{k}^r$, which is invertible as a 3×3 matrix, and we find

$$\lambda = -\omega_u \left[-i\bar{\epsilon}^i \zeta^A f^u{}_{iA} + \text{h.c.} \right] + \tfrac{1}{2}\theta^I \mathbf{r}_I . \tag{4.166}$$

The last case of (4.164) is

$$\delta^{\square} q^u = \delta q^u = \left[-i\bar{\epsilon}^i \zeta^A f^u{}_{iA} + \text{h.c.} \right] + \theta^I k_I^u . \tag{4.167}$$

Inserting this in (4.166), we obtain

$$\lambda = -\omega_u \delta q^u - \tfrac{1}{2}\kappa^2 \theta^I \mathbf{P}_I . \tag{4.168}$$

The second term determines that any gauge symmetry in the Poincaré theory has a contribution from the SU(2) R-symmetry in the conformal theory.

4.7 $D = 5$ and $D = 6$, $\mathcal{N} = 2$ Supergravities

So far we discussed $D = 4$ theories, but the main procedure is very similar for $D = 5$ and $D = 6$ theories. We review in the present section the main result for $D = 5$ and $D = 6$, $\mathcal{N} = 2$ supergravities.

4.7.1 $D = 5$

For $D = 5$ one can again use a vector multiplet and a hypermultiplet as compensating multiplets.[30] The difference from $D = 4$ is that now the vector

---

[30]For an overview of constructions with linear multiplets and other Weyl multiplets, see [49].

Table 4.1 Multiplets and fields in the superconformal construction for $D = 5$

spin	Weyl	vector	hyper	gauge fix	auxiliary
2	e_μ^a				
$\frac{3}{2}$	ψ_μ^i				
	$V_{\mu i}{}^j, T_{ab}$				auxiliary
1		$n+1$			
$\frac{1}{2}$	χ^i	$2(n+1)$	$2(n_H+1)$	2: S	χ^i with 2 others
0	D	$n+1$	$4(n_H+1)$	1: dilatations, 3: SU(2)	D and 1 other

multiplet has only a real scalar. This fits remarkably well with the fact that the R-symmetry group contains only SU(2), and no U(1), which we used in $D = 4$ to fix the phase of the scalar field of the vector multiplet. This is schematically represented in Table 4.1. We already gave the field equations in Sect. 4.4.1. The field equations in (4.43) constrain a doublet of spinors and a real scalar together with the gauge fixings of dilatation and S-supersymmetry, similar to how this happens in $D = 4$, see Sect. 4.4.2. Vice versa, the field equations of these components of the vector and hypermultiplet eliminate χ^i and D. We give here the bosonic gravity sector in more detail. The Riemann scalar appears in $f_a{}^a = -R/16 + \ldots$ in (2.100), which in its turn appears in the sum of (3.129) and (3.159)

$$\mathcal{L} = \mathcal{L}_g + \mathcal{L}_g = -\frac{1}{24} e\, R \left(C_{IJK} \sigma^I \sigma^J \sigma^K + k_{\mathrm{D}}{}^2 \right) + \cdots . \tag{4.169}$$

The dilatational gauge fixing is chosen such that this gives the canonical normalization of the Einstein–Hilbert action, i.e. $eR/(2\kappa^2)$. Combining with (4.43) leads to the analogue of (4.56):

$$C_{IJK} \sigma^I \sigma^J \sigma^K = -3\kappa^{-2} , \qquad k_{\mathrm{D}}{}^2 = -9\kappa^{-2} . \tag{4.170}$$

This is used to obtain the Poincaré theory as we will further illustrate in Sect. 6.2.

4.7.2 $D = 6$

As we discussed in Sect. 2.6, the $D = 6$ Weyl multiplet contains an antisymmetric tensor T_{abc}^-, which is anti-self-dual. If one wants to build an action with manifest Lorentz-invariance, this has to be combined with a self-dual tensor F_{abc}^+, which sits in a tensor multiplet. The sum of these two can then be considered as the field strength of a physical antisymmetric tensor $B_{\mu\nu}$. The added tensor multiplet acts also as first compensating multiplet. There are constructions with as second compensating multiplet a hypermultiplet or with a linear multiplet, see e.g. [50–53].

Importantly, the construction of $D = 6$ minimal supergravity differs slightly from that for $D = 4$ and $D = 5$, since the tensor multiplet involves constraints that define the field D of the Weyl multiplet in terms of fields of the tensor multiplet.

This has to be contrasted with the cases of $D = 4, 5$, in which the field equation of the field D in (4.47) combines with a dilatation gauge condition (4.55) to give a condition on a scalar of the first compensating multiplet and another condition on a scalar of the second multiplet, see (4.56). Indeed, in $D = 6$, the field D of the Weyl multiplet is not anymore independent from the tensor multiplet, and we thus have only one condition, which imply that there remains a physical scalar. The latter is in the Poincaré theory part of the tensor multiplet. This results matches with the expectation that, also in the Poincaré theory, we need one tensor multiplet in order to be able to construct Lorentz-invariant actions.[31]

Adding an arbitrary number of vector and hypermultiplets, one obtains super-Poincaré theories which all contain one tensor multiplet.[32] Such actions were also constructed independently in [55, 56]. The general super-Poincaré theory is given in [57]. It builds on earlier work, e.g. [58–60].

Famously, theories in $4n + 2$ dimensions can suffer from gravitational (and gauge) anomalies. These can be calculated with the methods of Alvarez-Gaumé and Witten [61]. Anomaly-free theories, using the Green–Schwarz mechanism, have been identified in [62, 63, 56, 64].

References

1. B. de Wit, P.G. Lauwers, A. Van Proeyen, Lagrangians of $N = 2$ supergravity–matter systems. Nucl. Phys. **B255**, 569–608 (1985). https://doi.org/10.1016/0550-3213(85)90154-3
2. B. de Wit, J.W. van Holten, Multiplets of linearized SO(2) supergravity. Nucl. Phys. **B155**, 530–542 (1979). https://doi.org/10.1016/0550-3213(79)90285-2
3. E.S. Fradkin, M.A. Vasiliev, Minimal set of auxiliary fields and S matrix for extended supergravity. Lett. Nuovo Cimento **25**, 79–90 (1979). https://doi.org/10.1007/BF02776267
4. B. de Wit, J.W. van Holten, A. Van Proeyen, Transformation rules of $N = 2$ supergravity multiplets. Nucl. Phys. **B167**, 186–204 (1980). https://doi.org/10.1016/0550-3213(80)90125-X
5. B. de Wit, R. Philippe, A. Van Proeyen, The improved tensor multiplet in $N = 2$ supergravity. Nucl. Phys. **B219**, 143–166 (1983). https://doi.org/10.1016/0550-3213(83)90432-7
6. M. Kaku, P.K. Townsend, Poincaré supergravity as broken superconformal gravity. Phys. Lett. **76B**, 54–58 (1978). https://doi.org/10.1016/0370-2693(78)90098-9
7. S. Ferrara, P. Van Nieuwenhuizen, Structure of supergravity. Phys. Lett. **B78**, 573 (1978) https://doi.org/10.1016/0370-2693(78)90642-1
8. D.Z. Freedman, A. Van Proeyen, *Supergravity* (Cambridge University Press, Cambridge, 2012). http://www.cambridge.org/mw/academic/subjects/physics/theoretical-physics-and-mathematical-physics/supergravity?format=AR
9. B. de Wit, S. Ferrara, On higher-order invariants in extended supergravity. Phys. Lett. **B81**, 317 (1979) https://doi.org/10.1016/0370-2693(79)90343-5

---

[31]Clearly the fermionic part is analogous: the field χ^i of the Weyl multiplet is determined by a constraint of the tensor multiplet. The S-gauge determines the fermion field of the second compensating multiplet, leaving the fermion of the tensor multiplet as physical field.

[32]If one is only interested in field equations, and not in Lorenz-invariant actions, one can add an arbitrary number of tensor multiplets, or even no tensor multiplet [54].

10. S. Ferrara, An overview on broken supergravity models, in *Proceedings of the Conference on 2nd Oxford Quantum Gravity* (1980)
11. B. de Wit, J.W. van Holten, A. Van Proeyen, Structure of $N = 2$ supergravity. Nucl. Phys. **B184**, 77–108 (1981). https://doi.org/10.1016/0550-3213(83)90548-5, https://doi.org/10.1016/0550-3213(81)90211-X [Erratum: Nucl. Phys. B **222**, 516 (1983)]
12. P. Fayet, Fermi–Bose hypersymmetry. Nucl. Phys. **B113**, 135 (1976). https://doi.org/10.1016/0550-3213(76)90458-2
13. P. Fayet, Spontaneous generation of massive multiplets and central charges in extended supersymmetric theories. Nucl. Phys. **B149**, 137 (1979). https://doi.org/10.1016/0550-3213(79)90162-7
14. A. Van Proeyen, $N = 2$ Supergravity multiplets, in *Proceedings of the 17th Winterschool Conference on Developments in the Theory of Fundamental Interactions*, Karpacz, ed. by L. Turko, A. Pekalski (Harwood Academic Publishers, Reading, 1981), pp. 57–93
15. B. de Wit, J.W. van Holten, A. Van Proeyen, Central charges and conformal supergravity. Phys. Lett. **95B**, 51–55 (1980). https://doi.org/10.1016/0370-2693(80)90397-4
16. B. de Wit, P.G. Lauwers, R. Philippe, A. Van Proeyen, Noncompact $N = 2$ supergravity. Phys. Lett. **135B**, 295 (1984). https://doi.org/10.1016/0370-2693(84)90395-2
17. M.F. Sohnius, P.C. West, An alternative minimal off-shell version of $N = 1$ supergravity. Phys. Lett. **B105**, 353 (1981). https://doi.org/10.1016/0370-2693(81)90778-4
18. T.L. Curtright, D.Z. Freedman, Nonlinear σ models with extended supersymmetry in four dimensions. Phys. Lett. **B90**, 71 (1980). https://doi.org/10.1016/0370-2693(80)90054-4, https://doi.org/10.1016/0370-2693(80)91028-X [Erratum: Phys. Lett. B **91**, 487 (1980)]
19. L. Andrianopoli, R. D'Auria, S. Ferrara, Supersymmetry reduction of N-extended supergravities in four dimensions. J. High Energy Phys. **3**, 025 (2002). https://doi.org/10.1088/1126-6708/2002/03/025, arXiv:hep-th/0110277 [hep-th]
20. L. Andrianopoli, R. D'Auria, S. Ferrara, Consistent reduction of $N = 2 \rightarrow N = 1$ four dimensional supergravity coupled to matter. Nucl. Phys. **B628**, 387–403 (2002). https://doi.org/10.1016/S0550-3213(02)00090-1, arXiv:hep-th/0112192 [hep-th]
21. Y. Yamada, Off-shell $N = 2 \rightarrow N = 1$ reduction in 4D conformal supergravity. J. High Energy Phys. **6**, 002 (2019) https://doi.org/10.1007/JHEP06(2019)002, arXiv:1902.00121 [hep-th]
22. T. Kugo, K. Ohashi, Off-shell $d = 5$ supergravity coupled to matter–Yang–Mills system. Prog. Theor. Phys. **105**, 323–353 (2001). https://doi.org/10.1143/PTP.105.323, arXiv:hep-ph/0010288 [hep-ph]
23. T. Kugo, K. Ohashi, Superconformal tensor calculus on an orbifold in $5D$. Prog. Theor. Phys. **108**, 203–228 (2002). https://doi.org/10.1143/PTP.108.203, arXiv:hep-th/0203276 [hep-th]
24. E. Bergshoeff, S. Cucu, T. de Wit, J. Gheerardyn, R. Halbersma, S. Vandoren, A. Van Proeyen, Superconformal $N = 2$, $D = 5$ matter with and without actions. J. High Energy Phys. **10**, 045 (2002). https://doi.org/10.1088/1126-6708/2002/10/045, arXiv:hep-th/0205230 [hep-th]
25. E. Bergshoeff, S. Cucu, T. de Wit, J. Gheerardyn, S. Vandoren, A. Van Proeyen, $N = 2$ supergravity in five dimensions revisited. Classical Quantum Gravity **21**, 3015–3041 (2004). https://doi.org/10.1088/0264-9381/23/23/C01, https://doi.org/10.1088/0264-9381/21/12/013, arXiv:hep-th/0403045[hep-th] [Erratum **23**, 7149 (2006)]
26. M. Günaydin, M. Zagermann, The gauging of five-dimensional, $N = 2$ Maxwell–Einstein supergravity theories coupled to tensor multiplets. Nucl. Phys. **B572**, 131–150 (2000). https://doi.org/10.1016/S0550-3213(99)00801-9, arXiv:hep-th/9912027 [hep-th]
27. A. Ceresole, G. Dall'Agata, General matter coupled $\mathcal{N} = 2$, $D = 5$ gauged supergravity. Nucl. Phys. **B585**, 143–170 (2000). https://doi.org/10.1016/S0550-3213(00)00339-4, arXiv:hep-th/0004111 [hep-th]
28. L. Andrianopoli, R. D'Auria, L. Sommovigo, On the coupling of tensors to gauge fields: $D = 5$, $N = 2$ supergravity revisited. arXiv:hep-th/0703188 [HEP-TH]
29. B. Vanhecke, A. Van Proeyen, Covariant field equations in supergravity. Fortsch. Phys. **65**(12), 1700071 (2017). https://doi.org/10.1002/prop.201700071, arXiv:1705.06675 [hep-th]
30. G. Sierra, P.K. Townsend, An introduction to $N = 2$ rigid supersymmetry, in *Supersymmetry and Supergravity 1983*, ed. by B. Milewski (World Scientific, Singapore, 1983)

31. B. Craps, F. Roose, W. Troost, A. Van Proeyen, What is special Kähler geometry? Nucl. Phys. **B503**, 565–613 (1997). https://doi.org/10.1016/S0550-3213(97)00408-2, arXiv:hep-th/9703082 [hep-th]

32. D.S. Freed, Special Kähler manifolds. Commun. Math. Phys. **203**, 31–52 (1999). https://doi.org/10.1007/s002200050604, arXiv:hep-th/9712042 [hep-th]

33. D.V. Alekseevsky, V. Cortés, C. Devchand, Special complex manifolds. J. Geom. Phys. **42**, 85–105 (2002). https://doi.org/10.1016/S0393-0440(01)00078-X, arXiv:math/9910091 [math.DG]

34. B. de Wit, P.G. Lauwers, R. Philippe, S.Q. Su, A. Van Proeyen, Gauge and matter fields coupled to $N = 2$ supergravity. Phys. Lett. **B134**, 37–43 (1984). https://doi.org/10.1016/0370-2693(84)90979-1

35. B. de Wit, A. Van Proeyen, Potentials and symmetries of general gauged $N = 2$ supergravity—Yang–Mills models. Nucl. Phys. **B245**, 89–117 (1984). https://doi.org/10.1016/0550-3213(84)90425-5

36. A. Strominger, Special geometry. Commun. Math. Phys. **133**, 163–180 (1990). https://doi.org/10.1007/BF02096559

37. G. Gibbons and P. Rychenkova, Cones, tri-Sasakian structures and superconformal invariance. Phys. Lett. **B443**, 138–142 (1998). https://doi.org/10.1016/S0370-2693(98)01287-8, arXiv:hep-th/9809158 [hep-th]

38. B. de Wit, B. Kleijn, S. Vandoren, Superconformal hypermultiplets. Nucl. Phys. **B568**, 475–502 (2000). https://doi.org/10.1016/S0550-3213(99)00726-9, arXiv:hep-th/9909228 [hep-th]

39. E. Cremmer, C. Kounnas, A. Van Proeyen, J. Derendinger, S. Ferrara, B. de Wit, L. Girardello, Vector multiplets coupled to $N = 2$ supergravity: super-Higgs effect, flat potentials and geometric structure, Nucl. Phys. **B250**, 385–426 (1985). https://doi.org/10.1016/0550-3213(85)90488-2

40. E. Cremmer, A. Van Proeyen, Classification of Kähler manifolds in $N = 2$ vector multiplet–supergravity couplings. Classical Quantum Gravity **2**, 445 (1985). https://doi.org/10.1088/0264-9381/2/4/010

41. M. Günaydin, G. Sierra, P.K. Townsend, The geometry of $N = 2$ Maxwell–Einstein supergravity and Jordan algebras. Nucl. Phys. **B242**, 244–268 (1984). https://doi.org/10.1016/0550-3213(84)90142-1

42. L. Castellani, R. D'Auria, S. Ferrara, Special Kähler geometry: an intrinsic formulation from $N = 2$ space-time supersymmetry. Phys. Lett. **B241**, 57–62 (1990). https://doi.org/10.1016/0370-2693(90)91486-U

43. L. Castellani, R. D'Auria, S. Ferrara, Special geometry without special coordinates. Classical Quantum Gravity **7**, 1767–1790 (1990). https://doi.org/10.1088/0264-9381/7/10/009

44. R. D'Auria, S. Ferrara, P. Frè, Special and quaternionic isometries: general couplings in $N = 2$ supergravity and the scalar potential. Nucl. Phys. **B359**, 705–740 (1991). https://doi.org/10.1016/0550-3213(91)90077-B

45. E. Witten, J. Bagger, Quantization of Newton's constant in certain supergravity theories. Phys. Lett. **B115**, 202–206 (1982). https://doi.org/10.1016/0370-2693(82)90644-X

46. J. Bagger, Supersymmetric sigma models, in *Supersymmetry*, ed. by K. Dietz et al. NATO Advanced Study Institute, Series B, Physics, vol. 125 (Plenum Press, New York, 1985)

47. D.Z. Freedman, D. Roest, A. Van Proeyen, Off-shell Poincaré supergravity. J. High Energy Phys. **02**, 102 (2017). https://doi.org/10.1007/JHEP02(2017)102, arXiv:1701.05216 [hep-th]

48. E. Bergshoeff, S. Cucu, T. de Wit, J. Gheerardyn, S. Vandoren, A. Van Proeyen, The map between conformal hypercomplex/hyper-Kähler and quaternionic(-Kähler) geometry. Commun. Math. Phys. **262**, 411–457 (2006). https://doi.org/10.1007/s00220-005-1475-6, arXiv:hep-th/0411209 [hep-th]

49. M. Ozkan, Off-shell $\mathcal{N} = 2$ linear multiplets in five dimensions. J. High Energy Phys. **11**, 157 (2016). https://doi.org/10.1007/JHEP11(2016)157, arXiv:1608.00349 [hep-th]

50. E. Bergshoeff, E. Sezgin, A. Van Proeyen, Superconformal tensor calculus and matter couplings in six dimensions. Nucl. Phys. **B264**, 653 (1986). https://doi.org/10.1016/0550-3213(86)90503-1 [Erratum: Nucl. Phys. B **598**, 667 (2001)]

51. A. Van Proeyen, $N = 2$ matter couplings in $d = 4$ and 6 from superconformal tensor calculus, in *Proceedings of the 1st Torino Meeting on Superunification and Extra Dimensions*, ed. R. D'Auria, P. Fré (World Scientific, Singapore, 1986), pp. 97–125

52. E. Bergshoeff, Superconformal invariance and the tensor multiplet in six dimensions, in *Proceedings of the 1st Torino Meeting on Superunification and Extra Dimensions*, ed. R. D'Auria, P. Fré (World Scientific, Singapore, 1986) pp. 126–137

53. F. Coomans, A. Van Proeyen, Off-shell $\mathcal{N} = (1, 0)$, $D = 6$ supergravity from superconformal methods. J. High Energy Phys. **1102**, 049 (2011). https://doi.org/10.1007/JHEP02(2011)049, arXiv:1101.2403 [hep-th]

54. N. Marcus, J.H. Schwarz, Field theories that have no manifestly Lorentz-invariant formulation. Phys. Lett. **115B**, 111 (1982). https://doi.org/10.1016/0370-2693(82)90807-3

55. H. Nishino, E. Sezgin, Matter and gauge couplings of $N = 2$ supergravity in six dimensions. Phys. Lett. **144B**, 187–192 (1984). https://doi.org/10.1016/0370-2693(84)91800-8

56. H. Nishino, E. Sezgin, The complete $N = 2$, $d = 6$ supergravity with matter and Yang-Mills couplings. Nucl. Phys. **B278**, 341, 353–379 (1986). https://doi.org/10.1016/0550-3213(86)90218-X

57. F. Riccioni, All couplings of minimal six-dimensional supergravity. Nucl. Phys. **B605**, 245–265 (2001). https://doi.org/10.1016/S0550-3213(01)00199-7, arXiv:hep-th/0101074 [hep-th]

58. L.J. Romans, Self-duality for interacting fields: covariant field equations for six-dimensional chiral supergravities. Nucl. Phys. **B276**, 71 (1986). https://doi.org/10.1016/0550-3213(86)90016-7,

59. H. Nishino, E. Sezgin, New couplings of six-dimensional supergravity. Nucl. Phys. **B505**, 497–516 (1997). https://doi.org/10.1016/S0550-3213(97)00357-X, arXiv:hep-th/9703075

60. S. Ferrara, F. Riccioni, A. Sagnotti, Tensor and vector multiplets in six-dimensional supergravity. Nucl. Phys. **B519**, 115–140 (1998). https://doi.org/10.1016/S0550-3213(97)00837-7, arXiv:hep-th/9711059

61. L. Alvarez-Gaumé, E. Witten, Gravitational Anomalies. Nucl. Phys. **B234**, 269, 269 (1984) https://doi.org/10.1016/0550-3213(84)90066-X,

62. M.B. Green, J.H. Schwarz, P.C. West, Anomaly-free chiral theories in six dimensions. Nucl. Phys. **B254**, 327–348 (1985). https://doi.org/10.1016/0550-3213(85)90222-6

63. S. Randjbar-Daemi, A. Salam, E. Sezgin, J.A. Strathdee, An anomaly-free model in six dimensions. Phys. Lett. **151B**, 351–356 (1985). https://doi.org/10.1016/0370-2693(85)91653-3

64. A. Sagnotti, A note on the Green-Schwarz mechanism in open string theories. Phys. Lett. **B294**, 196–203 (1992). https://doi.org/10.1016/0370-2693(92)90682-T, arXiv:hep-th/9210127 [hep-th]

Chapter 5
Special Geometries

Abstract In this chapter we will discuss the geometric concepts behind the special geometries. In terms of the supergravity theories, this means that we will restrict to the bosonic sector of these theories. In the first part of this chapter, we will examine the scalars of the vector multiplets, and we will see that the special Kähler geometry for the scalars is very much related to the symplectic transformations defined by the dualities of the vector fields. We will also discuss the attractor mechanism for black holes in this context. Then we will study the geometry of quaternionic-Kähler manifolds, related to the scalars of the hypermultiplets. Finally, we will present the relations between these manifolds and their versions in the different dimensions.

The *special geometries* are defined as the geometries of the scalars in $\mathcal{N} = 2$ supergravities in $D = 4, 5, 6$. As we will discuss, a very important role in these geometries is played by the R-symmetry group, namely SU(2) for $D = 5, 6$, and SU(2) × U(1) for $D = 4$. In particular:

- The SU(2) subgroup acts on the scalars of the hypermultiplets, leading to the three complex structures of the corresponding manifolds. The gauge connection of SU(2) promotes the hyper-Kähler manifold of hypermultiplets to a quaternionic manifold.
- In $D = 4$, the U(1) factor acts on the complex scalars of the vector multiplet, whose manifold therefore inherits one complex structure. The gauge connection of this U(1) will be the Kähler curvature.

There are three types of *special manifolds*. These are either associated to geometry of the real scalars of vector multiplets in $D = 5$, to the complex scalars of $D = 4$ vector multiplets or to the quaternionic scalars of hypermultiplets.[1] However, these spaces are not completely independent. An additional relation among them, called **c** map and **r** map (see Sect. 5.7), defines a precise correspondence.

---

[1] Since there are no scalars in the vector multiplets of $D = 6$ (see Table 3.1), there is no geometry in that case.

© Springer Nature Switzerland AG 2020

E. Lauria, A. Van Proeyen, $\mathcal{N} = 2$ *Supergravity in D = 4, 5, 6 Dimensions*,
Lecture Notes in Physics 966, https://doi.org/10.1007/978-3-030-33757-5_5

Most of this chapter will be concerned with the special Kähler geometry of the scalars in the $D = 4$ vector multiplet, which was obtained in the previous chapter. We will see how properties of this geometry, due to supersymmetry, are inherited from symplectic transformations acting on the gauge vectors in the multiplet. The latter generates dualities between $D = 4$, $\mathcal{N} = 2$ theories [1–4], which can be thought as generalizations of the well-known electric-magnetic dualities of Maxwell equations. Afterwards, we will discuss quaternionic-Kähler geometries (Sect. 5.6), and finally relations between all the manifolds (Sect. 5.7), where we will also consider special situations when the manifolds are homogeneous or symmetric.

5.1 $D = 4$, $\mathcal{N} = 2$ Bosonic Action

Before we start, let us rewrite the bosonic part of the action associated to the vector multiplets. The bosonic part of the matter-coupled supergravity action appears in the first 4 lines of (4.59). Omitting for now the hypermultiplet sector (q^X terms) and the Chern–Simons term in the fourth line (this will be included in the full action later, Sect. 6.1) we find

$$
e^{-1} \mathcal{L}_{\text{bos}} = \tfrac{1}{2} \kappa^{-2} R + \mathcal{L}_0 + \mathcal{L}_1 ,
$$

$$
\mathcal{L}_0 = - g_{\alpha\bar\beta} \nabla_\mu z^\alpha \nabla^\mu \bar z^\beta \; - \; V(z, \bar z), \qquad g_{\alpha\bar\beta} = \partial_\alpha \partial_{\bar\beta} \mathcal{K}(z, \bar z),
$$

$$
\mathcal{L}_1 = \tfrac{1}{2} \operatorname{Im} \left(\mathcal{N}_{IJ} \, F_{\mu\nu}^{+I} F^{+\mu\nu J} \right)
$$

$$
= \tfrac{1}{4} (\operatorname{Im} \mathcal{N}_{IJ}) F_{\mu\nu}^I F^{\mu\nu J} - \tfrac{1}{8} e^{-1} (\operatorname{Re} \mathcal{N}_{IJ}) \varepsilon^{\mu\nu\rho\sigma} F_{\mu\nu}^I F_{\rho\sigma}^J , \tag{5.1}
$$

with $\nabla_\mu z^\alpha$ defined in (4.63), and we have used the results of Sect. 4.5 to write the kinetic terms in $\mathcal{L}_0$. This is in fact the general form for a theory with scalar fields z^α in a Kähler manifold, and $n + 1$ non-abelian vector fields labelled by an index I. The scalar potential $V(z, \bar z)$ is very important for physical applications. Here we want to emphasize that $V(z, \bar z)$ is determined by the gauging as in the third line of (4.59) (see Sect. 6.1.1.2 for more details).

Crucially, the kinetic terms in $\mathcal{L}_1$ are controlled by the complex functions $\mathcal{N}_{IJ}$. The latter are defined by the second line of (4.59), upon using the equation of motion of T_{ab} (4.54),

$$
\mathcal{N}_{IJ}(z, \bar z) = \bar F_{IJ} + i \frac{(N_{IN} X^N)(N_{JK} X^K)}{N_{LM} X^L X^M} . \tag{5.2}
$$

These functions become the effective (complexified) gauge couplings, whenever the scalars have a non-zero vacuum expectation value (v.e.v.) on some classical background. In particular the v.e.v. of $\operatorname{Im} \mathcal{N}$ gives the effective gauge coupling, while that of $\operatorname{Re} \mathcal{N}$ gives the so-called theta angles.

Exercise 5.1 Consider the model of Exercise 4.1. Check that in this case, the kinetic terms of the two vectors are determined by the matrix (5.2)

$$
\mathcal{N}_{IJ} = \frac{1}{2} i \begin{pmatrix} 1 & 0 \\ 0 & -1 \end{pmatrix} + \frac{i}{z^2 - 1} \begin{pmatrix} 1 & -z \\ -z & z^2 \end{pmatrix} = -\frac{i}{1 - z^2} \begin{pmatrix} \frac{1}{2}(1 + z^2) & -z \\ -z & \frac{1}{2}(1 + z^2) \end{pmatrix}.
\tag{5.3}
$$

The imaginary part is negative definite as it should be for positive kinetic terms of the vectors. □

5.2 Symplectic Transformations

A prerequisite to understand the structure of special Kähler geometry is a study of symplectic transformations. As we will explain in this section, the latter first appear as generalized electric-magnetic dualities in the context of pure $D = 4$ abelian gauge theories. We will argue that symplectic transformations and consequent dualities can be extended to the full $\mathcal{N} = 2$ vector multiplets by means of supersymmetry.

5.2.1 Electric-Magnetic Dualities of Vector Fields in $D = 4$

A striking property of pure abelian gauge theories is that their equation of motions and Bianchi identities are left invariant by a group of symplectic transformations, which act on the (complexified) gauge couplings. In other words, symplectic transformations generate dualities of these theories.

Let us recall some key ingredients to understand symplectic transformations, following the main ideas discussed in [5, Sect. 4.2.4].[2] The starting point is to consider the kinetic terms of the vector fields, $\mathcal{L}_1$ in (5.1), which can be written as

$$
\mathcal{L}_1 = \frac{1}{2} \operatorname{Im} \left(\mathcal{N}_{IJ} F^{+I}_{\mu\nu} F^{+\mu\nu J} \right) = \frac{1}{2} \operatorname{Im} \left(F^{+I}_{\mu\nu} G^{\mu\nu}_{+I} \right),
\tag{5.4}
$$

with, viewing $\mathcal{L}_1$ as function of $F^{+I}_{\mu\nu}$ and $F^{-I}_{\mu\nu}$,

$$
G^{\mu\nu}_{+I} \equiv 2i \frac{\partial \mathcal{L}_1}{\partial F^{+I}_{\mu\nu}} = \mathcal{N}_{IJ} F^{+J\,\mu\nu},
\tag{5.5}
$$

---

[2]There the kinetic matrix $\mathcal{N}_{IJ}$ is written as $\mathcal{N}_{IJ} = -i \bar{f}_{IJ}$.

and the tensor $\mathcal{N}$ defined in (5.2). Importantly, in selecting the kinetic terms of (5.1), we are setting $f_{IJ}{}^K = 0$, such that (5.1) becomes effectively a $U(1)^{n+1}$ gauge theory. Non-abelian parts will be discussed later, in Sect. 5.4, when the symmetries of the kinetic terms will be identified and gauged.

The field equations and Bianchi identities are invariant under $Sp(2(n+1), \mathbb{R})$ transformations[3]

$$\begin{pmatrix} \widetilde{F}^+ \\ \widetilde{G}_+ \end{pmatrix} = \mathcal{S} \begin{pmatrix} F^+ \\ G_+ \end{pmatrix}, \qquad \mathcal{S} \in Sp(2(n+1), \mathbb{R}). \tag{5.6}$$

Specifically $\mathcal{S}$ is a symplectic transformation, namely a real matrix[4] satisfying

$$\mathcal{S} = \begin{pmatrix} A & B \\ C & D \end{pmatrix}, \qquad \mathcal{S}^T \Omega \mathcal{S} = \Omega, \qquad \text{where} \qquad \Omega = \begin{pmatrix} 0 & \mathbb{1} \\ -\mathbb{1} & 0 \end{pmatrix}. \tag{5.7}$$

This is equivalent to the requirements

$$A^T C - C^T A = 0, \qquad B^T D - D^T B = 0, \qquad A^T D - C^T B = \mathbb{1}. \tag{5.8}$$

It is not hard to verify that under (5.6) the kinetic matrix $\mathcal{N}$ transforms as

$$\widetilde{\mathcal{N}} = (C + D\mathcal{N})(A + B\mathcal{N})^{-1}. \tag{5.9}$$

Importantly, (5.9) preserves the condition $\text{Im}\,\mathcal{N} < 0$, namely the positivity of the kinetic terms for $\mathcal{L}_1$ (see Sect. 4.5.4). In the language of symplectic geometry, a $2(n+1)$-component column V transforming as $\widetilde{V} = \mathcal{S}V$. The prime example is (5.6). The invariant inner product of two symplectic vectors V and W is

$$\langle V, W \rangle \equiv V^T \Omega W. \tag{5.10}$$

Finally, observe that if we apply (5.6) to the last of (5.4), the action is invariant only if $B = 0$ (see that $\text{Im}\,F^{+I}F^{+J}$ is a total derivative). For this reason the subgroup of $Sp(2(n+1), \mathbb{R})$ with $B = 0$ are sometimes called 'perturbative symmetries'. It is possible to construct an action invariant under the more general transformations with $B \neq 0$, at the price of introducing magnetic duals of the vector fields W_μ^I.

---

[3] We use here and below the tilde to indicate the transformed fields. We hope that the reader does not confuse these with the duals, (A.8), which are contained in the (anti)self-dual $F^\pm$.
[4] Quantization of electric and magnetic charges breaks $Sp(2(n+1), \mathbb{R})$ to $Sp(2(n+1), \mathbb{Z})$. We do not discuss quantum effects in this section.

5.2.2 Symplectic Transformations in $\mathcal{N} = 2$

Let us apply the arguments of the previous section to $\mathcal{N} = 2$ supergravities. Our starting point is the tensor $\mathcal{N}$, which is determined as a function of F, as explained in Sect. 5.1. From its definition (5.2) we can immediately derive the following suggestive relations:

$$\mathcal{N}_{IJ} X^J = (\bar{F}_{IJ} + \mathrm{i} N_{IJ}) X^J = F_{IJ} X^J = F_I \,,$$

$$\mathcal{N}_{IJ} \nabla_{\bar{\alpha}} \bar{X}^J = \bar{F}_{IJ} \nabla_{\bar{\alpha}} \bar{X}^J = \nabla_{\bar{\alpha}} \bar{F}_I \,, \qquad (5.11)$$

where in the last line we used also (the complex conjugate of) (4.126) and we introduced the Kähler covariant derivatives of Sect. 4.5.7)

$$\nabla_{\bar{\alpha}} \bar{F}_I = \partial_{\bar{\alpha}} \bar{F}_I + \tfrac{1}{2} a^{-1} (\partial_{\bar{\alpha}} \mathcal{K}) \bar{F}_I \,, \qquad \nabla_{\bar{\alpha}} \bar{X}^J = \partial_{\bar{\alpha}} \bar{X}^J + \tfrac{1}{2} a^{-1} (\partial_{\bar{\alpha}} \mathcal{K}) \bar{X}^J \,. \qquad (5.12)$$

The identities (5.11) are of a similar form as (5.5) relating the lower components of a symplectic vector to its upper components by multiplication with $\mathcal{N}_{IJ}$. Hence we can identify two $2(n + 1)$-component symplectic vectors (and their complex conjugates)

$$V = \begin{pmatrix} X^I \\ F_I \end{pmatrix}, \qquad U_\alpha = \begin{pmatrix} \nabla_\alpha X^I \\ \nabla_\alpha F_I \end{pmatrix}, \qquad (5.13)$$

and check that they indeed transform as in (5.6). With this identification in mind, we can reconsider the kinetic terms of the scalars and note that the Kähler potential (4.93), and the constraint (4.82) are symplectic invariants. This is the starting point of the symplectic formulation of the special geometry, which we will discuss in Sect. 5.3. There we will see that (5.11) can be seen as an alternative definition of the matrix $\mathcal{N}$.

An immediate consequence of this construction is that a symplectic transformation on V induces a change of coordinates $X^I \mapsto \tilde{X}^I$ such that

$$\tilde{X}^I = A^I{}_J X^J + B^{IJ} F_J(X) \,, \qquad \tilde{F}_I = C_{IJ} X^J + D_I{}^J F_J(X) \,. \qquad (5.14)$$

If the first equation is invertible,[5] i.e. defines a relation $X^I(\tilde{X})$, then

$$\frac{\partial}{\partial \tilde{X}^J} \tilde{F}_I = \frac{\partial X^K}{\partial \tilde{X}^J} \frac{\partial \tilde{F}_I}{\partial X^K} \,. \qquad (5.15)$$

---

[5]The full symplectic matrix is always invertible due to (5.7), but this part may not be. In rigid supersymmetry, the invertibility of this transformation is necessary for the invertibility of $\mathcal{N}$ (due to the positive definiteness of the full metric), but in supergravity we may have that the $\tilde{X}^I$ do not form an independent set, and then $\tilde{F}$ cannot be defined. See below.

The inverse of the first factor and the last factor can be obtained from (5.14) and then (5.8) implies that (5.15) is symmetric in (IJ). This is the integrability condition for the (local) existence of a new *prepotential* $\tilde{F}(\tilde{X})$, such that

$$\tilde{F}_I(\tilde{X}) = \frac{\partial \tilde{F}(\tilde{X})}{\partial \tilde{X}^I} . \tag{5.16}$$

Invertibility of (5.14) thus guarantees the existence of a new formulation of the theory, and thus of the target space manifold, in terms of the prepotential $\tilde{F}(\tilde{X})$.

The attentive reader may have noticed that the prepotential has entered automatically our superconformal construction in Sect. 3.3.1. This is not surprising due to the homogeneity conditions (3.108), also valid for $\tilde{F}$, which allow one to determine $\tilde{F}$ from (5.14) as follows:

$$\tilde{F}(\tilde{X}(X)) = \frac{1}{2} V^T \begin{pmatrix} C^T A & C^T B \\ D^T A & D^T B \end{pmatrix} V . \tag{5.17}$$

However, as we will show in the following, there are also symplectic transformations such that the prepotential is not defined in the new symplectic basis, i.e. that the new $\tilde{F}_I$ cannot be obtained as a derivative of a function $\tilde{F}(\tilde{X})$. A concrete example exhibiting this feature will be given in Sect. 5.3. Therefore, one may expect new formulations without a prepotential to be generically available. We will show in the following that this is indeed the case.

Let us first consider an example [6], starting from the prepotential (4.98). If we apply the symplectic transformation

$$S = \begin{pmatrix} A & B \\ C & D \end{pmatrix} = \begin{pmatrix} 1 & 0 & 0 & 0 \\ 0 & 0 & 0 & 1/3 \\ 0 & 0 & 1 & 0 \\ 0 & -3 & 0 & 0 \end{pmatrix} , \tag{5.18}$$

using (5.17) one arrives at $\tilde{F}(\tilde{X}(X)) = -2(X^1)^3 X^0$, which is just (4.99) after the trivial field redefinition $X \to \tilde{X}$. Thus the two theories based on (4.98) and (4.99) are equivalent. We will call this a *pseudo symmetry* between the two formulations, and the transformation is called *symplectic reparameterization* .

On the other hand consider

$$S = \begin{pmatrix} 1+3\epsilon & \mu & 0 & 0 \\ \lambda & 1+\epsilon & 0 & 2\mu/9 \\ 0 & 0 & 1-3\epsilon & -\lambda \\ 0 & -6\lambda & -\mu & 1-\epsilon \end{pmatrix} , \tag{5.19}$$

for infinitesimal ϵ, μ, λ. Since F is invariant under this transformation, the latter defines a symmetry of the model. On the scalar field $z = X^1/X^0$, the transformations act as

$$\delta z = \lambda - 2\epsilon z - \mu z^2/3 . \tag{5.20}$$

All in all, the set of transformations (5.19) form an SU(1, 1) isometry group of the scalar manifold. The domain were the metric is positive definite is Im $z > 0$, hence proving the identification of the manifold as the coset space in (4.98), or equivalently (4.99).

In general, we have to distinguish two situations: [7]

1. The function $\tilde{F}(\tilde{X})$ is different from $F(\tilde{X})$. In that case $\tilde{F}$ provides a dual description of the classical field theory with prepotential F. The two functions describe equivalent physics and we have a *pseudo symmetry*. The transformations are called symplectic reparameterizations [8]. Hence we may find a variety of descriptions of the same theory in terms of different functions F.
2. If a symplectic transformation leads to the same function F, then we are dealing with a *proper symmetry*. This invariance reflects itself in an isometry of the target space manifold.

Note that a symplectic transformation with

$$S = \begin{pmatrix} \mathbb{1} & 0 \\ C & \mathbb{1} \end{pmatrix} , \tag{5.21}$$

does not change the X^I and gives $\tilde{F} = F + \frac{1}{2}C_{IJ}X^I X^J$. The matrix C must be symmetric as required for S to be symplectic. Indeed, the difference between $\tilde{F}$ and F is then of the form (3.111), and the action is invariant. Hence these are proper symmetries.

The two cases above are called 'duality symmetries', as they are generically accompanied by duality transformations on the field equations and the Bianchi identities. The question remains whether the duality symmetries comprise all the isometries of the target space, i.e. whether

$$Iso(\text{scalar manifold}) \subset Sp(2(n+1), \mathbb{R}) . \tag{5.22}$$

This question was investigated in [9] for the very special Kähler manifolds, and it was found that in that case one does obtain the complete set of isometries from the symplectic transformations. For generic special Kähler manifolds no isometries have been found that are not induced by symplectic transformations, but on the other hand we do not know a proof that these do not exist.

That the full supersymmetric theory allows such symplectic transformations can also be seen in another way. We mentioned before that the vector multiplets are chiral multiplets, with $w = 1$ that satisfy constraints (3.28). One of these constraints is the Bianchi identity for the vector. The functions $F_I(X)$ transform also in a chiral way under supersymmetry, and thus define also chiral multiplets, with Weyl weight $w = 1$. Now it turns out that the same constraints on these chiral multiplets are in fact the field equations. One of these is the field equation of the vectors. Thus the symplectic vectors V (5.13) are the lowest components of a symplectic vector of chiral multiplets. If the vector of chiral multiplets satisfies the mentioned constraints, then these imply as well that the multiplets are vector multiplets rather than general chiral multiplets, and that the fields of these vector multiplets satisfy the field equations. This is thus a supersymmetric generalization of the symplectic set-up of Sect. 5.2.1. For the supergravity case this has been worked out in [10]. In this way, even the equations for models in a parameterization without prepotential can be obtained. Such situations will be explained in the next section.

5.3   Characteristics of a Special Geometry

In Chap. 5 we defined the Special Kähler geometry [11] as a property of the scalar couplings in $\mathcal{N} = 2$ supergravities. However, one may wonder what is the definition of special geometry, independent of supersymmetry. A first step in that direction has been taken by Strominger [12]. He had in mind the moduli spaces of Calabi–Yau spaces. His definition is already based on the symplectic structure, which we also have emphasized. However, being already in the context of Calabi–Yau moduli spaces, his definition of special Kähler geometry omitted some ingredients that are automatically present in any Calabi–Yau moduli space, but have to be included as necessary ingredients in a generic definition. Another important step was made in [13]. Before, special geometry was necessarily connected to the existence of a holomorphic prepotential $F(X)$ or $\mathcal{F}(Z(z))$: it was recognized as those Kähler manifolds where the Kähler potential is of the form (4.93) with (4.92).

However, the authors of [13] found examples of $\mathcal{N} = 2$ supergravities coupled to Maxwell multiplets where the $\mathcal{F}_I$ in (4.93) is not a derivative of a prepotential as in (4.92). Crucially the latter were obtained from a symplectic transformation of a model admitting a prepotential. This surprising fact raised some natural questions: are all the models without prepotential symplectic dual to models with a prepotential? Can one still define special Kähler geometry as always symplectic dual to a formulation with a prepotential? And, of course, is there a more convenient definition that does not involve this prepotential? These questions have been answered in [14], and are reviewed here. A lot of this is based on formalism that has been developed using other methods in [15–18].

5.3.1   Symplectic Formulation of the Projective Kähler Geometry

The scope of this section is to reformulate the special Kähler geometry of Sect. 5.1 using the symplectic formalism. The basic building block is the symplectic vector V, which transforms as in (5.6),

$$V = \begin{pmatrix} X^I \\ F_I \end{pmatrix}. \tag{5.23}$$

It is important to realize that F_I is not necessarily a function of X^I, as is the case in the formulation with the prepotential. The dilatational gauge fixing condition (4.82) is chosen in order to decouple kinetic terms of the graviton from those of the scalars. It can be written as a condition on the symplectic inner product (5.10):

$$\langle V, \bar{V} \rangle = X^I \bar{F}_I - F_I \bar{X}^I = ia = i\kappa^{-2}. \tag{5.24}$$

To solve this condition, we define

$$V = y(z, \bar{z}) v(z), \qquad y(z, \bar{z}) = e^{\kappa^2 \mathcal{K}(z, \bar{z})/2}, \tag{5.25}$$

where $v(z)$ is a holomorphic symplectic vector

$$v(z) = \begin{pmatrix} Z^I(z) \\ \mathcal{F}_I(z) \end{pmatrix}, \tag{5.26}$$

and $\mathcal{K}(z, \bar{z})$ will be the Kähler potential. $Z^I(z)$ and $\mathcal{F}_I(z)$ are so far $2(n+1)$ arbitrary holomorphic functions in the n complex variables z^α, reflecting the freedom of choice of the coordinates z^α, up to a non-degeneracy condition that we will soon discover. In terms of (5.25) the condition (5.24) implies a value for the Kähler potential in terms of a symplectic invariant:

$$y^{-2} = e^{-\kappa^2 \mathcal{K}(z, \bar{z})} = -i\kappa^2 \langle v, \bar{v} \rangle. \tag{5.27}$$

We now act on (5.24) and (5.27) with the following Kähler covariant derivatives (see Sect. 4.5.7):

$$\nabla_\alpha v(z) \equiv \partial_\alpha v(z) + \kappa^2 v(z)(\partial_\alpha \mathcal{K}), \qquad \nabla_\alpha \bar{v} = 0,$$
$$\nabla_\alpha V \equiv \partial_\alpha V + \tfrac{1}{2}\kappa^2 V(\partial_\alpha \mathcal{K}), \qquad \nabla_\alpha \bar{V} = 0. \tag{5.28}$$

This implies

$$\langle \nabla_\alpha v, \bar{v} \rangle = 0 \,, \qquad \langle \nabla_\alpha V, \bar{V} \rangle = 0 \,. \tag{5.29}$$

A further derivation implies, using (4.123)

$$\langle \nabla_\alpha V, \nabla_{\bar{\beta}} \bar{V} \rangle = -\langle \nabla_{\bar{\beta}} \nabla_\alpha V, \bar{V} \rangle = -\mathrm{i} g_{\alpha\bar{\beta}} \,. \tag{5.30}$$

Hence the Kähler metric can be obtained as

$$g_{\alpha\bar{\beta}} = \mathrm{i} \langle \nabla_\alpha V, \nabla_{\bar{\beta}} \bar{V} \rangle \,. \tag{5.31}$$

There is an important extra condition on the inner product

$$\langle \nabla_\alpha V, V \rangle = 0 \,. \tag{5.32}$$

In most examples[6] the matrix (5.41) is invertible and we will prove at the end of Sect. 5.3.2 that then this condition implies the existence of a prepotential. Here we can already see that the condition is always satisfied for any $F_I = \partial_I F$. Indeed, in that case, due to the homogeneity of F_I

$$\langle \nabla_\alpha V, V \rangle = (\nabla_\alpha X^I) F_I - (\nabla_\alpha F_I) X^I = (\nabla_\alpha X^I) F_I - F_{IJ} (\nabla_\alpha X^J) X^I = 0 \,. \tag{5.33}$$

However, in full generality and without requiring a prepotential, we will impose (5.32) in a definition of special Kähler geometry, see Sect. 5.3.2. The conditions (5.24), (5.31) and (5.32) can be combined in the $(n+1) \times (n+1)$ matrix relation

$$G \equiv \mathrm{i} \left\langle \begin{pmatrix} \bar{V} \\ \nabla_\alpha V \end{pmatrix}, \begin{pmatrix} V & \nabla_{\bar{\beta}} \bar{V} \end{pmatrix} \right\rangle = \begin{pmatrix} \kappa^{-2} & 0 \\ 0 & g_{\alpha\bar{\beta}} \end{pmatrix} \,. \tag{5.34}$$

If we further require positivity of G, namely of the kinetic matrices $g_{\alpha\bar{\beta}}$ and $\kappa^{-2} > 0$,[7] this in turn implies that the $(n+1) \times (n+1)$ matrix

$$\left(\bar{X}^I \ \nabla_\alpha X^I \right) = y \left(\bar{Z}^I \ \nabla_\alpha Z^I \right) \,, \tag{5.35}$$

---

[6]It has been proven [14] that there always exist a symplectic frame where this is the case.

[7]These are thus the positivity requirements of the kinetic terms of the spin 0 and spin 2 fields. Recall from Sect. 4.5.4 that this in turn implies positivity for the kinetic terms of the spin 1 fields.

which is the upper part of $\left(\bar{V} \ \nabla_\alpha V\right)$, is invertible. For clarity: the matrix (5.35) is in full

$$
\begin{pmatrix}
\bar{X}^0 & \nabla_1 X^0 & \dots & \nabla_n X^0 \\
\bar{X}^1 & \nabla_1 X^1 & \dots & \nabla_n X^1 \\
& & \dots & \\
\bar{X}^n & \nabla_1 X^n & \dots & \nabla_n X^n
\end{pmatrix} . \tag{5.36}
$$

Such theorems on positivity and invertibility of symplectic matrices are collected in [14, Appendix B]. To prove the above statement, suppose that (5.35) is not invertible. Then there is a zero mode $a\bar{X}^I + b_\alpha \nabla_\alpha X^I = 0$. Also its complex conjugate is then zero, which leads to

$$
0 = \left\langle a\bar{V} + b_\alpha \nabla_\alpha V, \ \bar{a}V + \bar{b}_\alpha \nabla_\alpha \bar{V}\right\rangle = \begin{pmatrix} a & b^\alpha \end{pmatrix} \left\langle \begin{pmatrix} \bar{V} \\ \nabla_\alpha V \end{pmatrix}, \begin{pmatrix} V & \nabla_{\bar{\beta}} \bar{V} \end{pmatrix} \right\rangle \begin{pmatrix} \bar{a} \\ \bar{b}^\alpha \end{pmatrix}
$$

$$
= \begin{pmatrix} a & b^\alpha \end{pmatrix} (-iG) \begin{pmatrix} \bar{a} \\ \bar{b}^\alpha \end{pmatrix} . \tag{5.37}
$$

This is contradictory to the positivity of G. Invertibility of (5.35) is the non-degeneracy condition on the choice of (5.26) that we mentioned below that equation.

In terms of these quantities, and using the invertibility of (5.35), the kinetic matrix for the vectors is defined by

$$
\mathcal{N}_{IJ} = \left(F_I \ \nabla_{\bar{\alpha}} \bar{F}_I\right) \left(X^J \ \nabla_{\bar{\alpha}} \bar{X}^J\right)^{-1} , \tag{5.38}
$$

which is the product of two $(n+1) \times (n+1)$ matrices. This agrees with the definition of $\mathcal{N}_{IJ}$ starting from a prepotential, since the latter implied (5.11).

5.3.1.1 Models Without a Prepotential and Examples

In the previous sections, all the requirements listed above were satisfied with F_I defined via a prepotential, i.e. $F_I = \frac{\partial}{\partial X^I} F(X)$. The conclusion of the current analysis is that we can get rid completely of this last requirement. We can now appreciate how a symplectic transformation may relate symplectic vectors of a special Kähler manifold determined by a prepotential to new symplectic vectors where $\tilde{F}_I$ cannot be written as the derivative of a function $\tilde{F}(\tilde{X})$. A famous example [13] comes from the reduction to $\mathcal{N} = 2$ of two versions of $\mathcal{N} = 4$ supergravity, known, respectively, as the 'SO(4) formulation' [19–21] and the 'SU(4) formulation' of pure $\mathcal{N} = 4$ supergravity [22].[8] In the initial duality frame

---

[8]This was revisited in exercises 20.18 and 20.19 in [5].

the prepotential is

$$F = -i\kappa^{-2} X^0 X^1 . \tag{5.39}$$

After a symplectic transformation, one finds a new model based on the symplectic vector

$$v = \begin{pmatrix} 1 \\ i \\ -\kappa^{-2} z \\ -i\kappa^{-2} z \end{pmatrix} . \tag{5.40}$$

After this mapping, z does not appear anymore in $(\widetilde{Z}^0, \widetilde{Z}^1)$, the upper two components of the symplectic vector. It is then clear that from (5.26) we cannot define the function $\widetilde{F}(\widetilde{Z}^0, \widetilde{Z}^1)$, which would play the role of a prepotential. Another surprising phenomenon showing up in this model,[9] but being in fact very general, is that the set of perturbative symmetries (those with $B = 0$) is different before or after the transformation. In particular in the formulation with a prepotential there is only one 'perturbative' symmetry, while in the formulation without prepotential there are three.

The existence of a prepotential is intimately related to the invertibility of the matrix

$$\left(X^I \ \nabla_\alpha X^I \right) . \tag{5.41}$$

One can easily check that, as opposed to (5.35), the above is not always invertible—for example in the formulation (5.40). If that matrix is invertible, then a prepotential exists [14]. This will be proven below, related to (5.49). It turns out that the following 3 conditions are equivalent:

1. The matrix (5.41) is invertible;
2. Special coordinates are possible; these are coordinates defined in (4.79);
3. A prepotential $F(X)$ exists.

Another important example of the absence of a prepotential occurs when describing the manifold

$$\frac{\mathrm{SU}(1, 1)}{\mathrm{U}(1)} \otimes \frac{\mathrm{SO}(r, 2)}{\mathrm{SO}(r) \otimes \mathrm{SO}(2)} . \tag{5.42}$$

This is the only special Kähler manifold that is a product of two factors [23]. In physics, these manifold emerge in the classical limit of the compactified heterotic

---

[9]This is explicitly worked out for the above example in exercise 20.20 in [5].

string, where the dilaton does not mix with the scalars of the other vector multiplets and thus the target space should be a product of two factors. The first formulation of these spaces used a prepotential F of the form (4.101) [24]. As mentioned in Sect. 4.5.5 this is therefore an example of a 'very special Kähler manifold'.[10] In this case

$$d_{ABC} X^A X^B X^C = X^1 X^a \eta_{ab} X^b , \qquad a, b = 2, \ldots, r+1 , \qquad (5.43)$$

with η_{ab} a constant metric of signature $(+, -, \ldots, -)$.

In this description of (5.42) by a prepotential, only part of $SO(r, 2)$ belongs to the set of perturbative symmetries. In other words, one needs $B \neq 0$ in the duality group to generate the full $SO(2, r)$. This is somehow unexpected from string theory point of view, as from the superstring compactification one expects the full $SO(2, r; \mathbb{Z})$ to be explicitly realized as a perturbative (T-duality) symmetry group. The key point is that there exists another symplectic frame where the full $SO(r, 2)$ is realized as a perturbative symmetry [13]. After that symplectic transformation one finds a symplectic vector (X^I, F_I) satisfying

$$X^I \eta_{IJ} X^J = 0 ; \qquad F_I = S \eta_{IJ} X^J , \qquad (5.44)$$

where η_{IJ} is a metric for $SO(2, r)$ and the dilaton S is one of the variables $\{z^\alpha\}$. The first constraint comes on top of (5.24), and thus implies that the variables $\{z^\alpha\}$ cannot be chosen between the X^I only. Indeed, S occurs only in F_I. Therefore, in this new formulation, which realizes the full $SO(r, 2)$ as a symmetry of the action, special coordinates are not possible and we do not have a prepotential.

5.3.2 Definitions

After this extension of the formulation, the reader may wonder what is then really special Kähler geometry. This question has been addressed in [14] and leads to a few equivalent formulations of a definition. We will first give a definition using the prepotential, and then a second one using only the symplectic vectors. We will then show the equivalence. There is also a more mathematically inspired definition [25], which was summarized in [5, Appendix 20C] and we will not repeat here.

Definition 1 of (Local) Special Kähler Geometry
A special Kähler manifold is an n-dimensional Kähler–Hodge manifold with on any chart $n + 1$ holomorphic functions $Z^I(z)$ and a holomorphic function $F(Z)$, homogeneous of second degree, such that, with (5.26), the Kähler potential is given by (5.27) and the $v(z)$ are connected by symplectic transformations $Sp(2(n+1), \mathbb{R})$

---

[10]This example will reappear in Sect. 5.7.2: see the third line of Table 5.4.

and/or Kähler transformations.

$$v(z) \to e^{f(z)} S v(z),\tag{5.45}$$

on overlap of charts. The transition functions should satisfy the cocycle condition on overlap of regions of three charts.

We mentioned that there are formulations without a prepotential, e.g., the one based on the symplectic vector (5.40). In this case, the above definition turns out not to be applicable in an arbitrary symplectic frame. Therefore we will now give a second definition, and will comment about their equivalence.

Definition 2 of (Local) Special Kähler Geometry
A special Kähler manifold is an n-dimensional Kähler–Hodge manifold, that is the base manifold of a $\mathrm{Sp}(2(n+1)) \times \mathrm{U}(1)$ bundle. There should exist a holomorphic section $v(z)$ such that the Kähler potential can be written as (5.27) and it should satisfy the condition

$$\langle \nabla_\alpha v, \nabla_\beta v \rangle = 0.\tag{5.46}$$

Note that the latter condition guarantees the symmetry of $\mathcal{N}_{IJ}$. This condition did not appear in [12], where the author had in mind Calabi–Yau manifolds. As we will see below, in those applications, this condition is automatically fulfilled. For $n > 1$ the condition can be replaced by the equivalent condition

$$\langle \nabla_\alpha v, v \rangle = 0.\tag{5.47}$$

For $n = 1$, the condition (5.46) is empty, while (5.47) is not. In [10] it has been shown that models with $n = 1$ not satisfying (5.47) can be formulated.

Equivalence of the Two Definitions
It is thus legitimate to ask about the equivalence of the two definitions. Indeed, we saw that in some cases definition 2 is satisfied, but one cannot obtain a prepotential F. However, that example, as others in [13], was obtained by performing a symplectic transformation from a formulation where the prepotential does exist. In [14] it was shown that this is true in general. If definition 2 is applicable, then there exists a symplectic transformation to a basis such that $F(Z)$ exists. However, in the way physical problems are handled, the existence of formulations without prepotentials can play an important role. For example, these formulations were used to prove that one can break $\mathcal{N} = 2$ supersymmetry partially to $\mathcal{N} = 1$ [26] and not necessarily to $\mathcal{N} = 0$, as it was thought before. This is an extremely important property for phenomenological applications. Note that by introducing magnetic vectors the partial breaking can be obtained in the context of the theory with a prepotential, since this is just a dual formulation of the same theory. This is understood well using the embedding tensor formalism [27–31], as clarified in [32].

Recent investigations of the possibilities of partial breaking have been performed in [33, 34].[11]

On page 170, we gave equivalent conditions on the symplectic basis for the existence of a prepotential. The first and second conditions are clearly equivalent. Indeed, the invertibility of (5.41) is equivalent with the invertibility of $(X^I \, \partial_\alpha X^I)$, i.e. with ordinary derivatives, since the difference is proportional to the first column. Then one can define the coordinates as in (4.79), which clearly satisfy the invertibility condition. In these coordinates it can be shown that a prepotential exists. One first notices that (X^I / X^0) is independent of $\bar{z}$. Then also F_I / X^0 is by the homogeneity only function of z and one can define holomorphic functions $F_I(X)$, by replacing those z^α by their value in (4.79):

$$F_I(X) \equiv X^0 \frac{F_I}{X^0} \left(z \left(\frac{X^\alpha}{X^0} \right) \right). \tag{5.48}$$

The constraints (5.46) then imply

$$\begin{pmatrix} X^I \\ \partial_\alpha X^I \end{pmatrix} \partial_{[I} F_{J]} \left(X^J \, \partial_\alpha X^J \right) = 0, \tag{5.49}$$

and since the first and last factor is invertible, it follows that the middle factor should vanish, which is the integrability condition that in any patch $F_J = \frac{\partial}{\partial X^J} F(X)$ for some $F(X)$.

5.3.3 Symplectic Equations and the Curvature Tensor

Let us first summarize the symplectic inner products that we found. They can be simply written in terms of the $2(n + 1) \times 2(n + 1)$ matrix

$$\mathcal{V} = \left(\bar{V} \ U_\alpha \ V \ \kappa^{-2} \bar{U}^\alpha \right),$$

$$U_\alpha \equiv \nabla_\alpha V = \left[\partial_\alpha + \tfrac{1}{2} \kappa^2 (\partial_\alpha \mathcal{K}) \right] V,$$

$$\bar{U}_{\tilde{\alpha}} \equiv \nabla_{\tilde{\alpha}} \bar{V} = \left[\partial_{\tilde{\alpha}} + \tfrac{1}{2} \kappa^2 (\partial_{\tilde{\alpha}} \mathcal{K}) \right] \bar{V}, \qquad \bar{U}^\alpha = g^{\alpha \bar{\beta}} \bar{U}_{\bar{\beta}}, \tag{5.50}$$

as

$$\mathcal{V}^T \Omega \mathcal{V} = -\mathrm{i} \kappa^{-2} \Omega, \tag{5.51}$$

where Ω is the standard antisymmetric matrix in (5.7).

---

[11]Breaking by nonlinear terms has been investigated in [35].

Thus, the matrix $\mathcal{V}$ is also invertible. Covariant derivatives on these equations lead to new ones, like

$$\langle V, \nabla_\alpha U_\beta \rangle = \langle \bar{V}, \nabla_\alpha U_\beta \rangle = \langle \bar{U}_{\bar{\gamma}}, \nabla_\alpha U_\beta \rangle = 0,$$
$$\langle \nabla_\alpha U_\beta, U_\gamma \rangle + \langle U_\beta, \nabla_\alpha U_\gamma \rangle = 0. \tag{5.52}$$

Note that $\nabla_\alpha U_\beta$ contains also Levi-Civita connection.

$$\nabla_\alpha U_\beta = \left[\partial_\alpha + \tfrac{1}{2}\kappa^2 \left(\partial_\alpha \mathcal{K} \right) \right] \left[\partial_\beta + \tfrac{1}{2}\kappa^2 \left(\partial_\beta \mathcal{K} \right) \right] V - \Gamma_{\alpha\beta}^\gamma U_\gamma . \tag{5.53}$$

Due to these equations (5.52) and the invertibility of (5.50), $\nabla_\alpha U_\beta$ must be proportional to a $\bar{U}^\gamma$, defining a third rank tensor $C_{\alpha\beta\gamma}$:

$$\nabla_\alpha U_\beta = C_{\alpha\beta\gamma} \bar{U}^\gamma , \qquad C_{\alpha\beta\gamma} \equiv -\mathrm{i}\langle \nabla_\alpha U_\beta, U_\gamma \rangle. \tag{5.54}$$

Since (5.53) is already symmetric in $(\alpha\beta)$, (5.52) implies that $C_{\alpha\beta\gamma}$ is completely symmetric.

Exercise 5.2 Check that if there is a prepotential, then one can write

$$C_{\alpha\beta\gamma} = \mathrm{i}F_{IJK} \nabla_\alpha X^I \nabla_\beta X^J \nabla_\gamma X^K , \tag{5.55}$$

where the covariant derivatives may be replaced by ordinary derivatives due to (3.108).   □

The curvature of the projective manifold can be obtained from the commutator of covariant derivatives. The connections contain Levi-Civita and Kähler connections, see (5.53), so the commutator will lead to a sum of the curvature of the Kähler manifold as function of the Levi-Civita connection of $g_{\alpha\bar{\beta}}$ and the curvature of the Kähler connection. The latter is the original U(1) from the superconformal algebra (on Kähler-invariant quantities) when it is pulled back to the spacetime, see (4.116).

$$\left[\nabla_\alpha, \nabla_{\bar{\beta}} \right] V = -\mathrm{i}R_{\alpha\bar{\beta}}^{\mathrm{K}} V , \qquad R_{\alpha\bar{\beta}}^{\mathrm{K}} = \partial_\alpha \mathcal{A}_{\bar{\beta}} - \partial_{\bar{\beta}} \mathcal{A}_\alpha = -\mathrm{i}\kappa^2 g_{\alpha\bar{\beta}} . \tag{5.56}$$

Hence we have

$$\left[\nabla_\alpha, \nabla_{\bar{\beta}} \right] V = -\kappa^2 g_{\alpha\bar{\beta}} V , \qquad \rightarrow \qquad \nabla_{\bar{\beta}} \nabla_\alpha V = \nabla_{\bar{\beta}} U_\alpha = \kappa^2 g_{\alpha\bar{\beta}} V. \tag{5.57}$$

On U_γ we get

$$\left[\nabla_\alpha, \nabla_{\bar{\beta}} \right] U_\gamma = -\mathrm{i}R_{\alpha\bar{\beta}}^{\mathrm{K}} U_\gamma - R_{\alpha\bar{\beta}}{}^\delta{}_\gamma U_\delta . \tag{5.58}$$

To make the calculation of the left-hand side, one lemma that we still need is that $C_{\alpha\beta\gamma}$ is covariantly holomorphic. From the definition (5.54)

$$i\nabla_{\bar{\delta}}C_{\alpha\beta\gamma} = \langle\nabla_{\bar{\delta}}\nabla_{\alpha}U_{\beta}, U_{\gamma}\rangle + \langle\nabla_{\alpha}U_{\beta}, \nabla_{\bar{\delta}}U_{\gamma}\rangle = 0. \tag{5.59}$$

Both terms vanish separately. For the first term, use (5.58) and (5.57) and $\langle V, U_{\alpha}\rangle = 0$ and $\langle U_{\alpha}, U_{\beta}\rangle = 0$ (both part of (5.51)). The second term is by (5.57) and (5.52) immediately zero.

Now we can calculate the left-hand side of (5.58):

$$\nabla_{\alpha}\nabla_{\bar{\beta}}U_{\gamma} = \kappa^2 g_{\gamma\bar{\beta}}U_{\alpha},$$

$$\nabla_{\bar{\beta}}\nabla_{\alpha}U_{\gamma} = C_{\alpha\gamma\epsilon}g^{\epsilon\bar{\epsilon}}\bar{C}_{\bar{\epsilon}\bar{\beta}\bar{\delta}}g^{\bar{\delta}\delta}U_{\delta}. \tag{5.60}$$

Subtracting the two and insert the result in (5.58) gives

$$\left(\kappa^2 g_{\gamma\bar{\beta}}\delta^{\delta}_{\alpha} - C_{\alpha\gamma\epsilon}g^{\epsilon\bar{\epsilon}}\bar{C}_{\bar{\epsilon}\bar{\beta}\bar{\delta}}g^{\bar{\delta}\delta}\right)U_{\delta} = \left(-\kappa^2 g_{\alpha\bar{\beta}}\delta^{\delta}_{\gamma} - R_{\alpha\bar{\beta}}{}^{\delta}{}_{\gamma}\right)U_{\delta}. \tag{5.61}$$

We can drop the U_{δ}, e.g., by taking a symplectic product of this relation with $\bar{U}^{\phi}$. This establishes the form of the curvature tensor [24]

$$R_{\alpha\bar{\beta}\gamma\bar{\delta}} = \kappa^2\left(g_{\alpha\bar{\beta}}g_{\gamma\bar{\delta}} + g_{\alpha\bar{\delta}}g_{\gamma\bar{\beta}}\right) - C_{\alpha\gamma\epsilon}g^{\epsilon\bar{\epsilon}}\bar{C}_{\bar{\epsilon}\bar{\beta}\bar{\delta}}. \tag{5.62}$$

Hence, this has a contribution from the gauged Kähler symmetry and one from the Kähler curvature.

Having all this machinery, we want to derive a few more relations that are often used in special Kähler geometry. Define the $2(n+1) \times (n+1)$ matrix $\mathcal{U}$ as the left part of $\mathcal{V}$, (5.50),

$$\mathcal{U} = \left(\bar{V}\; U_{\alpha}\right) = \begin{pmatrix} \bar{X}^I & \nabla_{\alpha}X^I \\ \bar{F}_I & \nabla_{\alpha}F_I \end{pmatrix} = \begin{pmatrix} x^I \\ f_I \end{pmatrix}, \tag{5.63}$$

where the last equation defines $(n+1) \times (n+1)$ matrices. The matrix x^I (columns as in (5.63) and rows defined by I) is invertible. With these definitions, the matrix $\mathcal{N}$ in (5.38) is written as

$$\mathcal{N}_{IJ} = \bar{f}_I(\bar{x})^{-1}_J, \quad\text{or}\quad \mathcal{U} = \begin{pmatrix} 1 \\ \mathcal{N} \end{pmatrix}x. \tag{5.64}$$

The $(n+1) \times (n+1)$ matrix G, introduced in (5.34), can then also be written as

$$G = \begin{pmatrix} \kappa^{-2} & 0 \\ 0 & g_{\alpha\bar{\beta}} \end{pmatrix} = i\mathcal{U}^T\Omega\bar{\mathcal{U}} = -2\begin{pmatrix} \bar{X}^I \\ \nabla_{\alpha}X^I \end{pmatrix}\text{Im}\,\mathcal{N}_{IJ}\left(X^J\; \nabla_{\bar{\beta}}\bar{X}^J\right). \tag{5.65}$$

The above relation implies that $\operatorname{Im} \mathcal{N}_{IJ}$ is negative definite if the metric $g_{\alpha\bar\beta}$ is positive definite. The following consequences are often used[12]:

$$-\tfrac{1}{2}(\operatorname{Im}\mathcal{N})^{-1|IJ} = \nabla_\alpha X^I g^{\alpha\bar\beta}\nabla_{\bar\beta}\bar X^J + \kappa^2\bar X^I X^J ,$$

$$\mathcal{N}^{-1|IJ} = -\tfrac{1}{2}(\operatorname{Im}\mathcal{N})^{-1|IJ} - \kappa^2\left(X^I\bar X^J + \bar X^I X^J\right)$$

$$= \nabla_\alpha X^I g^{\alpha\bar\beta}\nabla_{\bar\beta}\bar X^J - \kappa^2 X^I\bar X^J . \tag{5.66}$$

Another way to write (5.65) is

$$\bar{\mathcal{U}}G^{-1}\mathcal{U}^T = \kappa^2 V\bar V^T + \bar U_{\bar\beta}g^{\bar\beta\alpha}U_\alpha^T = -\frac{1}{2}\begin{pmatrix}1\\\mathcal{N}\end{pmatrix}(\operatorname{Im}\mathcal{N})^{-1}\left(1\ \ \bar{\mathcal{N}}\right) = -\frac{1}{2}M + \frac{1}{2}i\Omega , \tag{5.67}$$

in terms of the real matrix M introduced in [16], that is often used for describing black hole solutions:

$$M \equiv \begin{pmatrix}\operatorname{Im}\mathcal{N}^{-1} & \operatorname{Im}\mathcal{N}^{-1}\operatorname{Re}\mathcal{N}\\\operatorname{Re}\mathcal{N}\operatorname{Im}\mathcal{N}^{-1} & \operatorname{Im}\mathcal{N}+\operatorname{Re}\mathcal{N}\operatorname{Im}\mathcal{N}^{-1}\operatorname{Re}\mathcal{N}\end{pmatrix} . \tag{5.68}$$

Exercise 5.3 Check from the definition (5.2) and the gauge fixing (4.56) the following useful relations:

$$\operatorname{Im}\mathcal{N}_{IJ}\bar X^J = -\frac{1}{2\kappa^2}\frac{\mathcal{N}_{IJ}X^J}{\mathcal{N}_{LM}X^L X^M} ,$$

$$X^I(\operatorname{Im}\mathcal{N}_{IJ})X^J = \frac{1}{2\kappa^4\bar X^I \mathcal{N}_{IJ}\bar X^J} ,$$

$$C_{\alpha\beta\gamma}g^{\gamma\bar\delta}\nabla_{\bar\delta}\bar X^I(\operatorname{Im}\mathcal{N}_{IJ})X^J = -\frac{i\kappa^{-2}}{2\bar X^I \mathcal{N}_{IJ}\bar X^J}F_{KLM}\nabla_\alpha X^K\nabla_\beta X^L\bar X^M . \tag{5.69}$$

Check also that (5.65) is consistent with the matrix equation

$$\begin{pmatrix}X^I\\\nabla_\alpha X^I\end{pmatrix}\mathcal{N}_{IJ}\left(\bar X^J\ \ \nabla_{\bar\beta}\bar X^J\right) = \begin{pmatrix}-\kappa^{-2} & 0\\0 & g_{\alpha\bar\beta}\end{pmatrix} . \tag{5.70}$$

The latter is similar to the one in $\mathcal{N} = 1$ supergravity, and illustrates that $\mathcal{N}_{IJ}$ has to be of indefinite signature.   □

---

[12]The first is the inverse of (5.65), while for the second, one proves first from the definition (5.2) that $\mathcal{N}^{-1|IJ} + \kappa^2(X^I\bar X^J + \bar X^I X^J)$ is the inverse of $-2\operatorname{Im}\mathcal{N}_{IJ}$.

Finally, we discuss the relation between the curvature in the projective mani-fold (5.62) and the geometry in the embedding manifold. The embedding manifold of the X^I in the formulation with a prepotential and metric (4.72) has affine connections (4.73) and therefore (see e.g. [5, (13.22)])

$$R_{I\bar{J}K\bar{L}} = G_{L\bar{L}}\bar{\partial}_{\bar{J}}\Gamma^L_{IK} = -F_{IKM}N^{-1|MN}\bar{F}_{\bar{N}\bar{J}\bar{L}}. \tag{5.71}$$

With (5.55) and the relation (5.66) for $N^{-1|MN}$ we can relate (5.62) to this curvature:

$$R_{\alpha\bar{\beta}\gamma\bar{\delta}} = \kappa^2\left(g_{\alpha\bar{\beta}}g_{\gamma\bar{\delta}} + g_{\alpha\bar{\delta}}g_{\gamma\bar{\beta}}\right) + \nabla_\alpha X^I \nabla_{\bar{\beta}}\bar{X}^J \nabla_\gamma X^K \nabla_{\bar{\delta}}\bar{X}^L R_{I\bar{J}K\bar{L}}. \tag{5.72}$$

This relation is also true in Kähler geometry in $\mathcal{N} = 1$ supergravity and can elegantly be shown when including Γ^I_{JK} connections (4.74) in a covariant derivative $\widehat{\nabla}$. First note that

$$\widehat{\nabla}_\alpha X^I \equiv (\partial_\alpha + \tfrac{1}{2}\kappa^{-2}(\partial_\alpha\mathcal{K}))X^I + \Gamma^I_{JK}X^K(\partial_\alpha X^J) = \nabla_\alpha X^I, \tag{5.73}$$

since the Γ term is vanishing due to (4.74). However, we then have

$$\widehat{\nabla}_\alpha\nabla_\beta X^I \equiv (\partial_\alpha + \tfrac{1}{2}\kappa^{-2}(\partial_\alpha\mathcal{K}))\nabla_\beta X^I - \Gamma^\gamma_{\alpha\beta}\nabla_\gamma X^I + \Gamma^I_{JK}\nabla_\beta X^K(\partial_\alpha X^J) = 0. \tag{5.74}$$

Indeed from (4.73) with (5.66) and the homogeneity relation $X^I F_{IJK} = 0$ we find

$$\Gamma^I_{JK}\nabla_\beta X^K \nabla_\alpha X^J = -\nabla_{\bar{\gamma}}X^I g^{\gamma\bar{\gamma}}C_{\alpha\beta\gamma} = -\mathrm{i}\nabla_\alpha\nabla_\beta X^I, \tag{5.75}$$

using (5.54). This vanishing of (5.74) can also be shown directly [36] from applying a $\widehat{\nabla}_\alpha$ derivative to the second line of (5.70) in the formulation where the last factor in that equation is invertible (existence of prepotential). Since $N_{IJ} = G_{I\bar{J}}$ it is inert under $\widehat{\nabla}_\alpha$.

The curvature relation (5.72) can then be obtained by acting with $\left[\widehat{\nabla}_\alpha, \widehat{\nabla}_{\bar{\beta}}\right]$ on $\nabla_\gamma X^I$ using the vanishing of (5.74). We summarize below the relevant equations for $\widehat{\nabla}$ on X^I:

$$\widehat{\nabla}_\alpha X^I = \nabla_\alpha X^I, \qquad \widehat{\nabla}_{\bar{\alpha}}X^I = \nabla_{\bar{\alpha}}X^I = 0,$$

$$\widehat{\nabla}_\alpha\nabla_\beta X^I = 0, \qquad \widehat{\nabla}_{\bar{\beta}}\nabla_\alpha X^I = \nabla_{\bar{\beta}}\nabla_\alpha X^I = \kappa^2 g_{\alpha\bar{\beta}}X^I. \tag{5.76}$$

However, we will only use $\widehat{\nabla}$ on objects where it is clear whether the I index is holomorphic or anti-holomorphic. The mixing of these two is in fact due to the constraints in chiral multiplets, in (3.23), which, applied to the gauge multiplet, connect Y^I_{ij} with its complex conjugate, and thus does not preserve a distinction of indices I for fields of the chiral multiplets and $\bar{I}$ for those of the antichiral multiplet. The same holds then also for the gauge field strengths.

5.4 Isometries and Symplectic Geometry

In considering the symmetries of special Kähler manifolds, there are a priori three different concepts, which we will relate:

- Isometries of the Kähler manifold,
- Symplectic transformations,
- Gauge group of the vector multiplets.

We will show that the group of transformations gauged by the vector multiplet belongs to the isometry group of the Kähler manifold and can be embedded in the group of symplectic transformations. On the other hand, not all isometries of the Kähler manifold are gauged. The gauged isometries can be selected from the general set of isometries by an embedding tensor formalism. We do not treat this here, and refer for the embedding tensor approach to [27–29] and to [30, 31] for reviews.

5.4.1 Isometries of a Kähler Metric

Isometries of a Kähler metric are defined as symmetries that preserve both the Hermitian metric and the covariantly constant complex structure. In the present section we will explain how isometries of the embedding manifold, with coordinates $\{X^I\}$, descend to isometries of the projective manifold. These steps apply to both $\mathcal{N} = 1$ and $\mathcal{N} = 2$ supergravity. For the convenience of the reader, we will first review the main results on isometries of a Kähler metric (more details can be found in [5, Sect. 13.4.1]).

We start with a generic Kähler metric $g_{\alpha\bar{\beta}} = \partial_\alpha \partial_{\bar{\beta}} \mathcal{K}(z, \bar{z})$ parameterized by scalars z^α. Let us consider general (not gauged) isometries labelled by an index Λ. Such isometries act on the scalars, defining a Killing vector $k_\Lambda{}^\alpha$ [13]:

$$\delta(\theta)z^\alpha = \theta^\Lambda k_\Lambda{}^\alpha(z) , \tag{5.77}$$

and $k_\Lambda{}^\alpha$ must be holomorphic in z in order to preserve the complex structure. Killing vectors are generated by a *moment map*, $\mathcal{P}_\Lambda(z, \bar{z})$, which is a real function defined

---

[13]The 'embedding tensor', mentioned above would be a tensor relating the index Λ to the index for gauge vectors, electric or magnetic. We will restrict here for gauging to identifying the index Λ, or part of its range, to I index of the gauge vectors W_μ^I.

as

$$k_\Lambda{}^\alpha(z) = -\mathrm{i}g^{\alpha\bar\beta}\partial_{\bar\beta}\mathcal{P}_\Lambda(z,\bar z)\,, \tag{5.78}$$

and satisfying

$$\nabla_\alpha\partial_\beta\mathcal{P}_\Lambda(z,\bar z) = 0\,. \tag{5.79}$$

The isometries (5.77) are not required to be symmetries of the Kähler potential. The latter can transform with a Kähler transformation depending on a holomorphic function $r_\Lambda(z)$:

$$\delta(\theta)\mathcal{K}(z,\bar z) = \theta^\Lambda[r_\Lambda(z) + \bar r_\Lambda(\bar z)]\,, \tag{5.80}$$

since the new potential then leads to the same Kähler metric. Due to this fact, one can find a general solution of (5.78) as

$$\mathcal{P}_\Lambda(z,\bar z) = \mathrm{i}[k_\Lambda{}^\alpha(z)\partial_\alpha\mathcal{K}(z,\bar z) - r_\Lambda(z)] = -\mathrm{i}[k_\Lambda{}^{\bar\alpha}(\bar z)\partial_{\bar\alpha}\mathcal{K}(z,\bar z) - \bar r_\Lambda(z)]\,. \tag{5.81}$$

Furthermore, when the isometries are non-abelian, the Killing vectors generate a Lie algebra

$$k_\Lambda{}^\beta\partial_\beta k_\Sigma{}^\alpha - k_\Sigma{}^\beta\partial_\beta k_\Lambda{}^\alpha = f_{\Lambda\Sigma}{}^\Gamma k_\Gamma{}^\alpha\,. \tag{5.82}$$

One can restrict the moment maps by a so-called 'equivariance relation' [18]

$$k_\Lambda{}^\alpha g_{\alpha\bar\beta}k_\Sigma{}^{\bar\beta} - k_\Sigma{}^\alpha g_{\alpha\bar\beta}k_\Lambda{}^{\bar\beta} = \mathrm{i}f_{\Lambda\Sigma}{}^\Gamma\mathcal{P}_\Gamma\,. \tag{5.83}$$

This relation is satisfied by the moment maps that appear in a supersymmetric Lagrangian.

The properties discussed above are general requirements that isometries for a general Kähler manifold must fulfill. In particular, they must apply for the isometries of the embedding manifold, which are further constrained by the conditions that the dilatational and R-symmetry U(1) structure have to be preserved. We can start with the general results written above using the following substitutions:

$$\begin{aligned} z^\alpha &\longrightarrow X^I\,, \\ \mathcal{K} &\longrightarrow N\,, \\ g_{\alpha\bar\beta} &\longrightarrow N_{IJ}\,, \\ r_\Lambda, \bar r_\Lambda &\longrightarrow 0\,. \end{aligned} \tag{5.84}$$

The first three rules follow directly from what was discussed in Sect. 4.5.2. The last rule in the previous equation follows from the fact that a transformation of the form (5.80) would now imply

$$\delta(\theta)N(X, \bar{X}) = \theta^\Lambda [r_\Lambda(X) + \bar{r}_\Lambda(\bar{X})], \tag{5.85}$$

which would be inconsistent with the (Weyl,chiral) weight (2,0) of the left-hand side. All together, the isometries of the embedding Kähler manifold act on the scalars X^I as follows:

$$\delta(\theta)X^I = \theta^\Lambda k_\Lambda{}^I(X). \tag{5.86}$$

The moment map (5.81) then becomes

$$\mathcal{P}_\Lambda(X, \bar{X}) = \mathrm{i}k_\Lambda{}^I N_{IJ} \bar{X}^J. \tag{5.87}$$

The goal is now to interpret these isometries of the embedding manifold as isometries in the projective manifold. We go to the projective manifold by splitting the X^I:

$$X^I = \{y, z^\alpha\}. \tag{5.88}$$

In these variables, only y carries the dilatational and U(1) weight. The requirement to preserve the dilatational and U(1) structure thus implies that y should transform linear in y and the transformation of z^α should not contain y. We write this as[14]

$$\delta(\theta)y = a^{-1} y\, r_\Lambda(z)\, \theta^\Lambda. \tag{5.89}$$

Note that we have introduced here a new holomorphic function $r_\Lambda(z)$. This is not to be confused with the function that is set to zero in (5.84). It will turn out that the latter $r_\Lambda(z)$ is the function that describes the non-invariance of the Kähler potential in the projective manifold. Imposing the commutation relations (5.82) for (5.89) leads to the condition

$$k_\Lambda{}^\alpha \partial_\alpha r_\Sigma - k_\Sigma{}^\alpha \partial_\alpha r_\Lambda = f_{\Lambda\Sigma}{}^\Gamma r_\Gamma. \tag{5.90}$$

The gauge fixing, i.e. restriction to the projective manifold, is performed by selecting the section (4.89). Since y transforms both under (4.78) and (5.89), there is a decomposition law for the invariance of the condition $y = \bar{y}$. This determines a dependence of λ_T on the parameters of the isometry θ^Λ. We indicate the remaining

---

[14]We keep here the notation $a = \kappa^{-2}$ since this part is also valid for $\mathcal{N} = 1$ with $a = 3\kappa^{-2}$.

transformation as $\delta^{\square}(\theta)$:

$$\delta^{\square}(\theta) = \delta(\theta) + \delta_T(\lambda_T(\theta)),$$

$$\lambda_T(\theta) = \frac{1}{2a}i\theta^{\Lambda}\left(r_{\Lambda}(z) - \bar{r}_{\Lambda}(\bar{z})\right) = \frac{1}{a}\theta^{\Lambda}\left(-\mathcal{P}_{\Lambda} + \frac{1}{2}i(k_{\Lambda}{}^{\alpha}\partial_{\alpha}\mathcal{K} - k_{\Lambda}{}^{\bar{\alpha}}\partial_{\bar{\alpha}}\mathcal{K})\right).$$

$$(5.91)$$

This implies for y

$$\delta^{\square}(\theta)y = \frac{1}{2a}y\,\theta^{\Lambda}\left(r_{\Lambda} + \bar{r}_{\Lambda}\right) = \frac{1}{2a}y\,\theta^{\Lambda}\left(k_{\Lambda}^{\alpha}\partial_{\alpha}\mathcal{K} + k_{\Lambda}^{\bar{\alpha}}\partial_{\bar{\alpha}}\mathcal{K}\right). \tag{5.92}$$

Note that, together with (4.89), (5.92) implies that $\mathcal{K}$ indeed transforms as in (5.80), by an amount precisely controlled by $r_{\Lambda}(z)$.[15] Equation (5.91) shows that, for any field that carries a non-trivial U(1) weight in the embedding manifold, a T transformation must be included in its effective symmetry transformation in the projective manifold, $\delta^{\square}(\theta)$. The new terms from λ_T have no effect on the scalars z^{α} since they are invariant under T transformations, while fermions do transform. In other words the scalars z^{α} transform as

$$\delta^{\square}(\theta)z^{\alpha} = \delta(\theta)z^{\alpha} = \theta^{\Lambda}k_{\Lambda}{}^{\alpha}(z), \tag{5.93}$$

and the Killing vectors in the projective manifold are related to those in the embedding manifold via

$$k_{\Lambda}{}^{I} = y\left(k_{\Lambda}{}^{\alpha}\nabla_{\alpha}Z^{I} + ia^{-1}\mathcal{P}_{\Lambda}Z^{I}\right), \tag{5.94}$$

using the Kähler covariant derivative (4.113). The moment map $\mathcal{P}_{\Lambda}$ is the same in the embedded and projected manifold. In the former it is written as (5.87). In the projected manifold an imaginary constant in $\mathcal{P}_{\Lambda}$ is undetermined from the transformations of the scalars. We can still define it as in (5.81), satisfying (5.78). The undetermined constant imaginary parts of r_{Λ} or real parts of $\mathcal{P}_{\Lambda}$ are the so-called Fayet–Iliopoulos terms. For a non-abelian theory these arbitrary constants are eliminated by the equivariance condition (5.83). The curious reader may have noticed already the similarity between (5.81), i.e.

$$k_{\Lambda}{}^{\alpha}\partial_{\alpha}\mathcal{K} = r_{\Lambda} - i\mathcal{P}_{\Lambda}, \qquad k_{\Lambda}{}^{\bar{\alpha}}\partial_{\bar{\alpha}}\mathcal{K} = \bar{r}_{\Lambda} + i\mathcal{P}_{\Lambda}, \tag{5.95}$$

and (4.157) for the projected quaternionic geometry, keeping in mind the identification of $\partial_{\alpha}\mathcal{K}$ as ω_{α} in (4.112). Note that the derivative of the first of (5.95) w.r.t. $\bar{z}^{\bar{\alpha}}$ brings us back to the definition of the moment map in (5.78).

---

[15]Note that this is consistent with (4.129).

5.4.2 Isometries in Symplectic Formulation

In this section we will show how the general isometries discussed in Sect. 5.4.1 can be embedded in the symplectic group. It will be useful to adopt the symplectic formulation (see Sect. 5.3.1) and work with covariant transformations (4.127) under the isometries:

$$\widehat{\delta}(\theta)v = \theta^\Lambda k_\Lambda{}^\alpha \nabla_\alpha v ,$$

$$\widehat{\delta}(\theta)V = \theta^\Lambda k_\Lambda{}^\alpha \nabla_\alpha V = y\widehat{\delta}(\theta)v . \tag{5.96}$$

Covariant transformations $\widehat{\delta}$ are related to the Poincaré transformations $\delta^\square$ (namely those induced by the dependence on z^α and $\bar{z}^{\bar{\alpha}}$) by (4.128)

$$\widehat{\delta}(\theta)v = \delta^\square(\theta)v + \kappa^2\theta^\Lambda(r_\Lambda - i\mathcal{P}_\Lambda)v ,$$

$$\widehat{\delta}(\theta)V = \delta^\square(\theta)V + \tfrac{1}{2}\kappa^2\theta^\Lambda(r_\Lambda - \bar{r}_\Lambda)V - i\kappa^2\theta^\Lambda\mathcal{P}_\Lambda V , \tag{5.97}$$

where we used the weights in the first and one but last row in (4.122) and (5.95). Furthermore, the Poincaré transformations are related to the transformations in the conformal setting δ by (5.91), and thus

$$\widehat{\delta}(\theta)v = \delta(\theta)v + \kappa^2\theta^\Lambda(r_\Lambda - i\mathcal{P}_\Lambda)v ,$$

$$\widehat{\delta}(\theta)V = \delta(\theta)V - i\kappa^2\theta^\Lambda\mathcal{P}_\Lambda V . \tag{5.98}$$

The correspondence between the two expressions above agrees with $V = yv$ and $\widehat{\delta}V = y\widehat{\delta}v$ using (5.89). We can re-express the second of (5.98) using (5.96) as

$$\delta(\theta)V = \theta^\Lambda\left(k_\Lambda{}^\alpha\nabla_\alpha V + i\kappa^2\mathcal{P}_\Lambda V\right) . \tag{5.99}$$

Therefore, upon contracting with $\bar{V}$ and using (5.29), the last equation implies a symplectic expression for the moment map associated to the isometries in the embedding manifold

$$\theta^\Lambda\mathcal{P}_\Lambda = -\langle\delta(\theta)V, \bar{V}\rangle . \tag{5.100}$$

From the expression above it is clear that $\delta(\theta)V$ must be part of the symplectic group. Indeed, being $\delta(\theta)V$ a transformation in the conformal setting, it must be holomorphic in X and consistent with dilatational symmetry. Moreover, since (5.100) is symplectic invariant, the only possibility is[16]

$$\delta(\theta)V = -\theta^\Lambda T_\Lambda V , \qquad T_\Lambda \in Sp(2(n+1), \mathbb{R}) , \tag{5.101}$$

---

[16]We use here the opposite sign convention for T_Λ as in [5, (20.147)].

where $\mathbb{1} + T_\Lambda$ is a symplectic matrix, such that the symplectic inner products discussed in Sect. 5.3 are preserved. The most general form of T_Λ, according to (5.8), is then

$$T_\Lambda = \begin{pmatrix} a_\Lambda{}^I{}_J & b_\Lambda{}^{IJ} \\ c_{\Lambda IJ} & -a_\Lambda{}^J{}_I \end{pmatrix}, \qquad b_\Lambda{}^{IJ} = b_\Lambda{}^{JI}, \qquad c_{\Lambda IJ} = c_{\Lambda JI}. \tag{5.102}$$

In particular for isometries we require the transformation (5.102) to be a symmetry of the action, which implies $b_\Lambda{}^{IJ} = 0$, according to the discussion at the end of 5.2.1. Isometries of the embedding manifold are therefore part of the symplectic group, as simply dictated by the following identity:

$$\mathcal{P}_\Lambda = \langle T_\Lambda V, \bar{V} \rangle = e^{\kappa^2 \mathcal{K}} \langle T_\Lambda v, \bar{v} \rangle. \tag{5.103}$$

Exercise 5.4 Here is an exercise leading to a formula that will be important in the context of the potential. First combine equations of this section to

$$k_\Lambda{}^\alpha \nabla_\alpha V = -T_\Lambda V - i\kappa^2 \theta^\Lambda \mathcal{P}_\Lambda V. \tag{5.104}$$

Then obtain

$$k_\Lambda{}^\alpha g_{\alpha\bar\beta} k_\Sigma{}^{\bar\beta} = i k_\Lambda{}^\alpha \langle \nabla_\alpha V, \nabla_{\bar\beta} \bar{V} \rangle k_\Sigma{}^{\bar\beta} = i\langle T_\Lambda V, T_\Sigma \bar{V} \rangle + \kappa^2 \mathcal{P}_\Lambda \mathcal{P}_\Sigma. \tag{5.105}$$

$\square$

5.4.3 Gauged Isometries as Symplectic Transformations

The symmetries gauged by the vector multiplet are of course a subgroup of the generic isometries Λ considered so far, and as such are expected to be represented by symplectic transformations. To specify such transformations we fix the index Λ to be in the range of I (which label the vector multiplets) as each symmetry is gauged by a vector. The moment map $\mathcal{P}_\Lambda$ is then identified with the P_I^0, introduced in (4.51).

The gauge transformations $\delta_G(\theta)$ of the scalars in the vector multiplet were written in (3.15)

$$\delta_G(\theta) X^I = -\theta^K f_{KJ}{}^I X^J, \tag{5.106}$$

where θ^K are the parameters. Notice that δ_G in (5.106) is understood to be a symmetry *before* the gauge fixing to the Poincaré theory. The transformations (5.106) fit in the general scheme of isometries defined by Killing vectors

$$k_J{}^I(X) = X^K f_{KJ}{}^I. \tag{5.107}$$

The associated moment map, from (5.87), is called the 'Kähler moment map' and it is identified with (4.51). The matrix T_Λ of (5.101) is thus of the form (5.102) and comparing with (5.106), we write

$$T_K V = \begin{pmatrix} f_{KJ}{}^I & 0 \\ 2C_{K,IJ} & -f_{KI}{}^J \end{pmatrix} \begin{pmatrix} X^J \\ F_J \end{pmatrix}, \tag{5.108}$$

where $C_{K,IJ}$ is symmetric in the last two indices according to (5.102). The transformation (5.101) with this matrix leads to (5.106) and

$$\delta_G(\theta)F_I = -2\theta^K C_{K,IJ} X^J + \theta^K f_{KI}{}^J F_J. \tag{5.109}$$

In particular

$$\delta_G(\theta)[X^I F_I] = -2\theta^K C_{K,IJ} X^I X^J. \tag{5.110}$$

In the tensor calculus, where F_I are functions of X^I and $X^I F_I = 2F(X)$, Eq. (5.110) tells us that the prepotential transforms in a real quadratic function, which does not contribute to the action. In fact, this corresponds to (3.112).[17] Furthermore, the matrices T_K should satisfy the algebra of gauge transformations (5.106):

$$[T_I, T_J] = f_{IJ}{}^K T_K, \tag{5.111}$$

which imposes again the condition (3.117).

In the more general case without a prepotential, we can proceed from comparing (5.99) and (5.101) replacing the θ^Λ by X^J, leading to

$$\delta_G(X)X^I = X^J k_J{}^\alpha \nabla_\alpha X^I + i\kappa^2 P_J^0 X^J X^I = 0,$$
$$\delta_G(X)F_I = X^J k_J{}^\alpha \nabla_\alpha F_I + i\kappa^2 P_J^0 X^J F_I = -2X^K C_{K,IJ} X^J + X^K f_{KI}{}^J F_J. \tag{5.112}$$

Multiplying the latter with X^I and using the first one gives

$$-2X^I X^J C_{J,IK} X^K = X^I X^J k_J^\alpha \nabla_\alpha F_I - X^J k_J^\alpha F_I \nabla_\alpha X^I = -X^J k_J^\alpha \langle V, \nabla_\alpha V \rangle = 0. \tag{5.113}$$

We thus re-obtain (3.115) in this more general setting. This is the equation that says that if all the X^I are independent, $C_{I,JK}$ does not have a completely symmetric part.

For the gauged symmetries, we already found other relations for the moment maps in (4.51)–(4.53). The relation between the embedding and projective gauge

---

[17] This is the reason why we put the factor 2 in (5.108).

transformations (5.94) is now written as

$$k_J{}^I = y(k_J{}^\alpha \nabla_\alpha Z^I + i\kappa^2 P_J^0 Z^I) = k_J{}^\alpha \nabla_\alpha X^I + i\kappa^2 P_J^0 X^I = X^K f_{KJ}{}^I , \quad (5.114)$$

where we repeat at the end also the expression (5.107) for easy reference below.

Exercise 5.5 Prove the following interesting relation from the definition (5.78) and (4.53):

$$X^I k_I{}^\alpha = -i g^{\alpha\bar\beta} X^I \partial_{\bar\beta} P_I^0 = i g^{\alpha\bar\beta} P_I^0 \nabla_{\bar\beta} X^I . \quad (5.115)$$

Further, with (5.114) rewrite the gauge terms in the covariant derivative (4.63) as

$$D_\mu X^I = \nabla_\mu X^I - W_\mu^K k_K{}^\alpha \nabla_\alpha X^I = \nabla_\alpha X^I \left(\partial_\mu z^\alpha - W_\mu^K k_K{}^\alpha \right) . \quad (5.116)$$

$\square$

5.5 Electric-Magnetic Charges: Attractor Phenomenon

In this section we put all the ingredients of special geometry together to study some universal property of charged black hole solutions in $N = 2$ supergravity.[18] It turns out that scalars in the background of charged black holes take fixed values at the horizon. These values are universally determined by (conserved) electric and magnetic charges, and independent of the initial configuration of the scalars (i.e. their value at infinite distance from the black hole). This phenomenon is called 'attractor mechanism' in special geometry [38, 39]. Even though we will restrict here to $D = 4$ for simplicity, the phenomenon has a $D = 5, 6$ counterpart in $N = 2$ theories[40–42].

This section is structured as follows. First, we obtain from the supergravity action (5.1) an effective action for the extremal black holes. In the usual way to proceed, at least in the case of static and spherically symmetric black holes, one eliminates the vectors through their equation of motion and then integrates out the angular dependence of the supergravity action to obtain an effective theory that governs the dynamics of the scalars. Crucially, the integration introduces a charge-dependent black hole potential along the way. For completeness, a Hamiltonian constraint has to be imposed on the system. Instead, we will employ an alternative derivation of such an effective action using only Einstein's equations, together with the equation of motion for the scalars and those on the vectors [43]. The Einstein equations that cannot be derived from this effective action, will become constraints that have to be imposed on the solution. Along the way we will show

---

[18] A recent review on black hole solutions in these theories can be found in [37].

that the black hole potential governing the effective action has a simple expression in the symplectic language. We will conclude this section by showing the universal solution corresponding to the attractor mechanism for the scalars.

5.5.1 The Spacetime Ansatz and an Effective Action

Consider a static spacetime metric[19]:

$$ds^2 = -e^{2U}dtdt + e^{-2U}\gamma_{mn}dx^m dx^n,$$

$$\text{i.e. } g_{00} = -e^{2U}, \quad g_{mn} = e^{-2U}\gamma_{mn}, \quad \sqrt{g} = e^{-2U}\sqrt{\gamma},$$

$$\partial_t U = \partial_t \gamma_{mn} = 0.$$

$$(5.117)$$

We are interested in the theory of a complex scalar z coupled to Maxwell fields and gravity.[20] The relevant action can be recovered from the more general (5.1) and takes the schematic form

$$S = S_{\text{Einstein}} + S^{(0)} + S^{(1)},$$

$$S_{\text{Einstein}} = \int d^4x \tfrac{1}{2}\sqrt{g}R(g),$$

$$S^{(0)} = -\int d^4x \sqrt{g}g^{\mu\nu}\partial_\mu z \partial_\nu \bar{z},$$

$$S^{(1)} = \int d^4x \left[\tfrac{1}{4}\sqrt{g}(\text{Im}\,\mathcal{N}_{IJ})F^I_{\mu\nu}F^{\mu\nu J} - \tfrac{1}{8}(\text{Re}\,\mathcal{N}_{IJ})\varepsilon^{\mu\nu\rho\sigma}F^I_{\mu\nu}F^J_{\rho\sigma} \right],$$

$$(5.118)$$

where $F^I_{\mu\nu} = 2\partial_{[\mu}W^I_{\nu]}$. Note that the metric appears in $S^{(1)}$ only in the term with Im $\mathcal{N}$.

As explained at the beginning of this section, we want to obtain an effective action starting from the Einstein equations. Let us consider the Einstein tensor:

$$G_{\mu\nu} = 2(\sqrt{g})^{-1}\frac{\delta S_{\text{Einstein}}}{\delta g^{\mu\nu}} = R_{\mu\nu} - \tfrac{1}{2}g_{\mu\nu}R.$$

$$(5.119)$$

---

[19]'Static' means that it admits a global, nowhere zero, timelike hypersurface orthogonal Killing vector field. A generalization are the 'stationary' spacetimes, which admit a global, nowhere zero timelike Killing vector field. In that case the components g_{0m} could be non-zero. For simplicity we look here to the static spacetimes.

[20]The generalization to non-trivial Kähler manifolds is obvious.

For the static metric (5.117), one obtains

$$G_{00} = e^{4U}\left(\tfrac{1}{2}R(\gamma) + 2D_m\partial^m U - \partial^m U \partial_m U\right),$$

$$G_{mn} = \left(\gamma_{mn}\gamma^{rs} - 2\delta^r_m\delta^s_n\right)\left(-\tfrac{1}{2}R_{rs}(\gamma) + \partial_r U \partial_s U\right). \qquad (5.120)$$

Here γ_{mn} is used to raise and lower indices, and to define the covariant derivative D_m. For the bosonic sector, the Einstein equations are

$$G_{\mu\nu} = T_{\mu\nu}, \qquad (5.121)$$

where $T_{\mu\nu}$ is the energy–momentum tensor. We split it in the scalar part and the spin 1 part:

$$T_{\mu\nu} = T^{(0)}_{\mu\nu} + T^{(1)}_{\mu\nu},$$

$$T^{(0)}_{\mu\nu} = -2(\sqrt{g})^{-1}\frac{\delta S^{(0)}}{\delta g^{\mu\nu}}, \qquad T^{(1)}_{\mu\nu} = -2(\sqrt{g})^{-1}\frac{\delta S^{(1)}}{\delta g^{\mu\nu}}. \qquad (5.122)$$

The energy–momentum tensor induced from $S^{(0)}$ is

$$T^{(0)}_{\mu\nu} = -g_{\mu\nu}g^{\rho\sigma}\partial_\rho z \partial_\sigma \bar{z} + 2\partial_\mu z \partial_\nu \bar{z}. \qquad (5.123)$$

For our metric ansatz (5.117), this gives

$$T^{(0)}_{00} = e^{4U}\gamma^{mn}\partial_m z \partial_n \bar{z}, \qquad T^{(0)}_{mn} = \left(-\gamma_{mn}\gamma^{rs} + 2\delta^r_{(m}\delta^s_{n)}\right)\partial_r z \partial_s \bar{z}. \qquad (5.124)$$

The energy–momentum tensor induced from $S^{(1)}$ is

$$T^{(1)}_{\mu\nu} = -\operatorname{Im}\mathcal{N}_{IJ}\left(-\tfrac{1}{4}g_{\mu\nu}F^I_{\rho\sigma}F^{J\rho\sigma} + F^I_{\mu\rho}F^J_{\nu\sigma}g^{\rho\sigma}\right). \qquad (5.125)$$

If we now use the spacetime metric (5.117), the non-zero terms are

$$T^{(1)}_{00} = -\operatorname{Im}\mathcal{N}_{IJ}\left(\tfrac{1}{2}e^{2U}F^I_{0m}\gamma^{mn}F^J_{0n} + \tfrac{1}{4}e^{6U}F^I_{mn}\gamma^{mp}\gamma^{nq}F^J_{pq}\right),$$

$$T^{(1)}_{mn} = -\operatorname{Im}\mathcal{N}_{IJ}\left(\tfrac{1}{2}e^{-2U}\gamma_{mn}F^I_{0p}\gamma^{pq}F^J_{0q} - \tfrac{1}{4}e^{2U}\gamma_{mn}F^I_{pq}\gamma^{pp'}\gamma^{qq'}F^J_{p'q'}\right.$$

$$\left. -e^{-2U}F^I_{0m}F^J_{0n} + e^{2U}F^I_{mp}\gamma^{pq}F^J_{nq}\right). \qquad (5.126)$$

We now introduce the magnetic vectors

$$F^I_m = \tfrac{1}{2}\gamma_{mn}(\sqrt{\gamma})^{-1}\varepsilon^{npq}F^I_{pq}, \qquad (5.127)$$

such that

$$F_{mn}^I = \sqrt{\gamma}\,\varepsilon_{mnp}\gamma^{pq}F_q^I,$$

$$F_{mn}^I\gamma^{mp}\gamma^{nq}F_{pq}^J = 2F_r^I\gamma^{rs}F_s^J, \qquad F_{mp}^I\gamma^{pq}F_{nq}^J = \gamma_{mn}F_r^I\gamma^{rs}F_s^J - F_n^I F_m^J.$$

$$(5.128)$$

Using these, we can again write the energy–momentum tensor in a similar form as for the gravity field and for the scalars. We find

$$T_{00}^{(1)} = e^{6U}\gamma^{mn}V_{mn}, \qquad T_{mn}^{(1)} = e^{2U}\left(\gamma_{mn}\gamma^{rs} - 2\delta_m^r\delta_n^s\right)V_{rs}, \qquad (5.129)$$

where

$$V_{mn} = -\tfrac{1}{2}\,\mathrm{Im}\,\mathcal{N}_{IJ}\left(e^{-4U}F_{0m}^I F_{0n}^J + F_m^I F_n^J\right). \qquad (5.130)$$

The Einstein equations thus reduce to the following two equations:

$$-\tfrac{1}{2}R_{mn}(\gamma) + \partial_m U\,\partial_n U + \partial_{(m}z\partial_{n)}\bar{z} - e^{2U}V_{mn} = 0, \qquad (5.131)$$

$$D_m\partial^m U - e^{2U}\gamma^{mn}V_{mn} = 0. \qquad (5.132)$$

We can obtain (5.132) using the following effective action:

$$S_{\text{eff}} = \int d^3x\,\sqrt{\gamma}\left[-\partial_m U\,\gamma^{mn}\partial_n U - \partial_m z\,\gamma^{mn}\partial_n\bar{z} - e^{2U}\gamma^{mn}V_{mn}\right]. \qquad (5.133)$$

Indeed, the field equation of this action for U is (5.132) if we keep V_{mn} fixed during the variation. Of course, as U is never involved in the z-dependent part of the action, the latter is not determined by this requirement. We will prove that the field equation for the scalar z can also be derived from this action for a specific parameterization of V_{mn}. This will be clarified in Sect. 5.5.2. Only then it will be clear how to use this effective action. We remark here that γ^{mn} should not be seen as a dynamical variable in this action. Instead of its field equations as following from this effective action, we impose the extra constraint (5.131), which is not derivable from (5.133).

5.5.2 Maxwell Equations and the Black Hole Potential

The quantity (5.130) is expressed in components of the field strengths $F_{\mu\nu}^I$. However, we can write it in terms of the symplectic vectors of field strengths and field equations. To do so, we start from the real form of (5.5), which can be rewritten using the Hodge duality relation (A.8) as follows:

$$G_{I\mu\nu} = \operatorname{Re} \mathcal{N}_{IJ} F_{\mu\nu}^J + \tfrac{1}{2} \operatorname{Im} \mathcal{N}_{IJ} g_{\mu\mu'} g_{\nu\nu'} (\sqrt{g})^{-1} \varepsilon^{\mu'\nu'\rho\sigma} F_{\rho\sigma}^J . \qquad (5.134)$$

Then $F_{\mu\nu}^I$, $G_{I\mu\nu}$ are components of a symplectic vector. Introducing now the three-dimensional duals as a generalization of (5.127):

$$F_m^I = \tfrac{1}{2} \gamma_{mn} (\sqrt{\gamma})^{-1} \varepsilon^{npq} F_{pq}^I, \qquad G_{Im} = \tfrac{1}{2} \gamma_{mn} (\sqrt{\gamma})^{-1} \varepsilon^{npq} G_{Ipq}, \qquad (5.135)$$

we obtain from (5.134) with $\varepsilon^{0npq} = -\varepsilon^{npq}$ (as we use $\varepsilon^{0123} = -1$, see (A.5))

$$\begin{pmatrix} F_{0m}^I \\ G_{I0m} \end{pmatrix} = -e^{2U} \mathcal{M}\Omega \begin{pmatrix} F_m^J \\ G_{Jm} \end{pmatrix}, \qquad (5.136)$$

where $\mathcal{M}$ was given in (5.68) and Ω is the symplectic metric (5.7). These matrices contain indices I and J at appropriate positions automatically for (5.136) to make sense. Equation (5.130) then becomes

$$V_{mn} = \tfrac{1}{2} \left(F_m^I \ G_{Im} \right) \Omega \mathcal{M}\Omega \begin{pmatrix} F_n^J \\ G_{Jn} \end{pmatrix}, \qquad (5.137)$$

where

$$\Omega \mathcal{M}\Omega = \begin{pmatrix} -I - RI^{-1}R & RI^{-1} \\ I^{-1}R & -I^{-1} \end{pmatrix}, \qquad R = \operatorname{Re}\mathcal{N}, \quad I = \operatorname{Im}\mathcal{N}. \qquad (5.138)$$

Note that U does not appear in this expression for V_{mn}. This implies that if we consider V_{mn} as a function of F_m^I, G_{Im} and the scalars implicitly present in (5.138) and we insert it as such in the effective action (5.133) then this action still generates the field equation (5.132) for U. We now check that in this way it also generates the same scalar field equations as those obtained from the original action $S^{(0)}$ and $S^{(1)}$, where the vector fields W_μ^I were the other independent variables. Hence these field equations that should be reproduced are

$$0 = \partial_\mu \sqrt{g} g^{\mu\nu} \partial_\nu \bar{z} + \tfrac{1}{4} \sqrt{g} \partial_{\bar{z}} (\operatorname{Im} \mathcal{N}_{IJ}) F_{\mu\nu}^I F^{\mu\nu J} - \tfrac{1}{8} \partial_{\bar{z}} (\operatorname{Re} \mathcal{N}_{IJ}) \varepsilon^{\mu\nu\rho\sigma} F_{\mu\nu}^I F_{\rho\sigma}^J . \qquad (5.139)$$

Specifying the metric (5.117) and the expressions for the field strengths in terms of F_m^I and G_{Im}, this becomes

$$0 = \partial_m \sqrt{\gamma} \gamma^{mn} \partial_n \bar{z} + \tfrac{1}{2} \sqrt{\gamma} e^{2U} \gamma^{mn} \left(F_m^I \ G_{mI} \right) \Omega \partial_{\bar{z}} \mathcal{M}\Omega \begin{pmatrix} F_n^J \\ G_{nJ} \end{pmatrix}, \qquad (5.140)$$

where the indices I and J appear again in appropriate positions on the submatrices of $\Omega \partial_z \mathcal{M}\Omega$. The latter is indeed the field equation obtained from the effective action

$$S_{\text{eff}}(U, z) = \int d^3 x \sqrt{\gamma}\, \gamma^{mn} \left[-\partial_m U \, \partial_n U - \partial_m z \, \partial_n \bar{z} \right.$$

$$\left. - \tfrac{1}{2} e^{2U} \left(F_m^I \ G_{Im} \right) \Omega \mathcal{M}\Omega \begin{pmatrix} F_n^J \\ G_{Jn} \end{pmatrix} \right]. \tag{5.141}$$

The (U, z) in the left-hand side indicates that S_{eff} should be considered as an effective action for varying with respect to $\{U, z\}$, while γ^{mn}, F_m^I and G_{Im} should be considered as background. We saw already that the field equations of the original action for γ^{mn} lead to the constraint (5.131). We will now check what the field equations of the vector sector impose.

The field equations from $S^{(1)}$ in (5.118) with independent vectors W_μ^I are equivalent to the field equations and Bianchi identities

$$\varepsilon^{\mu\nu\rho\sigma} \partial_\nu \begin{pmatrix} F_{\rho\sigma}^I \\ G_{I\rho\sigma} \end{pmatrix} = 0. \tag{5.142}$$

Using our preferred variables, this gives

$$\partial_m \sqrt{\gamma}\, \gamma^{mn} \begin{pmatrix} F_n^I \\ G_{In} \end{pmatrix} = 0, \qquad \partial_{[m} e^{2U} \mathcal{M}\Omega \begin{pmatrix} F_{n]}^I \\ G_{In]} \end{pmatrix} = 0. \tag{5.143}$$

One way of solving these equations is to define $F_m^I = \partial_m H^I$ and $G_{Im} = \partial_m H_I$, such that the first of (5.143) becomes

$$\partial_m \sqrt{\gamma}\, \gamma^{mn} \partial_n H^I = \partial_m \sqrt{\gamma}\, \gamma^{mn} \partial_n H_I = 0. \tag{5.144}$$

We remain then with Bianchi identities of the form

$$\left(\partial_{[m} e^{2U} \mathcal{M} \right) \Omega \partial_{n]} \begin{pmatrix} H^I \\ H_I \end{pmatrix} = 0, \tag{5.145}$$

which can be solved by assuming that all fields (U, the scalars z and the harmonic H^I and H_I) depend only on one coordinate. This ensures that the ∂_m and ∂_n for $m \neq n$ in the above equation cannot both be nonvanishing. We denote this one coordinate as τ. Thus $U(\tau)$, $z(\tau)$, $H^I(\tau)$ and $H_I(\tau)$.

A convenient metric is, e.g., [44]

$$\gamma_{mn} dx^m dx^n = \frac{c^4}{\sinh^4 c\tau} d\tau^2 + \frac{c^2}{\sinh^2 c\tau} (d\theta^2 + \sin^2\theta d\phi^2). \tag{5.146}$$

Details on this parameterization are given in an appendix of [45]. This parameterization has the property $\sqrt{\gamma} \gamma^{\tau\tau} = \sin\theta$, which will be useful. In this basis, condition (5.144) is just $H'' = 0$ (where a prime is now a derivative w.r.t. τ), so we can take

$$H = \begin{pmatrix} H^I \\ H_I \end{pmatrix} = \Gamma\tau + h, \qquad h = \begin{pmatrix} h^I \\ h_I \end{pmatrix}, \qquad \Gamma = \begin{pmatrix} p^I \\ q_I \end{pmatrix}. \qquad (5.147)$$

We have here introduced the magnetic and electric charges in the symplectic vector Γ. We come back to their meaning in Sect. 5.5.3. The effective Lagrangian for the scalars is obtained upon plugging these solutions into (5.141). Up to a constant we get

$$\mathcal{L}_{\text{eff}} = U'^2 + e^{2U} V_{\text{BH}} + z'\bar{z}', \qquad (5.148)$$

where the 'black hole potential' is now

$$V_{\text{BH}} = V_{\tau\tau} = \tfrac{1}{2}\Gamma^T \Omega\mathcal{M}\Omega\Gamma. \qquad (5.149)$$

The one-dimensional effective Lagrangian (5.148) does not reproduce all the Einstein equations. Indeed the Einstein equations (5.131)–(5.132) lead to 2 independent equations:

$$c^2 - U'^2 - z'\bar{z}' + e^{2U} V_{\text{BH}} = 0, \qquad -U'' + e^{2U} V_{\text{BH}} = 0. \qquad (5.150)$$

The second one is the one that can be obtained from the effective action, while the first one cannot. It must be considered as an extra constraint.

5.5.3 Field Strengths and Charges

In (5.147) we already wrote p^I and q_I for the components of Γ, suggesting that they are charges. Indeed, when we consider field configurations with electric and/or magnetic charges in 4 dimensions, this means that there are 2-cycles S^2 surrounding the sources such that

$$\int_{S^2} F^I_{\mu\nu} dx^\mu \wedge dx^\nu = 8\pi p^I, \qquad \int_{S^2} G_{I\mu\nu} dx^\mu \wedge dx^\nu = 8\pi q_I. \qquad (5.151)$$

Exercise 5.6 Check that the solution that we gave above, leads indeed to the identification of the charges here and in (5.147). □

There is also the field strength that occurs in the gravitino transformation, see e.g. (2.90), which is the value of the auxiliary field $T_{\mu\nu}$ of the Weyl multiplet. When we restrict to the bosonic part of its value, determined in (4.54), we obtain

$$T_{\mu\nu}^- = 2T_I F_{\mu\nu}^{-I} \,, \tag{5.152}$$

$$T_I = \frac{N_{IJ}\bar{X}^J}{\bar{X}^L N_{LM}\bar{X}^M} = -2\kappa^2 \operatorname{Im} N_{IJ} X^J = \mathrm{i}\kappa^2 \left(F_I - \bar{N}_{IJ} X^J \right), \qquad \bar{X}^I T_I = 1,$$

where use has been made of (5.69).

Assuming that F_I and X^I are sufficiently constant in the integration region (such that they can be taken in and out of the integral), the integral of this quantity gives [46]

$$\begin{aligned}
\mathcal{Z} &\equiv \frac{\mathrm{i}}{16\pi\kappa^2} \int_{S^2} T_{\mu\nu}^- \mathrm{d}x^\mu \wedge \mathrm{d}x^\nu = \frac{1}{8\pi} \int_{S^2} \left(X^I G_{I\mu\nu}^- - F_I F_{\mu\nu}^{-I} \right) \mathrm{d}x^\mu \wedge \mathrm{d}x^\nu \\
&= \frac{1}{8\pi} \int_{S^2} \left(X^I G_{I\mu\nu} - F_I F_{\mu\nu}^I \right) \mathrm{d}x^\mu \wedge \mathrm{d}x^\nu \\
&= X^I q_I - F_I p^I \,.
\end{aligned} \tag{5.153}$$

Between the first and the second line we used that the combination with the self-dual field strengths vanishes due to $F_I = N_{IJ} X^J$. The object $\mathcal{Z}$ is called the central charge, because its value appears in the commutator of two supersymmetries, as can be seen from (2.96)–(2.97).

When we take the holomorphic covariant derivatives of the final expression, then we have to use $\nabla_\alpha F_I = \bar{N}_{IJ}\nabla_\alpha X^J$, and therefore only the self-dual parts remain. This gives thus

$$\begin{aligned}
\nabla_\alpha \mathcal{Z} &= \nabla_\alpha X^I q_I - \nabla_\alpha F_I p^I \\
&= \frac{1}{8\pi} \int_{S^2} \left(\nabla_\alpha X^I G_{I\mu\nu}^+ - \nabla_\alpha F_I F_{\mu\nu}^{+I} \right) \mathrm{d}x^\mu \wedge \mathrm{d}x^\nu \\
&= \frac{2\mathrm{i}}{8\pi} \int_{S^2} \mathcal{D}_\alpha X^I \operatorname{Im} N_{IJ} F_{\mu\nu}^{+J} \mathrm{d}x^\mu \wedge \mathrm{d}x^\nu \,.
\end{aligned} \tag{5.154}$$

The latter quantities $\nabla_\alpha \mathcal{Z}$ are the objects that appear also in the transformation laws of the physical gauginos. Indeed, the fermions of the conformal multiplets transform according to (3.15) to quantities $\mathcal{F}_{\mu\nu}^-$, whose bosonic part is

$$\mathcal{F}_{\mu\nu}^{-I} = \left(\delta_J^I - \bar{X}^I T_J \right) F_{\mu\nu}^{-J} \,. \tag{5.155}$$

The physical fermions are the ones that satisfy the S-gauge condition (4.58), which means that they vanish under projection with T_I. We find indeed $T_I \mathcal{F}_{\mu\nu}^{-I} = 0$.

Using the vector Γ of (5.147), we can write in a symplectic notation

$$\mathcal{Z} = \langle V, \Gamma \rangle = V^T \Omega \Gamma, \qquad \nabla_\alpha \mathcal{Z} = \langle U_\alpha, \Gamma \rangle, \tag{5.156}$$

and can derive from (5.67) a simple expression for the 'black hole potential' [16, 44, 47]

$$V_{BH} \equiv \mathcal{Z}\bar{\mathcal{Z}} + \nabla_\alpha \mathcal{Z} g^{\alpha\bar{\beta}} \nabla_{\bar{\beta}} \bar{\mathcal{Z}} = \tfrac{1}{2} \Gamma^T \Omega \mathcal{M} \Omega \Gamma. \tag{5.157}$$

Similarly, by using the same identity, one derives

$$V\bar{\mathcal{Z}} + \bar{U}_{\bar{\beta}} g^{\bar{\beta}\alpha} \nabla_\alpha \mathcal{Z} = -\tfrac{1}{2}(\mathcal{M}\Omega + i\mathbb{1})\Gamma. \tag{5.158}$$

5.5.4 Attractors

The attractor solution [38, 39, 47] is the solution near the horizon. This is the large τ behaviour. In that case supersymmetry is preserved, which is expressed as $\nabla_\alpha \mathcal{Z} = 0$. This extremizes the black hole potential. So it is consistent with constant z^α as solution of the field equation for the scalars. In this case (5.158) simplifies. The imaginary part is

$$-2\operatorname{Im}(V\bar{\mathcal{Z}}) = \Gamma. \tag{5.159}$$

These are the attractor equations. The BH potential reduces to

$$V_{BH,BPS} = |\mathcal{Z}|^2, \tag{5.160}$$

where the indication 'BPS' is used because we started from an extremal black hole solution. Then we determine U by the constraint

$$\dot{U}^2 = e^{2U} V_{BH,BPS}, \qquad \text{i.e.} \qquad \dot{U} = \pm e^U \sqrt{V_{BH,BPS}}. \tag{5.161}$$

The $V_{BH,BPS}$ being constant, this automatically implies the other field equation

$$\ddot{U} = e^{2U} V_{BH,BPS}. \tag{5.162}$$

Finally, the solution is

$$e^{-U} = \mp |\mathcal{Z}|\tau + \text{constant}. \tag{5.163}$$

Since $\mathcal{Z}$ is completely determined by the charges via (5.153), the solution above shows that the near-horizon ($\tau \rightarrow \infty$) value of the scalars is universal and independent of the initial conditions.

5.6 Quaternionic-Kähler Manifolds

Quaternionic-Kähler manifolds entered supergravity research first in the seminal work [48]. In [49] a lot of interesting properties were already discussed. Workshops on quaternionic geometry have been organized where mathematics and physics results were brought together [50]. Other important papers that reviewed the properties of quaternionic manifolds are [51–53].

As a result of Swann [54], every quaternionic manifold can be obtained as a cone with an SU(2) gauging. The condition that a hyper-Kähler manifold can be formulated in a conformal way is equivalent to the condition that there is a cone. Therefore, those hyper-Kähler that satisfy this condition are one-to-one related with the hyper-Kähler manifolds that can be made quaternionic by an SU(2) gauging. The result of Swann has been made explicit in [55] by the construction with the embedding of the manifold in a conformal manifold, as written in Sect. 4.6. There it has been shown how any quaternionic-Kähler metric can be obtained in this way. The procedure is the same whether applied for $D = 4$, $D = 5$ or $D = 6$.

5.6.1 Supersymmetry and Quaternionic Geometry

The quaternionic geometry is a bi-product of the supersymmetry algebra, which leads to the definition of the almost quaternionic structures $\mathbf{J}_X{}^Y$, see (3.37). Furthermore, in order to build an action, we require this manifold to possess an invariant metric g_{XY}, which was defined in (3.145). If the almost quaternionic structures are covariantly constant (up to a rotation among them) with the Levi-Civita connection associated to g_{XY} (see (3.39)), the manifold is promoted to a hyper-Kähler manifold. The requirement of conformal symmetry further restricts the manifold. The gauge fixing then leads from a $4(n_H + 1)$-dimensional embedding manifold to a projection on a $4n_H$-dimensional submanifold, which turns out to be quaternionic-Kähler (see Sect. 4.6). The latter is parameterized by scalar fields q^u, while its tangent space contains vectors labelled by indices (i, A). Using the basis (4.130), a quaternionic-Kähler metric $h_{uv}(q)$, is obtained from the embedding metric in (4.153). Similarly all other geometrical quantities are obtained by a projection from the embedding to the quaternionic-Kähler space.[21]

Considering the differential equation on the frame fields in the quaternionic-Kähler space, (4.146), we see that it contains SU(2) connections $\omega_{uj}{}^i$ as well as USp($2n_H$) connections[22] $\omega_{uB}{}^A$. The latter is similar to the condition in the embedding manifold, (3.32), but the SU(2) connection was absent in the hyper-Kähler manifold. This SU(2) connection ω_u, promotes the almost quaternionic structure to a quaternionic structure, such that the resulting manifolds are 'quaternionic'. If the SU(2) connection is zero, they are called 'hypercomplex', which is the case before the gauge fixing (and in rigid supersymmetry). Note that the same SU(2) connection also appears in the differential equation on the hypercomplex structures in the quaternionic-Kähler manifold, (4.143), while it was absent in that of the hyper-Kähler structure (3.39). For physics applications we further require a positive definite energy in the Einstein–Hilbert action, which results in additional constraints on the geometry.

5.6.2 Quaternionic Manifolds

As mentioned above, a main ingredient is the equation that states the covariant constancy of the frame fields (4.146), which can be considered in terms of a connection on the tangent space

$$\Omega_{u\,jB}{}^{iA} \equiv f^v{}_{jB} \left(\partial_u f_v^{iA} - \Gamma_{uv}^w f^{iA}{}_w \right) = -\omega_{uj}{}^i \delta_B{}^A - \omega_{uB}{}^A \delta_j{}^i . \qquad (5.164)$$

If this $\Omega_{u\,jB}{}^{iA}$, for each u, would be a general $4n_H \times 4n_H$ matrix, then we would say that the holonomy is not restricted (or sits in $G\ell(4n_H, \mathbb{R})$). The splitting as in the right-hand side of this equation implies that the holonomy group is restricted to SU(2) × $G\ell(n_H, \mathbb{H})$.

The integrability condition of (4.146) leads to an expression for the curvature of quaternionic manifolds:

$$R^w{}_{xuv} = f^w{}_{iA} f^{jA}{}_x \mathcal{R}_{uvj}{}^i + f^w{}_{iA} f^{iB}{}_x \mathcal{R}_{uvB}{}^A = -\mathbf{J}_x{}^w \mathcal{R}_{uv} + f^w{}_{iA} f^{iB}{}_x \mathcal{R}_{uvB}{}^A , \qquad (5.165)$$

---

[21] The terminology 'quaternionic' and 'quaternionic-Kähler' is used in mathematics in this sense: the first requires a quaternionic structure on the manifold, and the second demands moreover compatibility with a metric structure.

[22] We demanded the preservation of the metric g_{XY} (see (3.145)), therefore the $G\ell(n_H, \mathbb{H})$ is restricted to USp($2n_H$).

where

$$R^w{}_{xuv} \equiv 2\partial_{[u}\Gamma^w_{v]x} + 2\Gamma^w_{y[u}\Gamma^y_{v]x},$$

$$R_{uv} \equiv 2\partial_{[u}\boldsymbol{\omega}_{v]} + 2\boldsymbol{\omega}_{[u} \times \boldsymbol{\omega}_{v]},$$

$$\mathcal{R}_{uvB}{}^A \equiv 2\partial_{[u}\omega_{v]B}{}^A + 2\omega_{[u|C|}{}^A\omega_{v]B}{}^C. \tag{5.166}$$

Thus the curvature of the affine connection is split in an SU(2) part and a $G\ell(n_H, \mathbb{H})$ part.

5.6.3 *Quaternionic-Kähler Manifolds*

Quaternionic-Kähler manifolds (which include 'hyper-Kähler manifolds' in the limiting case that the SU(2) curvature vanishes) by definition have a metric, h_{uv}. This corresponds to the requirement in the matter couplings that there is an action. There is the request that the connections preserve the metric, which restricts the holonomy group to $SU(2) \times USp(2n_H)$. The affine connection Γ^w_{uv} is now the metric connection, and the first line of (5.166) is the metric curvature, while the last line is the $USp(2n_H)$ curvature.

For $n_H > 1$ one can prove that these manifolds are Einstein, and that the SU(2) curvatures are proportional to the complex structures[23]

$$R_{uv} \equiv R^w{}_{uwv} = \frac{1}{4n_H}h_{uv}R, \qquad \mathcal{R}_{uv} = \tfrac{1}{2}\nu\mathbf{J}_{uv}, \qquad \nu = \frac{1}{4n_H(n_H+2)}R.$$
$$\tag{5.167}$$

For $n_H = 1$ this is part of the definition of quaternionic-Kähler manifolds. Hyper-Kähler manifolds are those where the SU(2) curvature is zero, and these are thus also Ricci-flat.

5.6.4 *Quaternionic-Kähler Manifolds in Supergravity*

In supergravity we find all these constraints from requiring a supersymmetric action. Moreover, we need for the invariance of the action that the last equation of (5.167) is satisfied with $\nu = -\kappa^2$, see (4.147). This implies that the scalar curvature is $R = -4n_H(n_H + 2)\kappa^2$. The fact that this is negative excludes, e.g., the compact symmetric spaces. In Sect. 5.7.2 all the symmetric spaces in this class will be mentioned, and these are thus all noncompact.

---

[23]In the conventions of this book, such proofs are given in [53, Appendix B].

5.7 Relations Between Special Manifolds

So far we discussed separately the Kähler manifolds (associated to the vector multiplets) and the quaternionic-Kähler manifolds (hypermultiplets), even though we paid attention, along the way, to various similarities among these spaces. Although our discussion has been mostly limited to $D = 4$, as we pointed out at the beginning of Chap. 5, there exists an analogous special geometry associated to the real scalars in the $D = 5$ vector multiplets. Perhaps not surprisingly, all these 'special manifolds' are related by maps, which are denoted as **c** map and **r** map.

5.7.1 c-Map and r-Map

The **c** map [8] connects a special Kähler to a quaternionic-Kähler manifold. It is induced by dimensional reduction of an $\mathcal{N} = 2$ supergravity theory from $D = 4$ to $D = 3$, by suppressing the dependence on one of the (spatial) coordinates. The resulting $D = 3$ supergravity theory can be written in terms of $D = 3$ fields and this rearranges the original fields such that the number of scalar fields increases from $2n$ to $4(n + 1)$. This is shown in Table 5.1. Essential in this map is that $D = 4$ vectors leave first a scalar component in $D = 3$, but also the other part, vectors in $D = 3$, are dual to scalars. This map is also obtained in string theory context by changing from the reduction of a type IIA theory to the reduction of the T-dual type IIB theory or vice versa.

This leads to the notion of '*special quaternionic manifolds*', which are those manifolds appearing in the image of the **c** map. They are a subclass of the quaternionic manifolds.

The **r** map [9] is determined by the reduction of vector multiplets in $D = 5$ to vector multiplets in $D = 4$. Starting with $n - 1$ physical vector multiplets in $D = 5$, one ends up with n vector multiplets in $D = 4$ ($2n$ real scalars) as schematically shown in Table 5.2. The manifolds defined by coupling (real) scalars to vector multiplets in 5 dimensions are called 'very special real manifolds'. They are determined by the symmetric tensor C_{IJK} in Sect. 3.3.2. The corresponding $D = 4$

Table 5.1 The **c** map as dimensional reduction from $D = 4$ to $D = 3$ supergravity

$D = 4$ spins	2	1	0
numbers	1	$n + 1$	$2n$

$D = 3$ spins 2 0	1 2		
	1		
	2	$2(n + 1)$	$2n$

The number of fields of various spins is indicated

Table 5.2 The **r** map
induced by dimensional
reduction from $D = 5$ to
$D = 4$ supergravity

$D = 5$ spins	2	1	0
numbers	1	n	$n - 1$
$D = 4$ spins			
2	1		
1	1	n	
0	1	n	$n - 1$

The number of fields of integer spins is
indicated

couplings will then also be determined by such a tensor, and one parameterization
is in terms of the prepotential (4.101). As already indicated there, the special Kähler
manifolds that are in the image of this **r** map are then denoted as 'very special
Kähler manifolds' . Their further image under the c∘**r** map are then called 'very
special quaternionic manifolds'.

The **c** map has been studied in superspace in [56, 57] and the mathematical
structure has been clarified in terms of the so-called Swann bundle in [58]. An off-
shell extension has been found [59] in view of applications with higher derivatives.
It is then useful to formulate the hypermultiplets in the form of tensor multiplets
[60].

5.7.2   Homogeneous and Symmetric Spaces

Homogeneous and symmetric spaces are the most known manifolds. These are
spaces of the form G/H, where G is the isometry group and H is its isotropy
subgroup. The group G is not necessarily a semi-simple group, and thus not all the
homogeneous spaces have a clear name. The symmetric spaces are those for which
the algebra splits as $g = h + k$ and all commutators $[k, k] \subset h$. The homogeneous
special manifolds are classified in [61].

It turns out that homogeneous special manifolds are in one-to-one correspondence to realizations of real Clifford algebras with signature $(q + 1, 1)$ for real,
$(q + 2, 2)$ for Kähler, and $(q + 3, 3)$ for quaternionic manifolds. Thus, the spaces
are identified by giving the number q, which specifies the Clifford algebra, and by
specifying its representation. If q is not a multiple of 4, then these Clifford algebras
have only one irreducible representations, and we thus just have to mention the
multiplicity P of this representation. The spaces are denoted as $L(q, P)$. If $q = 4m$
then there are two inequivalent representations, chiral and antichiral, and the spaces
are denoted as $L(q, P, \dot{P})$. The fact that the chiral and the antichiral representations
are conjugate implies $L(4m, P, \dot{P}) = L(4m, \dot{P}, P)$. A special case is $q = 0$ for
which $L(0, n) = L(n, 0)$. These manifolds are listed in Table 5.3. If we use n as the

complex dimension of the special Kähler space, the dimension of these manifolds is ($\dot{P} = 0$ if $q \neq 4m$)

$$n = 3 + q + (P + \dot{P})\mathcal{D}_{q+1}, \quad \begin{cases} \dim_{\mathbb{R}}[\text{very special real } L(q, P, \dot{P})] = n - 1 \\ \dim_{\mathbb{R}}[\text{special Kähler } L(q, P, \dot{P})] = 2n \\ \dim_{\mathbb{R}}[\text{quaternionic-Kähler } L(q, P, \dot{P})] = 4(n + 1). \end{cases}$$
(5.168)

where $\mathcal{D}_{q+1}$ is the dimension of the irreducible representation of the Clifford algebra in $q + 1$ dimensions with positive signature, i.e.

$$\mathcal{D}_{q+1} = 1 \text{ for } q = -1, 0, \quad \mathcal{D}_{q+1} = 2 \text{ for } q = 1, \quad \mathcal{D}_{q+1} = 4 \text{ for } q = 2,$$

$$\mathcal{D}_{q+1} = 8 \text{ for } q = 3, 4, \quad \mathcal{D}_{q+1} = 16 \text{ for } q = 5, 6, 7, 8, \quad \mathcal{D}_{q+8} = 16 \mathcal{D}_q.$$
(5.169)

The very special manifolds are defined by coefficients C_{IJK}. For the homogeneous ones, we can write them as

$$C_{IJK} h^I h^J h^K = 3 \left\{ h^1 \left(h^2 \right)^2 - h^1 \left(h^\mu \right)^2 - h^2 \left(h^i \right)^2 + \gamma_{\mu ij} h^\mu h^i h^j \right\}.$$
(5.170)

We decomposed the indices $I = 1, \ldots, n$ into $I = 1, 2, \mu, i$, with $\mu = 1, \ldots, q+1$ and $i = 1, \ldots, (P + \dot{P})\mathcal{D}_{q+1}$. Here, $\gamma_{\mu ij}$ is the $(q + 1, 0)$ Clifford algebra representation that we mentioned. Note that these models have predecessors in 6 dimensions, with $q + 1$ tensor multiplets and $(P + \dot{P})\mathcal{D}_{q+1}$ vector multiplets. The gamma matrices are then the corresponding coupling constants between the vector and tensor multiplets.

Considering further Table 5.3, we find in the quaternionic spaces the homogeneous ones that were found in [62], together with those that were discovered in [61] (the ones with a $\star$ except for the series $L(0, P, \dot{P})$, which were already in [62], and denoted there as $W(P, \dot{P})$). A new overview of the properties of these homogeneous manifolds can be found in [63], where these (apart from the pure $D = 4$ and pure $D = 5$ theories) are also constructed as double copies of Yang–Mills theories.

Observe that the classification of homogeneous spaces exhibits that the quaternionic projective spaces have no predecessor in special geometry, and that the complex projective spaces have no predecessor in very special real manifolds. Similarly, only those with $q \geq -1$ can be obtained from 6 dimensions. $L(-1, 0)$ corresponds to pure supergravity in 6 dimensions. In general, the scalars of the tensor multiplets in $D = 6$ describe a SO$(1, q + 1)$/SO$(q + 1)$ manifold.

In the range $q \geq -1$, some of these manifolds are in fact symmetric manifolds. These are collected in Table 5.4. For the symmetric special Kähler spaces, this reproduces the classification obtained in [64]. A study of the full set of isometries could be done systematically in these models. All this has been summarized in [65].

Table 5.3 Homogeneous manifolds

	n	real	Kähler	quaternionic
$L(-3,P)$	P			$\frac{USp(2P+2,2)}{USp(2P+2)\otimes SU(2)}$
SG_4	0		SG	$\frac{U(1,2)}{U(1)\otimes U(2)}$
$L(-2,P)$	$1+P$		$\frac{U(P+1,1)}{U(P+1)\otimes U(1)}$	$\frac{SU(P+2,2)}{SU(P+2)\otimes SU(2)\otimes U(1)}$
SG_5	1	SG	$\frac{SU(1,1)}{U(1)}$	$\frac{G_2}{SU(2)\otimes SU(2)}$
$L(-1,P)$	$2+P$	$\frac{SO(P+1,1)}{SO(P+1)}$	$\star$	$\star$
$L(4m,P,\dot P)$		$\star$	$\star$	$\star$
$L(q,P)$		$X(P,q)$	$H(P,q)$	$V(P,q)$

In this table, q, P, $\dot P$ and m denote positive integers or zero, and $q \neq 4m$. SG denotes an empty space, which corresponds to supergravity models without scalars. The horizontal lines separate spaces of different rank. The first non-empty space in each column has rank 1. Going to the right or down a line increases the rank by 1. The manifolds indicated by a $\star$ did not get a name. The number n is the complex dimension of the Kähler space as in (5.168), which simplifies in many cases

Table 5.4 Symmetric very special manifolds

	n	Real	Kähler	Quaternionic
$L(-1,0)$	2	$SO(1,1)$	$\left[\frac{SU(1,1)}{U(1)}\right]^2$	$\frac{SO(3,4)}{(SU(2))^3}$
$L(-1,P)$	$2+P$	$\frac{SO(P+1,1)}{SO(P+1)}$		
$L(0,P)$	$3+P$	$SO(1,1)\otimes\frac{SO(P+1,1)}{SO(P+1)}$	$\frac{SU(1,1)}{U(1)}\otimes\frac{SO(P+2,2)}{SO(P+2)\otimes SO(2)}$	$\frac{SO(P+4,4)}{SO(P+4)\otimes SO(4)}$
$L(1,1)$	6	$\frac{S\ell(3,\mathbb{R})}{SO(3)}$	$\frac{Sp(6)}{U(3)}$	$\frac{F_4}{USp(6)\otimes SU(2)}$
$L(2,1)$	9	$\frac{S\ell(3,C)}{SU(3)}$	$\frac{SU(3,3)}{SU(3)\otimes SU(3)\otimes U(1)}$	$\frac{E_6}{SU(6)\otimes SU(2)}$
$L(4,1)$	15	$\frac{SU^*(6)}{USp(6)}$	$\frac{SO^*(12)}{SU(6)\otimes U(1)}$	$\frac{E_7}{SO(12)\otimes SU(2)}$
$L(8,1)$	27	$\frac{E_6}{F_4}$	$\frac{E_7}{E_6\otimes U(1)}$	$\frac{E_8}{E_7\otimes SU(2)}$

Note that the very special real manifolds $L(-1,P)$ are symmetric, but not their images under the **r** map. The number n is the dimension as in Table 5.3

References

1. S. Ferrara, J. Scherk, B. Zumino, Algebraic properties of extended supergravity theories. Nucl. Phys. **B121**, 393–402 (1977). http://dx.doi.org/10.1016/0550-3213(77)90161-4
2. B. de Wit, Properties of SO(8) extended supergravity. Nucl. Phys. **B158**, 189–212 (1979). http://dx.doi.org/10.1016/0550-3213(79)90195-0
3. E. Cremmer, B. Julia, The SO(8) supergravity. Nucl. Phys. **B159**, 141–212 (1979). http://dx.doi.org/10.1016/0550-3213(79)90331-6

4. M.K. Gaillard, B. Zumino, Duality rotations for interacting fields. Nucl. Phys. **B193**, 221 (1981). http://dx.doi.org/10.1016/0550-3213(81)90527-7

5. D.Z. Freedman, A. Van Proeyen, *Supergravity*. (Cambridge University Press, Cambridge, 2012). http://www.cambridge.org/mw/academic/subjects/physics/theoretical-physics-and-mathematical-physics/supergravity?format=AR

6. B. de Wit, A. Van Proeyen, Special geometry and symplectic transformations. Nucl. Phys. Proc. Suppl. **45BC**, 196–206(1996). http://dx.doi.org/10.1016/0920-5632(95)00637-0. arXiv:hep-th/9510186 [hep-th]

7. C.M. Hull, A. Van Proeyen, Pseudoduality. Phys. Lett. **B351**, 188–193 (1995). http://dx.doi.org/10.1016/0370-2693(95)00408-D. arXiv:hep-th/9503022 [hep-th]

8. S. Cecotti, S. Ferrara, L. Girardello, Geometry of type II superstrings and the moduli of superconformal field theories. Int. J. Mod. Phys. **A4**, 2475 (1989). http://dx.doi.org/10.1142/S0217751X89000972

9. B. de Wit, A. Van Proeyen, Broken sigma model isometries in very special geometry. Phys. Lett. **B293**, 94–99 (1992). http://dx.doi.org/10.1016/0370-2693(92)91485-R. arXiv:hep-th/9207091 [hep-th]

10. P. Claus, K. Van Hoof, A. Van Proeyen, A symplectic covariant formulation of special Kähler geometry in superconformal calculus. Class. Quant. Grav. **16**, 2625–2649 (1999). http://dx.doi.org/10.1088/0264-9381/16/8/305. arXiv:hep-th/9904066 [hep-th]

11. B. de Wit, A. Van Proeyen, Potentials and symmetries of general gauged $N = 2$ supergravity–Yang–Mills models. Nucl. Phys. **B245**, 89–117 (1984). http://dx.doi.org/10.1016/0550-3213(84)90425-5

12. A. Strominger, Special geometry. Commun. Math. Phys. **133**, 163–180 (1990). http://dx.doi.org/10.1007/BF02096559

13. A. Ceresole, R. D'Auria, S. Ferrara, A. Van Proeyen, Duality transformations in supersymmetric Yang–Mills theories coupled to supergravity. Nucl. Phys. **B444**, 92–124 (1995). http://dx.doi.org/10.1016/0550-3213(95)00175-R. arXiv:hep-th/9502072 [hep-th]

14. B. Craps, F. Roose, W. Troost, A. Van Proeyen, What is special Kähler geometry? Nucl. Phys. **B503**, 565–613 (1997). http://dx.doi.org/10.1016/S0550-3213(97)00408-2. arXiv:hep-th/9703082 [hep-th]

15. L. Castellani, R. D'Auria, S. Ferrara, Special Kähler geometry: an intrinsic formulation from $N = 2$ spacetime supersymmetry. Phys. Lett. **B241**, 57–62 (1990). http://dx.doi.org/10.1016/0370-2693(90)91486-U

16. A. Ceresole, R. D'Auria, S. Ferrara, The symplectic structure of $N = 2$ supergravity and its central extension. Nucl. Phys. Proc. Suppl. **46**, 67–74 (1996). http://dx.doi.org/10.1016/0920-5632(96)00008-4. arXiv:hep-th/9509160 [hep-th]

17. L. Andrianopoli, M. Bertolini, A. Ceresole, R. D'Auria, S. Ferrara, P. Frè, General matter-coupled $N = 2$ supergravity. Nucl. Phys. **B476**, 397–417 (1996). http://dx.doi.org/10.1016/0550-3213(96)00344-6. arXiv:hep-th/9603004 [hep-th]

18. L. Andrianopoli, M. Bertolini, A. Ceresole, R. D'Auria, S. Ferrara, P. Frè, T. Magri, $N = 2$ supergravity and $N = 2$ super Yang–Mills theory on general scalar manifolds: symplectic covariance, gaugings and the momentum map. J. Geom. Phys. **23**, 111–189 (1997). http://dx.doi.org/10.1016/S0393-0440(97)00002-8. arXiv:hep-th/9605032 [hep-th]

19. A. Das, SO(4) invariant extended supergravity. Phys. Rev. **D15**, 2805 (1977). http://dx.doi.org/10.1103/PhysRevD.15.2805

20. E. Cremmer, J. Scherk, S. Ferrara, U(N) invariance in extended supergravity. Phys. Lett. **68B**, 234–238 (1977). http://dx.doi.org/10.1016/0370-2693(77)90277-5

21. E. Cremmer, J. Scherk, Algebraic simplifications in supergravity theories. Nucl. Phys. **B127**, 259–268 (1977). http://dx.doi.org/10.1016/0550-3213(77)90214-0

22. E. Cremmer, J. Scherk, S. Ferrara, SU(4) invariant supergravity theory. Phys. Lett. **74B**, 61–64 (1978). http://dx.doi.org/10.1016/0370-2693(78)90060-6

23. S. Ferrara, A. Van Proeyen, A theorem on $N = 2$ special Kähler product manifolds. Class. Quant. Grav. **6**, L243 (1989). http://dx.doi.org/10.1088/0264-9381/6/12/002

24. E. Cremmer, C. Kounnas, A. Van Proeyen, J. Derendinger, S. Ferrara, B. de Wit, L. Girardello, Vector multiplets coupled to $N = 2$ supergravity: super-Higgs effect, flat potentials and geometric structure. Nucl. Phys. **B250**, 385–426 (1985). http://dx.doi.org/10.1016/0550-3213(85)90488-2

25. D.S. Freed, Special Kähler manifolds. Commun. Math. Phys. **203**, 31–52 (1999). http://dx.doi.org/10.1007/s002200050604. arXiv:hep-th/9712042 [hep-th]

26. S. Ferrara, L. Girardello, M. Porrati, Minimal Higgs branch for the breaking of half of the supersymmetries in $N = 2$ supergravity. Phys. Lett. **B366**, 155–159 (1996). http://dx.doi.org/10.1016/0370-2693(95)01378-4. arXiv:hep-th/9510074 [hep-th]

27. F. Cordaro, P. Frè, L. Gualtieri, P. Termonia, M. Trigiante, $N = 8$ gaugings revisited: an exhaustive classification. Nucl. Phys. **B532**, 245–279 (1998). http://dx.doi.org/10.1016/S0550-3213(98)00449-0. arXiv:hep-th/9804056

28. H. Nicolai, H. Samtleben, Compact and noncompact gauged maximal supergravities in three dimensions. J. High Energy Phys. **4**, 022 (2001). http://dx.doi.org/10.1088/1126-6708/2001/04/022. arXiv:hep-th/0103032 [hep-th]

29. B. de Wit, H. Samtleben, M. Trigiante, Magnetic charges in local field theory. J. High Energy Phys. **9**, 016 (2005). http://dx.doi.org/10.1088/1126-6708/2005/09/016. arXiv:hep-th/0507289 [hep-th]

30. H. Samtleben, Lectures on gauged supergravity and flux compactifications. Class. Quant. Grav. **25**, 214002 (2008). http://dx.doi.org/10.1088/0264-9381/25/21/214002. arXiv:0808.4076 [hep-th]

31. M. Trigiante, Gauged Supergravities. Phys. Rept. **680**, 1–175 (2017). http://dx.doi.org/10.1016/j.physrep.2017.03.001. arXiv:1609.09745 [hep-th]

32. J. Louis, P. Smyth, H. Triendl, The $N = 1$ low-energy effective action of spontaneously broken $N = 2$ supergravities. J. High Energy Phys. **1010**, 017 (2010). http://dx.doi.org/10.1007/JHEP10(2010)017. arXiv:1008.1214 [hep-th]

33. I. Antoniadis, J.-P. Derendinger, P.M. Petropoulos, K. Siampos, All partial breakings in $N = 2$ supergravity with a single hypermultiplet. J. High Energy Phys. **08**, 045 (2018). http://dx.doi.org/10.1007/JHEP08(2018)045. arXiv:1806.09639 [hep-th]

34. H. Abe, S. Aoki, S. Imai, Y. Sakamura, Interpolation of partial and full supersymmetry breakings in $N = 2$ supergravity. Nucl. Phys. **B**, 114690 (2019). http://dx.doi.org/10.1016/j.nuclphysb.2019.114690. arXiv:1901.05679 [hep-th]

35. F. Farakos, P. Kočí, G. Tartaglino-Mazzucchelli, R. von Unge, Partial $N = 2$ supersymmetry breaking and deformed hypermultiplets. J. High Energy Phys. **03**, 037 (2019). http://dx.doi.org/10.1007/JHEP03(2019)037. arXiv:1807.03715 [hep-th]

36. D.Z. Freedman, D. Roest, A. Van Proeyen, Off-shell Poincaré Supergravity. J. High Energy Phys. **2**, 102 (2017). http://dx.doi.org/10.1007/JHEP02(2017)102. arXiv:1701.05216 [hep-th]

37. A. Gallerati, *Constructing Black Hole Solutions in Supergravity Theories* (2019). arXiv:1905.04104 [hep-th]

38. S. Ferrara, R. Kallosh and A. Strominger, $N = 2$ extremal black holes. Phys. Rev. **D52**, R5412–R5416 (1995). http://dx.doi.org/10.1103/PhysRevD.52.R5412. arXiv:hep-th/9508072 [hep-th]

39. A. Strominger, Macroscopic entropy of $N = 2$ extremal black holes. Phys. Lett. **B383**, 39–43 (1996). http://dx.doi.org/10.1016/0370-2693(96)00711-3. arXiv:hep-th/9602111 [hep-th]

40. P. Kraus, F. Larsen, Attractors and black rings. Phys. Rev. **D72**, 024010 (2005). http://dx.doi.org/10.1103/PhysRevD.72.024010. arXiv:hep-th/0503219 [hep-th]

41. S. Ferrara, M. Günaydin, Orbits and attractors for $N = 2$ Maxwell-Einstein supergravity theories in five dimensions. Nucl. Phys. **B759**, 1–19 (2006). http://dx.doi.org/10.1016/j.nuclphysb.2006.09.016. arXiv:hep-th/0606108 [hep-th]

42. F. Larsen, The attractor mechanism in five dimensions. Lect. Notes Phys. **755**, 249–281 (2008). arXiv:hep-th/0608191 [hep-th]

43. A. Van Proeyen, B. Vercnocke, Effective action for the field equations of charged black holes. Class. Quant. Grav. **25**, 035010 (2008). http://dx.doi.org/10.1088/0264-9381/25/3/035010. arXiv:0708.2829 [hep-th]

44. S. Ferrara, G.W. Gibbons, R. Kallosh, Black holes and critical points in moduli space. Nucl. Phys. **B500**, 75–93 (1997). http://dx.doi.org/10.1016/S0550-3213(97)00324-6. arXiv:hep-th/9702103

45. R. Kallosh, N. Sivanandam, M. Soroush, The Non-BPS black hole attractor equation. J. High Energy Phys. **03**, 060 (2006). http://dx.doi.org/10.1088/1126-6708/2006/03/060. arXiv:hep-th/0602005 [hep-th]

46. M. Billó, A. Ceresole, R. D'Auria, S. Ferrara, P. Fré, T. Regge, P. Soriani, A. Van Proeyen, A search for nonperturbative dualities of local $N = 2$ Yang-Mills theories from Calabi-Yau threefolds. Class. Quant. Grav. **13**, 831–864 (1996). http://dx.doi.org/10.1088/0264-9381/13/5/007. arXiv:hep-th/9506075 [hep-th]

47. S. Ferrara, R. Kallosh, Supersymmetry and attractors. Phys. Rev. **D54**, 1514–1524 (1996). http://dx.doi.org/10.1103/PhysRevD.54.1514. arXiv:hep-th/9602136

48. J. Bagger, E. Witten, Matter couplings in $N = 2$ supergravity. Nucl. Phys. **B222**, 1–10 (1983). http://dx.doi.org/10.1016/0550-3213(83)90605-3

49. K. Galicki, A generalization of the momentum mapping construction for quaternionic Kähler manifolds. Commun. Math. Phys. **108**, 117 (1987). http://dx.doi.org/10.1007/BF01210705

50. S. Marchiafava, P. Piccinni, M. Pontecorvo (eds.), Quaternionic structures in mathematics and physics, in *Proceedings of Workshop in Roma*, September 1999 (World Scientific, 2001). Available on http://www.univie.ac.at/EMIS/proceedings/QSMP99/

51. P. Frè, Gaugings and other supergravity tools of *p*-brane physics, in *Proceedings of the Workshop on Latest Development in M-Theory*, Paris, 1–9 February 2001. hep-th/0102114

52. R. D'Auria, S. Ferrara, On fermion masses, gradient flows and potential in supersymmetric theories. J. High Energy Phys. **05**, 034 (2001). http://dx.doi.org/10.1088/1126-6708/2001/05/034. arXiv:hep-th/0103153 [hep-th]

53. E. Bergshoeff, S. Cucu, T. de Wit, J. Gheerardyn, R. Halbersma, S. Vandoren, A. Van Proeyen, Superconformal $N = 2$, $D = 5$ matter with and without actions. J. High Energy Phys. **10**, 045 (2002). http://dx.doi.org/10.1088/1126-6708/2002/10/045. arXiv:hep-th/0205230 [hep-th]

54. A. Swann, HyperKähler and quaternionic Kähler geometry. Math. Ann. **289**, 421–450 (1991)

55. E. Bergshoeff, S. Cucu, T. de Wit, J. Gheerardyn, S. Vandoren, A. Van Proeyen, The map between conformal hypercomplex/hyper-Kähler and quaternionic(-Kähler) geometry. Commun. Math. Phys. **262**, 411–457 (2006) http://dx.doi.org/10.1007/s00220-005-1475-6. arXiv:hep-th/0411209 [hep-th]

56. S. J. Gates, Jr., T. Hübsch, S.M. Kuzenko, CNM models, holomorphic functions and projective superspace *C*-maps. Nucl. Phys. **B557**, 443–458 (1999). http://dx.doi.org/10.1016/S0550-3213(99)00370-3. arXiv:hep-th/9902211 [hep-th]

57. M. Roček, C. Vafa, S. Vandoren, Hypermultiplets and topological strings. J. High Energy Phys. **02**, 062 (2006). http://dx.doi.org/10.1088/1126-6708/2006/02/062. arXiv:hep-th/0512206 [hep-th]

58. M. Roček, C. Vafa, S. Vandoren, *Quaternion-Kähler Spaces, Hyperkahler Cones, and the c-Map* (2006). arXiv:math/0603048 [math-dg]

59. N. Banerjee, B. de Wit, S. Katmadas, The off-shell c-map. J. High Energy Phys. **01**, 156 (2016). http://dx.doi.org/10.1007/JHEP01(2016)156. arXiv:1512.06686 [hep-th]

60. B. de Wit, F. Saueressig, Off-shell $N = 2$ tensor supermultiplets. J. High Energy Phys. **9**, 062 (2006). http://dx.doi.org/10.1088/1126-6708/2006/09/062. arXiv:hep-th/0606148 [hep-th]

61. B. de Wit, A. Van Proeyen, Special geometry, cubic polynomials and homogeneous quaternionic spaces. Commun. Math. Phys. **149**, 307–334 (1992). http://dx.doi.org/10.1007/BF02097627. arXiv:hep-th/9112027 [hep-th]

62. D.V. Alekseevsky, Classification of quaternionic spaces with a transitive solvable group of motions. Math. USSR Izv. **9**, 297–339 (1975)

63. A. Anastasiou, L. Borsten, M. J. Duff, A. Marrani, S. Nagy, M. Zoccali, Are all supergravity theories Yang–Mills squared? Nucl. Phys. **B934**, 606–633 (2018). http://dx.doi.org/10.1016/j. nuclphysb.2018.07.023. arXiv:1707.03234 [hep-th]
64. E. Cremmer, A. Van Proeyen, Classification of Kähler manifolds in $N = 2$ vector multiplet–supergravity couplings. Class. Quant. Grav. **2**, 445 (1985). http://dx.doi.org/10.1088/0264-9381/2/4/010
65. B. de Wit, A. Van Proeyen, Isometries of special manifolds, in *Proceedings of the Meeting on Quaternionic Structures in Mathematics and Physics*, Trieste, September 1994. Available on http://www.emis.de/proceedings/QSMP94/. hep-th/9505097

Chapter 6
Final Results

Abstract The previous part of this book first considered conformal couplings, and finally these were projected to a subspace after gauge choices. We saw how then special and quaternionic-Kähler geometries emerge. In this chapter we obtain the final action and transformation laws for $\mathcal{N} = 2$, $D = 4$ (Sect. 6.1) and $D = 5$ (Sect. 6.2) Poincaré supergravity coupled to n_V vector multiplets and n_H hypermultiplets after elimination of auxiliary fields and in terms of the variables of Poincaré supergravity. We end with final remarks indicating also future directions.

6.1 Final $D = 4$ Poincaré Supergravity Results

This chapter repeats all definitions and is readable by itself for what concerns the final results. For the origin of the equations and more detailed explanations we refer to the previous chapters. Here we will first collect the already obtained relevant results for the bosonic sector, and then complete it for the fermionic sector. These final results could also have been obtained in other ways. In particular a geometric and rheonomic approach has been used in [1], where the complete results found in this chapter have also been obtained.

We consider the theory with

- Supergravity including the frame field e_μ^a, the doublet gravitino $\psi_\mu^i = P_L\psi_\mu^i$ and the graviphoton. The description of the graviphoton is unified with the description of the vector multiplets.
- $n = n_V$ vector multiplets with n_V complex scalars z^α, n_V physical fermions, described as $\chi_i^\alpha = P_L\chi_i^\alpha$ (and their right-handed components $\chi^{i\bar\alpha}$) and n_V vectors. The unified description of these vectors and the graviphoton is in W_μ^I, $I = 0, \ldots, n_V$, with the field strengths $F_{\mu\nu}^I$

$$F_{\mu\nu}^I = 2\partial_{[\mu} W_{\nu]}^I + f_{JK}{}^I W_\mu^J W_\nu^K . \tag{6.1}$$

The gauge algebra has structure constants $f_{JK}{}^I$. We will say more about the definition of the physical fermions in Sect. 6.1.2. Here we concentrate on the

© Springer Nature Switzerland AG 2020   205
E. Lauria, A. Van Proeyen, $\mathcal{N} = 2$ *Supergravity in D = 4, 5, 6 Dimensions*,
Lecture Notes in Physics 966, https://doi.org/10.1007/978-3-030-33757-5_6

description of the scalars. The latter are described as independent variables in a holomorphic $2(n_V + 1)$ component symplectic vector,

$$v(z) = \begin{pmatrix} Z^I(z) \\ \mathcal{F}_I(z) \end{pmatrix}. \tag{6.2}$$

This symplectic vector satisfies constraints such that the independent physical scalars are z^α with $\alpha = 1, \ldots, n_V$. The constraint is expressed using a symplectic inner product defined by (5.10):

$$\langle \nabla_\alpha v, \nabla_\beta v \rangle = 2 \left(\nabla_{[\alpha} Z^I \right) \left(\nabla_{\beta]} \mathcal{F}_I \right) = 0, \qquad \nabla_\alpha v \equiv \partial_\alpha v + \kappa^2 (\partial_\alpha \mathcal{K}) v, \tag{6.3}$$

where the Kähler potential $\mathcal{K}$ is determined by

$$e^{-\kappa^2 \mathcal{K}(z, \bar{z})} = -i\kappa^2 \langle v, \bar{v} \rangle = -i\kappa^2 \left(Z^I \bar{\mathcal{F}}_I - \mathcal{F}_I \bar{Z}^I \right). \tag{6.4}$$

This Kähler potential is related to the metric of the scalar manifold via:

$$g_{\alpha \bar{\beta}} = \partial_\alpha \partial_{\bar{\beta}} \mathcal{K} = i e^{\kappa^2 \mathcal{K}} \langle \nabla_\alpha v, \nabla_{\bar{\beta}} \bar{v} \rangle. \tag{6.5}$$

We require this metric[1] to be positive definite in the physical domain of the scalars z.

The geometric structure of the manifold is most visible in the fundamental equation (5.34):

$$i \left\langle \left(\bar{V} \, \nabla_\alpha V \right), \begin{pmatrix} V \\ \nabla_{\bar{\beta}} \bar{V} \end{pmatrix} \right\rangle = G \equiv \begin{pmatrix} \kappa^{-2} & 0 \\ 0 & g_{\alpha \bar{\beta}} \end{pmatrix}, \tag{6.6}$$

where we use the symplectic sections

$$V(z, \bar{z}) = y(z, \bar{z}) v(z) = \begin{pmatrix} X^I \\ F_I \end{pmatrix}, \qquad y = e^{\kappa^2 \mathcal{K}/2}. \tag{6.7}$$

Demanding positive kinetic terms, the matrix G should be positive definite, and thus the vector

$$\left(Z^I \, \nabla_{\bar{\alpha}} \bar{Z}^I \right) \qquad \text{or} \qquad \left(\bar{Z}^I \, \nabla_\alpha Z^I \right) \tag{6.8}$$

---

[1]There are further cohomological restrictions concerning the global structure of the metric, i.e. it should be Kähler manifolds of restricted type or 'Hodge manifolds', but these global restrictions are not discussed here. See footnote 24 page 143.

should be invertible (see proof in (5.37)). This then allows to define the kinetic matrix for the vectors (see the action below in (6.15)):

$$\mathcal{N}_{IJ} = \left(\mathcal{F}_I \, \nabla_{\tilde{\alpha}} \overline{\mathcal{F}}_I\right)\left(Z^J \, \nabla_{\tilde{\alpha}} \bar{Z}^J\right)^{-1}. \tag{6.9}$$

The positivity properties of G imply that the imaginary part of $\mathcal{N}_{IJ}$ is negative definite.

The 'usual case' is when the $(n_V + 1) \times (n_V + 1)$ matrix

$$\left(Z^I \, \nabla_\alpha Z^I\right) \tag{6.10}$$

is invertible. This can always be obtained by symplectic transformations (but one might sometimes prefer not to use such a symplectic basis). In this 'usual case' the condition (5.46) implies the existence of a holomorphic function $\mathcal{F}(Z)$, homogeneous of second order in Z, such that

$$\mathcal{F}_I(z) = \frac{\partial}{\partial Z^I} \mathcal{F}(Z(z)). \tag{6.11}$$

In the basis where a prepotential $\mathcal{F}(Z)$ exists, (6.9) can be expressed as in (5.2)

$$\mathcal{N}_{IJ}(z, \bar{z}) = \overline{\mathcal{F}}_{IJ} + i\frac{N_{IN}N_{JK} \, Z^N Z^K}{N_{LM} \, Z^L Z^M},$$

$$N_{IJ} \equiv 2\,\mathrm{Im}\,\mathcal{F}_{IJ} = -i\mathcal{F}_{IJ} + i\overline{\mathcal{F}}_{IJ}, \qquad \mathcal{F}_{IJ} = \partial_I \partial_J \mathcal{F}, \tag{6.12}$$

where ∂_I are derivatives w.r.t. Z^I. Due to the homogeneity of $F(X)$, we have $\mathcal{F}_{IJ} = F_{IJ}$, where the latter is the second derivative of $F(X)$ w.r.t. X^I.

- n_H hypermultiplets with scalars q^u, $u = 1, \ldots, 4n_H$ and spinors $\zeta^A = P_L \zeta^A$ (their right components are $\zeta_{\bar{A}}$) with $A, \bar{A} = 1, \ldots, 2n_H$. We introduced the fermions already in (4.138) and give more details in Sect. 6.1.2. The scalar manifold has a quaternionic-Kähler property, which has also been obtained from a projection in (4.153). The main ingredients are the metric h_{uv}, the triplet of hypercomplex structures $\mathbf{J}_u{}^v$ and a triplet connection $\boldsymbol{\omega}_u(q)$. They are related by (4.143), which extends the Levi-Civita covariant derivatives ∇_u, defined from the metric h_{uv}, to covariant derivatives $\tilde{\nabla}_u$ that preserve the complex structures:

$$\tilde{\nabla}_w \mathbf{J}_u{}^v \equiv \nabla_w \mathbf{J}_u{}^v + 2\,\boldsymbol{\omega}_w \times \mathbf{J}_u{}^v = 0. \tag{6.13}$$

The interactions are determined by the frame fields, $f^{iA}{}_u$, invertible as $4n_H \times 4n_H$ matrices, which satisfy a covariant constancy condition that contains also a $\mathrm{USp}(2n_H)$ connection $\omega_{uA}{}^B$

$$\tilde{\nabla}_v f^{iA}{}_u \equiv \partial_v f^{iA}{}_u + f^{jA}{}_u \omega_{vj}{}^i + f^{iB}{}_u \omega_{vB}{}^A - \Gamma^w_{vu} f^{iA}{}_w = 0. \tag{6.14}$$

Table 6.1 Multiplets and fields of the super-Poincaré theories for $D = 4$

field	spin	pure SG	vector mult.	hypermult.	indices
e_μ^a	2	1			$\mu, a = 0, \ldots, 3$
ψ_μ^i	$\frac{3}{2}$	2			$i = 1, 2$
W_μ^I	1	1	n_V		$I = 0, \ldots, n_V$
χ_α^i, ζ^A	$\frac{1}{2}$		$2n_V$	$2n_H$	$A = 1, \ldots, 2n_H$
z^α, q^u	0		$2n_V$	$4n_H$	$\alpha = 1, \ldots, n_V; u = 1, \ldots, 4n_H$

Notice that the connection ω_u is written here as a traceless anti-hermitian $\omega_{ui}{}^j$. We often switch between one and the other notation using (A.19). There are also reality conditions on these objects, for which we refer to (4.142).

Summarizing, Table 6.1 gives the physical fields in the Poincaré theory, multiplets in Poincaré language and the corresponding range of indices.

6.1.1   The Bosonic Action

We start by summarizing the results for the bosonic sector. The bosonic action is

$$
e^{-1}\mathcal{L}_{\text{bos}} = \tfrac{1}{2}\kappa^{-2}R - g_{\alpha\bar\beta}\hat\partial_\mu z^\alpha \hat\partial^\mu \bar z^\beta - \tfrac{1}{2}h_{uv}\hat\partial_\mu q^u \hat\partial^\mu q^v - V(z, \bar z, q)
$$
$$
+ \tfrac{1}{4}(\text{Im}\,\mathcal{N}_{IJ})F_{\mu\nu}^I F^{\mu\nu J} - \tfrac{1}{8}(\text{Re}\,\mathcal{N}_{IJ})e^{-1}\varepsilon^{\mu\nu\rho\sigma}F_{\mu\nu}^I F_{\rho\sigma}^J
$$
$$
+ \tfrac{2}{3}C_{I,JK}e^{-1}\varepsilon^{\mu\nu\rho\sigma}W_\mu^I W_\nu^J \left(\partial_\rho W_\sigma^K + \tfrac{3}{8}f_{LM}{}^K W_\rho^L W_\sigma^M\right). \quad (6.15)
$$

The first line starts with the Einstein–Hilbert term, which is the kinetic term for the graviton. It has been normalized to this form by our choice of dilatational gauge (4.55). Then appear the kinetic terms for the scalars related to the geometric structure. The main feature visible in these kinetic terms is that the scalar field space is divided in a special Kähler manifold, parametrized by $\{z^\alpha, \bar z^\alpha\}$ (and was obtained in (5.1)) and a quaternionic-Kähler manifold parametrized by $\{q^u\}$, with a metric obtained in (4.152). The hats on the kinetic terms are related to the gauging, see Sect. 6.1.1.1. The potential $V(z, \bar z, q)$ will be discussed in Sect. 6.1.1.2, since it is determined by this gauging. The second line contains the kinetic terms of the vectors using the matrix (6.9) or (6.12). The last line contains a Chern–Simons term that is also related to the embedding of the gauge group in the symplectic group.

6.1.1.1   The Gauging

The scalars have covariant kinetic terms for possible gaugings by the gauge vectors W_μ^I, with $I = 0, \ldots, n$:

$$\hat{\partial}_\mu z^\alpha = \partial_\mu z^\alpha - W_\mu{}^I k_I{}^\alpha , \qquad \hat{\partial}_\mu q^u = \partial_\mu q^u - W_\mu{}^I k_I{}^u , \tag{6.16}$$

First note that in these covariant derivatives the U(1), respectively SU(2), connections are absent, since the physical scalars z^α do not transform under the U(1) transformation and q^u do not transform under the SU(2) (see (4.130)).

The covariant derivatives are covariant for the gauge transformations, which are for the bosons:

$$\delta_G(\theta) W_\mu^I = \partial_\mu \theta^I - \theta^J f_{JK}{}^I W_\mu^K ,$$

$$\delta_G(\theta) z^\alpha = \theta^I k_I{}^\alpha ,$$

$$\delta_G(\theta) q^u = \theta^I k_I{}^u . \tag{6.17}$$

The Killing vectors $k_I{}^\alpha(z)$ and $k_I{}^u(q)$ separately satisfy the algebra with structure constants $f_{JK}{}^I$. They are projections of the Killing vectors of the embedding manifolds, see (5.114) and (4.155). They, respectively, preserve the complex and quaternionic structures, which can be seen in that they can be related to a real moment map P_I^0 and triplet real moment map $\mathbf{P}_I$, see (5.78) and (4.156):

$$g_{\alpha\bar{\beta}} k_I^{\bar{\beta}} = i\partial_\alpha P_I^0 ,$$

$$h_{uv} k_I^v = -\tfrac{1}{3} \mathbf{J}_u{}^v \cdot \tilde{\nabla}_v \mathbf{P}_I . \tag{6.18}$$

The gauge transformations do not necessarily leave the Kähler potential invariant

$$\delta_G(\theta) \mathcal{K}(z, \bar{z}) = \theta^I \left(r_I(z) + \bar{r}_I(\bar{z}) \right) , \qquad r_I(z) = k_I^\alpha \partial_\alpha \mathcal{K} + i P_I^0 . \tag{6.19}$$

Supersymmetry implies that the gauge transformations of the scalars are related to those of the vectors, and thus related to the structure constants.

This can best be seen in the symplectic formulation, where the vectors V transform in the conformal setting with a matrix

$$T_I = \begin{pmatrix} f_{IK}{}^J & 0 \\ 2C_{I,JK} & -f_{IJ}{}^K \end{pmatrix} , \tag{6.20}$$

where $C_{I,JK}$ are real coefficients, symmetric in the last two indices and $Z^I Z^J Z^K C_{I,JK} = 0$. They appear in the last line of (6.15).

Following Sect. 5.4.2 the gauge transformations in the Poincaré frame and the covariant transformations are of the form

$$\delta_G^\square(\theta)v(z) = \theta^I(-T_I - \kappa^2 r_I)v(z) , \qquad \widehat{\delta}_G(\theta)v(z) = \theta^I(-T_I - \kappa^2 P_I^0)v(z) . \tag{6.21}$$

The moment map can be written as a symplectic inner product

$$P_I^0 = e^{\kappa^2 \mathcal{K}} \langle T_I v(z), \bar{v}(\bar{z})\rangle = -e^{\kappa^2 \mathcal{K}} \langle v(z), T_I \bar{v}(\bar{z})\rangle . \tag{6.22}$$

Other ways to write P_I^0 are given in (4.52).

On the quaternionic side, the moment maps $\mathbf{P}_I$ are defined from the Killing vectors in (4.156). We use here h_{uv} to raise and lower indices, and thus the main equations are

$$\widetilde{\nabla}_u \mathbf{P}_I \equiv \partial_u \mathbf{P}_I + 2\omega_u \times \mathbf{P}_I = \mathbf{J}_{uv} k_I{}^v ,$$

$$k_I{}^u \mathbf{J}_{uv} k_J{}^v + \kappa^2 \mathbf{P}_I \times \mathbf{P}_J = f_{IJ}{}^K \mathbf{P}_K . \tag{6.23}$$

As we discussed around (4.160), the solution for $\mathbf{P}_I$ is unique when $n_H \neq 0$, and for $n_H = 0$ there are two types of FI terms possible: for SU(2) and for $U(1)$ factors.

From the derivatives of the quaternionic Killing vectors, one can define matrices

$$t_{IA}{}^B \equiv \tfrac{1}{2} f^v{}_{iA} \nabla_v k_I{}^u f^{iB}{}_u , \tag{6.24}$$

which satisfy the gauge algebra and are relevant for the gauge transformation of the fermions of the hypermultiplets.

6.1.1.2 The Potential

The remaining bosonic part of the action (6.15) is the scalar potential. It originates in the conformal setting from the third line in (4.59). We can use the bosonic part of the field equations for $\mathbf{Y}$ (4.54) to obtain

$$V = N^{-1|IJ} \mathbf{P}_I \cdot \mathbf{P}_J + N^{-1|IJ} P_I^0 P_J^0 + 2\bar{X}^I X^J k_I{}^X k_J{}^Y g_{XY} . \tag{6.25}$$

We rewrite this now in the Poincaré frame. For the first two terms we use (5.66) and (4.53). For the last term, we use the decomposition of the metric (4.134) and Killing vector (4.155) with the conditions (4.147), (4.157), (4.136):

$$k_I{}^X k_J{}^Y g_{XY} = k_I{}^u k_J{}^v h_{uv} - \kappa^2 \mathbf{P}_I \cdot \mathbf{P}_J . \tag{6.26}$$

We obtain

$$V = -4\kappa^2 X^I \bar{X}^J \mathbf{P}_I \cdot \mathbf{P}_J - \frac{1}{2}(\text{Im}\,\mathcal{N})^{-1|IJ} \left[\mathbf{P}_I \cdot \mathbf{P}_J + P_I^0 P_J^0\right] + 2\bar{X}^I X^J k_I{}^u k_J{}^v h_{uv}. \tag{6.27}$$

Since $\text{Im}\,\mathcal{N}$ is negative definite, only the first term gives a negative contribution to the potential. If there are no physical hypermultiplets ($n_H = 0$), this term can still be there for constant vectors $\mathbf{P}_I$, which are then the 'Fayet–Iliopoulos terms'.

Since the potential is important for applications, it is worthwhile to give alternative expressions. Therefore, let us first introduce [1–3]

$$U^{IJ} \equiv g^{\alpha\bar{\beta}} \nabla_\alpha X^I \nabla_{\bar{\beta}} \bar{X}^J = -\frac{1}{2}(\text{Im}\,\mathcal{N})^{-1|IJ} - \kappa^2 \bar{X}^I X^J \tag{6.28}$$

and using (4.53), we can write

$$V = -3\kappa^2 \bar{X}^I X^J \mathbf{P}_I \cdot \mathbf{P}_J + U^{IJ} (\mathbf{P}_I \cdot \mathbf{P}_J + P_I^0 P_J^0) + 2h_{uv} k_I^u k_J^v \bar{X}^I X^J . \tag{6.29}$$

Only the first term is negative. After we have discussed the supersymmetry transformations, see (6.88), we will understand the three terms in this expression as squares of supersymmetry transformations of, respectively, the gravitino, the gauginos (split in the SU(2) triplet and SU(2) singlet part) and the hyperinos. The fact that the scalar potential can be written in such a way is a general feature in supergravity (see, e.g., [4]).

Finally, let us still give alternative expressions for the part

$$V_1 = U^{IJ} P_I^0 P_J^0 = g^{\alpha\bar{\beta}} \nabla_\alpha X^I \nabla_{\bar{\beta}} \bar{X}^J P_I^0 P_J^0 = g_{\alpha\bar{\beta}} k_I{}^\alpha k_J{}^{\bar{\beta}} , \tag{6.30}$$

where for the last expression we used (5.115). Another way uses (5.105) on the last expression (remember that for gauged symmetries $\mathcal{P}_\Lambda$ is P_I^0), leading to

$$V_1 = i\bar{X}^I X^J \langle T_I V, T_J \bar{V}\rangle . \tag{6.31}$$

Using the explicit form of the matrices in (5.108) we get

$$V_1 = -i\bar{X}^I X^J f_{IJ}{}^K X^L \left(f_{LK}{}^M \bar{F}_M + 2C_{L,KM} \bar{X}^M\right) + \text{h.c.} \tag{6.32}$$

Exercise 6.1 As a simple example, consider the special Kähler manifold that we have discussed in example 4.1, which has $n = 1$, i.e. 2 vectors. The algebra should be abelian to leave F invariant. We consider no physical hypermultiplet, i.e. in the superconformal setup there is just a trivial one that is needed for compensation. Then the potential can only originate in the gauging of the compensating hypermultiplet, which is equivalent to having constant moment maps $\mathbf{P}_I$. In the equivariance condition, second of (6.23), only the second term survives, which says that $\mathbf{P}_1$ and $\mathbf{P}_2$ should point in the same direction in SU(2) space. Hence as in (4.163) we take

$$\mathbf{P}_I = \xi_I \mathbf{e}, \qquad \mathbf{e} \cdot \mathbf{e} = 1. \tag{6.33}$$

To calculate the potential we first obtain U^{IJ}, which can most easily be determined using $\nabla_z Z^I$. This gives, using (4.105),

$$\nabla_z Z^I = \partial_z Z^I + \kappa^2 Z^I \partial_z \mathcal{K} = \frac{1}{1 - z\bar{z}} \begin{pmatrix} \bar{z} \\ 1 \end{pmatrix}, \qquad \nabla_z X^I = e^{\kappa^2 \mathcal{K}/2} \nabla_z Z^I,$$

$$e^{\kappa^2 \mathcal{K}} = \frac{1}{1 - z\bar{z}}, \qquad g^{z\bar{z}} = \kappa^2 (1 - z\bar{z})^2,$$

$$U^{IJ} = \frac{\kappa^2}{1 - z\bar{z}} \begin{pmatrix} z\bar{z} & \bar{z} \\ z & 1 \end{pmatrix}, \qquad \bar{X}^I X^J = \frac{1}{1 - z\bar{z}} \begin{pmatrix} 1 & \bar{z} \\ z & z\bar{z} \end{pmatrix}. \tag{6.34}$$

Thus the potential, originating from the first two terms of (6.29) leads to

$$V = \frac{\kappa^2}{1 - z\bar{z}} \left[\xi_0^2 (z\bar{z} - 3) - 2\xi_0 \xi_1 (z + \bar{z}) + \xi_1^2 (1 - 3z\bar{z}) \right]. \tag{6.35}$$

Before the gauging the model has an SU(1, 1) rigid symmetry. The properties of the potential depend on the SU(1, 1)-invariant $\xi_0^2 - \xi_1^2$. There are thus 3 relevant cases, whether this invariant is positive, negative or zero [5]. In each case, we can take a standard choice by reparametrization, and we find, respectively, the following extrema

- Take $\xi_0 = \xi$ and $\xi_1 = 0$. There is an extremum at $z = 0$ with negative $V(z = 0)$, i.e. anti-de Sitter. $\xi_I U^{IJ} \xi_J$, which is the contribution from the supersymmetry of the gaugini, vanishes. Therefore this vacuum preserves supersymmetry.
- $\xi_0 = \xi_1$. In this case there is no extremum in the positivity domain $|z| < 1$.
- $\xi_0 = 0$ and $\xi_1 = \xi$. There is an extremum with positive V, i.e. de Sitter, with non-vanishing $\xi_I U^{IJ} \xi_J$, i.e. (spontaneously) broken supersymmetry.

Note that in the first case we can omit the scalars, and this case thus occurs in pure $\mathcal{N} = 2$ supergravity with possible gauging leading to the anti-de Sitter $\mathcal{N} = 2$ supergravity.   □

6.1.2   Physical Fermions

In the bosonic sector we have defined appropriate coordinates for implementing the conditions (4.56). We saw in the vector multiplet sector that it was often useful to use the invertible matrix (6.8) to split equations between the compensating part and physical part, as is most clear from (6.6). For usual symplectic bases where a prepotential is defined, we can simplify it to (6.10). We will use such a basis in this part, since the tensor calculus on which we base ourselves is developed in this setting.[2] For the hypermultiplets we defined an appropriate basis (4.130) where the first quaternion $\{q^0, q^r\}$ is put to a constant using the conformal gauge conditions, and the physical fields are q^u.

Similarly, we will now split the fermions in the parts that are put equal to zero by (4.58) and the physical fermions of the matter-coupled Poincaré supergravity theory.

6.1.2.1 Fermions of the $\mathcal{N} = 2$ Vector Multiplets

For the fermions of the conformal vector multiplet we consider the following splitting in $1 + n$ doublets:

$$\Omega_i^I = \chi_i^0 X^I + \chi_i^\alpha \nabla_\alpha X^I, \qquad \Omega^{iI} = \chi^{i0} \bar{X}^I + \chi^{i\bar{\alpha}} \nabla_{\bar{\alpha}} \bar{X}^I. \qquad (6.36)$$

Since the first condition in (4.56) leads to (5.70), the first of (4.58) imposes

$$\chi_i^0 = 0, \qquad \rightarrow \qquad \Omega_i^I = \chi_i^\alpha \nabla_\alpha X^I. \qquad (6.37)$$

The inverse can be obtained using (5.70) or (5.65):

$$\chi_i^\alpha = g^{\alpha\bar{\beta}} \nabla_{\bar{\beta}} \bar{X}^I N_{IJ} \Omega_i^J = -2g^{\alpha\bar{\beta}} \nabla_{\bar{\beta}} \bar{X}^I \operatorname{Im} N_{IJ} \Omega_i^J. \qquad (6.38)$$

Thus χ_i^α will be the n physical spin 1/2 fields, and their conjugates are defined as

$$\Omega^{iI} = \chi^{i\bar{\alpha}} \bar{\nabla}_{\bar{\alpha}} \bar{X}^I. \qquad (6.39)$$

In the conformal setting the transformations of X^I are

$$\delta X^I = \tfrac{1}{2} \bar{\epsilon}^i \Omega_i^I + (\lambda_{\rm D} + {\rm i}\lambda_T) X^I + \theta^K k_K{}^I. \qquad (6.40)$$

If we use (6.37) and (5.114) for the gauged isometries, we can write this as

$$\delta X^I = \tfrac{1}{2} \bar{\epsilon}^i \chi_i^\alpha \nabla_\alpha X^I + (\lambda_{\rm D} + {\rm i}\lambda_T) X^I + \theta^K \left(k_K{}^\alpha \nabla_\alpha X^I + {\rm i}\kappa^2 P_K^0 X^I \right). \qquad (6.41)$$

In the Poincaré theory, only supersymmetry and the gauge symmetry survive, and the other parameters $\lambda_{\rm D}$ and λ_T become functions of ϵ and θ^K by the decomposition laws.

---

[2]For a more general formulations where the fermions are also in a symplectic vectors, see [6].

On the other hand, after these gauge transformations X^I is only function of z^α and $\bar{z}^{\bar\alpha}$, and thus these transformations should also be obtainable as

$$\delta^\Box X^I = \delta z^\alpha \partial_\alpha X^I + \delta\bar{z}^{\bar\alpha} \partial_{\bar\alpha} X^I$$

$$= \delta z^\alpha \left(\nabla_\alpha - \tfrac{1}{2}\kappa^2 \partial_\alpha \mathcal{K} \right) X^I + \tfrac{1}{2}\kappa^2 \delta\bar{z}^{\bar\alpha} \partial_{\bar\alpha} \mathcal{K} X^I , \tag{6.42}$$

where we used (4.125) and the definition of the covariant derivatives in that section. Identifying this with (6.41) we find

$$\delta(\epsilon,\theta)z^\alpha = \tfrac{1}{2}\bar{\epsilon}^i \chi_i^\alpha + \theta^K k_K{}^\alpha ,$$

$$\lambda_{\mathrm{D}}(\epsilon,\theta) = 0 , \qquad \lambda_T(\epsilon,\theta) = \tfrac{1}{2}\mathrm{i}\kappa^2(\delta z^\alpha \partial_\alpha \mathcal{K} - \delta\bar{z}^{\bar\alpha}\partial_{\bar\alpha}\mathcal{K}) - \kappa^2\theta^K P_K^0 . \tag{6.43}$$

Thus our parametrization (6.36) is convenient for defining the physical fermions χ_i^α as supersymmetry partners of the z^α.

On the other hand, when we use covariant transformations (4.127), the U(1) compensation is much simpler. Indeed, the covariant transformation is

$$\hat{\delta} X^I = \nabla_\alpha X^I \, \delta z^\alpha , \tag{6.44}$$

which is then identified with (6.41) by the same z^α transformation and

$$\lambda_{\mathrm{D}}(\epsilon,\theta) = 0 , \qquad \lambda_T(\epsilon,\theta) = -\kappa^2\theta^K P_K^0 . \tag{6.45}$$

One can also check that the difference between (6.44) and (6.42) is in agreement with (4.128).

6.1.2.2 Fermions of the $\mathcal{N} = 2$ Hypermultiplets

For the hypermultiplet we use the coordinates introduced in Sect. 4.6:

$$\{\zeta^{\mathcal{A}}\} = \{\zeta^i, \zeta^A\} . \tag{6.46}$$

In these coordinates the second part of (4.58) with (4.141) reduces to

$$\zeta^i = 0 , \tag{6.47}$$

and the physical fermions are ζ^A. The reduction of (4.64), including (6.17), is

$$\delta q^u = -\mathrm{i}\bar{\epsilon}^i \zeta^A f^u{}_{iA} + \mathrm{i}\varepsilon^{ij}\rho^{\bar{A}B}\bar{\epsilon}_i\zeta_{\bar{A}} f^u{}_{jB} + \theta^I k_I{}^u . \tag{6.48}$$

The Poincaré supersymmetry transformations have a contribution from the SU(2) symmetry as in (4.168):

$$\lambda(\epsilon, \theta) = -\omega_u \delta q^u - \tfrac{1}{2}\kappa^2 \theta^I \mathbf{P}_I \,. \tag{6.49}$$

One may notice the similarity with (6.43). However, with the coordinate choice (4.131), these transformations (3.69) do not act on the physical scalars q^u.

For objects that transform under the SU(2), we can define covariant transformations that contain the geometric part of the gauge field (4.149), which then, similarly to the case for the T-transformations above, reduces the remaining compensation of SU(2) transformations to a gauge transformation:

$$\lambda(\epsilon, \theta) = -\tfrac{1}{2}\kappa^2 \theta^I \mathbf{P}_I \,, \tag{6.50}$$

similar to (6.45).

6.1.3 The Fermionic Part of the Poincaré Action

All this allows us to obtain the action for the physical fields. In the language of conformal fields, this action was given in (4.59). The first four lines gave rise to the bosonic action (6.15), and the gravitino kinetic term

$$\mathcal{L}_{\text{kin,gravitino}} = -e\bar{\psi}_{i\mu}\gamma^{\mu\nu\rho}D_\nu\psi^i_\rho \,,$$

$$D_\mu\psi_{vi} = \left(\partial_\mu + \tfrac{1}{4}\omega_\mu{}^{ab}(e)\gamma_{ab} + \tfrac{1}{2}i\mathcal{A}_\mu\right)\psi_{vi} + \mathcal{V}_{\mu i}{}^j\psi_{vj}, \,. \tag{6.51}$$

Here and below appear the (bosonic part of the) effective U(1) and SU(2) composite gauge fields:

$$\mathcal{A}_\mu = \tfrac{1}{2}\kappa^2 i\left[(\partial_\alpha \mathcal{K})\partial_\mu z^\alpha - (\partial_{\bar\alpha}\mathcal{K})\partial_\mu \bar{z}^{\bar\alpha}\right] - \kappa^2 W_\mu^I P_I^0$$

$$= \tfrac{1}{2}\kappa^2 i\left[(\partial_\alpha \mathcal{K})\hat\partial_\mu z^\alpha - (\partial_{\bar\alpha}\mathcal{K})\hat\partial_\mu \bar{z}^{\bar\alpha}\right] + \tfrac{1}{2}i W_\mu^I(r_I - \bar{r}_I),$$

$$\mathcal{V}_\mu = -\omega_u \partial_\mu q^u - \tfrac{1}{2}\kappa^2 W_\mu^I \mathbf{P}_I$$

$$= -\omega_u \hat\partial_\mu q^u + \tfrac{1}{2}W_\mu^I \mathbf{r}_I \,, \tag{6.52}$$

where ω_u, which is in the conformal variables $-\tfrac{1}{2}\kappa^2 \mathbf{k}_u$, is in the Poincaré variables the connection defined by (6.13).

Moreover these first four lines of (4.59) contain auxiliary fields, whose expressions in the conformal terminology have been given in (4.54) and (4.61) and contain also fermionic terms.

To translate to Poincaré language, it is first useful to use the expression of $D_\mu X^I$ obtained in (5.116)

$$D_\mu X^I = \nabla_\alpha X^I \, \widehat{\partial}_\mu z^\alpha \,, \tag{6.53}$$

with $\widehat{\partial}_\mu z^\alpha$ defined in (6.16). It is then convenient to repeatedly use (5.70), which then e.g. shows that the first term on the fifth line of (4.59) vanishes. For the second term, we note that the bosonic part of (4.48) is $D_\mu q^X \mathbf{k}_X$, which thus vanishes when we use field equations.

The sixth line of (4.59) contains the kinetic terms of the gauginos and hyperinos. We obtain for the covariant derivatives of the gauginos

$$\widehat{D}_\mu \Omega_i^I = (\widehat{D}_\mu \chi_i^\alpha) \nabla_\alpha X^I + \chi_i^\alpha \kappa^2 g_{\alpha\bar{\beta}} X^I \widehat{\partial}_\mu \bar{z}^{\bar{\beta}} \,, \tag{6.54}$$

$$\widehat{D}_\mu \chi_i^\alpha = \left(\partial_\mu + \tfrac{1}{4} \omega_\mu{}^{ab}(e) \gamma_{ab} + \tfrac{1}{2} i \mathcal{A}_\mu \right) \chi_i^\alpha + \mathcal{V}_{\mu i}{}^j \chi_j^\alpha - W_\mu^I \chi_i^\beta \nabla_\beta k_I{}^\alpha$$
$$+ \Gamma_{\beta\gamma}^\alpha \chi_i^\gamma \partial_\mu z^\beta \,. \tag{6.55}$$

Let us take some time to explain this in more detail, since this is also relevant to understand the gauge transformations of the χ_α^i in Sect. 6.1.5. The covariant derivative that is used here is given in (4.60). Then (6.37) should be used. The chiral weight of Ω_i^I is 1/2. The chiral weight of $\nabla_\alpha X^I$ is the same of that of X^I, i.e. +1. Thus the chiral weight of χ_i^α is $-1/2$. We will have to care then also for the difference between the term $-\tfrac{1}{2} i \mathcal{A}_\mu \Omega_I^i$ in $D_\mu \Omega_i^I$ and the one included in the term with $D_\mu \chi_i^\alpha$, i.e. this remaining term is

$$(D_\mu \Omega_i^I - D_\mu \chi_i^\alpha) \nabla_\alpha X^I = \ldots - i \mathcal{A}_\mu \chi_i^\alpha \nabla_\alpha X^I \,. \tag{6.56}$$

For the geometric part, we write

$$\widehat{D}_\mu \Omega_i^I = \widehat{D}_\mu \left(\chi_i^\alpha \nabla_\alpha X^I \right) = \left(\widehat{D}_\mu \chi_i^\alpha \right) \nabla_\alpha X^I + \chi_i^\alpha \widehat{\nabla}_\beta \nabla_\alpha X^I \partial_\mu z^\beta + \chi_i^\alpha \widehat{\nabla}_{\bar{\beta}} \nabla_\alpha X^I \partial_\mu \bar{z}^{\bar{\beta}} \,, \tag{6.57}$$

and with (5.76) this leads to (6.54) (and the covariant derivative $\widehat{\nabla}_\beta \nabla_\alpha X^I$ takes into account (6.56) for what concerns the non-gauge part). The remaining question concerns the gauge part: how it reduces to the W_μ^I terms mentioned in (6.55). For that, we see that the explicit W_μ^I terms in (4.60) lead to the terms

$$\widehat{D}_\mu \Omega_i^I = \ldots + \chi_i^\alpha \widehat{\nabla}_\alpha (W_\mu^J X^K f_{JK}{}^I) \,. \tag{6.58}$$

Note that W_μ^J appears here inside the bracket in order that the covariantization is respected. In this equation we then also treat the index J as holomorphic. The interpretation is similar to the discussion after (3.46): we treat the W_μ^J as independent of the z^α, but then $\widehat{\nabla}_\alpha W_\mu^J = \Gamma_{KL}^J W_\mu^L \partial_\alpha X^K$. Then we insert (5.114) and obtain

$$\widehat{D}_\mu \Omega_i^I = \ldots - \chi_i^\alpha \widehat{\nabla}_\alpha \left(W_\mu^J k_J{}^I \right)$$

$$= \ldots - \chi_i^\alpha \nabla_\alpha \left(W_\mu^J k_J{}^\beta \right) \nabla_\beta X^I - i\kappa^2 \chi_i^\alpha \nabla_\alpha (W_\mu^J P_J^0) X^I - \chi_i^\alpha W_\mu^J P_J^0 \nabla_\alpha X^I \,,$$

$$(6.59)$$

using once more (5.76). Note that in the last line we removed the hats on the covariant derivatives either because the J index is anyway contracted or the fact that $\widehat{\nabla}_\alpha X^I = \nabla_\alpha X^I$. This then allows to extract the W_μ^J out of the covariant derivative. In the last term we use that the gauge part of $\mathcal{A}_\mu$ is $-\kappa^2 W_\mu^J P_J^0$ and this term thus cancels the gauge part in (6.56). The second term can be identified with the gauge term included in $\hat{\partial}_\mu \bar{z}^{\bar\beta}$ in (6.54) using (5.78). This finally leaves the explicit W_μ^I term in (6.55).

Using (5.70), the parts proportional to X^I in $\widehat{D}_\mu \Omega_i^I$ do not contribute to the kinetic terms, which reduce to

$$e^{-1} \mathcal{L}_{\text{kin, matter ferm}} = -\tfrac{1}{4} g_{\alpha\bar\beta} \bar\chi_i^\alpha \widehat{\slashed{D}} \chi^{i\bar\beta} - \bar\zeta_{\bar A} \widehat{\slashed{D}} \zeta^B d^{\bar A}{}_B + \text{h.c.} \,, \qquad (6.60)$$

where our task at the hypermultiplets side was simple since the gauge choices just restrict $\zeta^{\mathcal{A}}$ to ζ^A. The covariant derivative is

$$\widehat{D}_\mu \zeta^A = \left(\partial_\mu + \tfrac{1}{4} \omega_\mu{}^{ab}(e) \gamma_{ab} + \tfrac{1}{2} i \mathcal{A}_\mu \right) \zeta^A - W_\mu^I t_{IB}{}^A \zeta^B + \partial_\mu q^u \omega_{uB}{}^A \zeta^B \,,$$

$$(6.61)$$

where $\omega_{uB}{}^A$ appears in (6.14). Note that since we now are restricted to the physical fermions we could replace $d^{\bar A}{}_{\mathcal{B}}$ by $\delta^{\mathcal{A}}{}_{\mathcal{B}}$, thus reducing also $\zeta^{\mathcal{A}}$ to ζ_A. We will keep here the covariant way of writing.

For the terms bilinear in fermions and derivatives of bosons, which are Noether terms (seventh line), we use (6.53) for the gaugino part, and for the hyperini the restriction of $\zeta^{\mathcal{A}}$ to ζ^A implies that we can use

$$D_\mu q^X f^{iA}{}_X = D_\mu q^u f^{iA}{}_u = \widehat{\partial}_\mu q^u f^{iA}{}_u \,, \qquad \hat{\partial}_\mu q^u = \partial_\mu q^u - W_\mu^I k_I^u \,. \qquad (6.62)$$

Therefore, these Noether terms are

$$e^{-1} \mathcal{L}_{\text{Noether}} = \bar\psi_{i\mu} \left[\tfrac{1}{2} g_{\alpha\bar\beta} (\widehat{\slashed{\partial}} z^\alpha) \gamma^\mu \chi^{i\bar\beta} + i (\widehat{\slashed{\partial}} q^u) \gamma^\mu f^{iA}{}_u \zeta_A \right] + \text{h.c.} \qquad (6.63)$$

For the coupling of the gauge field strengths to the fermions, we first have to take into account that the graviphoton field strength (last line of (4.54)) contains also a fermionic part, which is thus included in the TT term in the second line of (4.59). We can rewrite the expression for T_{ab} using the special geometry relations (5.69) and (5.54) as follows:

$$T_{ab}^- = -\kappa^2 \operatorname{Im} \mathcal{N}_{IJ} X^J \left[4\widehat{F}_{ab}^{-I} + \tfrac{1}{2} C_{\alpha\beta\gamma} g^{\gamma\bar{\delta}} \nabla_{\bar{\delta}} \bar{X}^I \, \bar{\chi}_i^\alpha \gamma_{ab} \chi_j^\beta \varepsilon^{ij} \right.$$
$$\left. +2\kappa^2 X^I \bar{\zeta}^A \gamma_{ab} \zeta^B C_{AB} \right] . \tag{6.64}$$

The square of this expression contains 2-fermion terms, which combine with the terms on the eighth line in (4.59) to the so called 'Pauli terms'[3]

$$e^{-1} \mathcal{L}_{\text{Pauli}} = F_{ab}^{-I} \operatorname{Im} \mathcal{N}_{IJ} Q^{ab-J} + \text{h.c.},$$
$$Q^{ab-J} \equiv \nabla_{\bar{\alpha}} \bar{X}^J \left(\tfrac{1}{8} g^{\beta\bar{\alpha}} C_{\beta\gamma\delta} \bar{\chi}_i^\gamma \gamma^{ab} \chi_j^\delta \varepsilon^{ij} + \bar{\chi}^{\bar{\alpha}i} \gamma^a \psi^{bj} \varepsilon_{ij} \right)$$
$$+ X^J \left(\bar{\psi}_i^a \psi_j^b \varepsilon^{ij} + \tfrac{1}{2} \kappa^2 \bar{\zeta}^A \gamma^{ab} \zeta^B C_{AB} \right) . \tag{6.65}$$

Notice that the above expression contains the coupling between fermions and gauge field strengths. The precise form of this term is governed by symplectic geometry. In particular, by similarity with (5.38) we can identify

$$\begin{pmatrix} Q^{ab-J} \\ \mathcal{N}_{IJ} Q^{ab-J} \end{pmatrix}, \tag{6.66}$$

as a symplectic vector [7]. The Pauli term is then i/2 times the symplectic inner product of this with the symplectic vector of F_{ab}^{-I} and $G_{ab\,I}^-$ defined in Sect. 5.2.1.

The ninth line in (4.59) are Goldstino terms:

$$e^{-1} \mathcal{L}_{\text{goldstino}} = -(\bar{\psi}_i \cdot \gamma \upsilon^i + \text{h.c.}), \qquad \upsilon^i = \tfrac{1}{2} W_\alpha{}^{ij} \chi_j^\alpha + 2N^i{}_A \zeta^A ,$$
$$W_\alpha{}^{ij} \equiv \left(i\varepsilon^{ij} P_I^0 - P_I{}^{ij} \right) \nabla_\alpha X^I = -\varepsilon^{ij} g_{\alpha\bar{\beta}} k_I^{\bar{\beta}} \bar{X}^I - P_I{}^{ij} \nabla_\alpha X^I ,$$
$$N^i{}_A \equiv i X^I k_I{}^u f^{iB}{}_u C_{BA} , \tag{6.67}$$

where we used (4.52).

The remaining terms in (4.59) are in the last two lines, and fermionic terms originating from the auxiliary field term $-\mathcal{N}_{IJ} \mathbf{Y}^I \cdot \mathbf{Y}^J$, using the expression

---

[3]In the expression of Q^{ab-J} we did not explicitly write that one should take the anti-self-dual combination in $[ab]$ since anyway this is multiplied by an anti-self-dual field strength in the Pauli terms.

in (4.54). The latter we can write using (5.66) and (5.55) as

$$Y_{ij}^I = -N^{-1|IJ} P_{Jij} - \tfrac{1}{4}\left[\bar{\chi}_i^\alpha \chi_j^\beta C_{\alpha\beta\gamma} g^{\gamma\bar{\gamma}} \nabla_{\bar{\gamma}} \bar{X}^I + \bar{\chi}^{i\bar{\alpha}} \chi^{j\bar{\beta}} \bar{C}_{\bar{\alpha}\bar{\beta}\bar{\gamma}} g^{\gamma\bar{\gamma}} \nabla_\gamma X^I\right].$$
$$(6.68)$$

These are thus fermion mass terms, which we can parametrize as

$$\mathcal{L}_m = \tfrac{1}{2} S_{ij} \bar{\psi}_\mu^i \gamma^{\mu\nu} \psi_\nu^j - \tfrac{1}{2} m^{ij}{}_{\alpha\beta} \bar{\chi}_i^\alpha \chi_j^\beta - m^i_{\alpha A} \bar{\chi}_i^\alpha \zeta^A - \tfrac{1}{2} m_{AB} \bar{\zeta}^A \zeta^B + \text{h.c.}$$
$$(6.69)$$

All these expressions are determined by the gauging. We obtain

$$S_{ij} = P_{Iij} \bar{X}^I,$$

$$m^{ij}{}_{\alpha\beta} = \tfrac{1}{2} P_I^{ij} C_{\alpha\beta\gamma} g^{\gamma\bar{\delta}} \nabla_{\bar{\delta}} \bar{X}^I + \varepsilon^{ij} \nabla_\alpha X^I k_I{}^{\bar{\gamma}} g_{\beta\bar{\gamma}},$$

$$m^{\bar{A}}_{i\bar{\alpha}} = 2i k_I^u \varepsilon_{ij} f^{jA}{}_u \nabla_{\bar{\alpha}} \bar{X}^I,$$

$$m^i_{\alpha A} = -2i k_I^u f^{iB}{}_u C_{BA} \nabla_\alpha X^I,$$

$$m_{AB} = -4 X^I t_{IAB}.$$
$$(6.70)$$

For the $\chi\chi$ mass term we used (5.114).

6.1.4   Total Action

We have rewritten the action in Poincaré variables. The result is

$$\mathcal{L} = \mathcal{L}_{\text{bos}} \qquad\qquad (6.15)$$
$$+ \mathcal{L}_{\text{kin,gravitino}} \qquad\qquad (6.51)$$
$$+ \mathcal{L}_{\text{kin, matter ferm}} \qquad\qquad (6.60)$$
$$+ \mathcal{L}_{\text{Noether}} \qquad\qquad (6.63)$$
$$+ \mathcal{L}_{\text{Pauli}} \qquad\qquad (6.65)$$
$$+ \mathcal{L}_{\text{goldstino}} \qquad\qquad (6.67)$$
$$+ \mathcal{L}_m \qquad\qquad (6.69)$$
$$+ \text{4-fermion terms,} \qquad\qquad (6.71)$$

where we indicate where these parts can be found.

6.1.5   Supersymmetry and Gauge Transformations

We still have to obtain the transformation rules under which this action is invariant. After all the gauge choices, the remaining local symmetries are (apart from the obvious general coordinate transformations and Lorentz symmetry): supersymmetry and the gauge transformations parametrized by θ^I.

In Sect. 6.1.2 we already obtained the transformations of the scalar fields in the Poincaré theory[4]:

$$\delta(\epsilon, \theta) z^\alpha = \tfrac{1}{2}\bar{\epsilon}^i \chi_i^\alpha + \theta^K k_K{}^\alpha ,$$

$$\delta(\epsilon, \theta) q^u = -i\bar{\epsilon}^i \zeta^A f^u{}_{iA} + i\varepsilon^{ij} \rho^{\bar{A}B} \bar{\epsilon}_i \zeta_{\bar{A}} f^u{}_{jB} + \theta^I k_I{}^u . \tag{6.72}$$

The other bosonic physical fields are the gravitons and the gauge fields W_μ^I, for which we can immediately obtain from (4.64) and (4.65)

$$\delta(\epsilon, \theta) e_\mu{}^a = \tfrac{1}{2}\bar{\epsilon}^i \gamma^a \psi_{\mu i} + \text{h.c.},$$

$$\delta(\epsilon, \theta) W_\mu^I = \tfrac{1}{2}\varepsilon^{ij} \bar{\epsilon}_i \gamma_\mu \chi_j^\alpha \nabla_\alpha X^I + \varepsilon^{ij} \bar{\epsilon}_i \psi_{\mu j} X^I + \text{h.c.} + \partial_\mu \theta^I + \theta^J W_\mu^K f_{KJ}{}^I . \tag{6.73}$$

We can read off the gauge transformations of the fermions from their covariant derivatives (6.55) and (6.61). We explained the derivation there in full. The explanation of the gauge transformations is identical by replacing W_μ^I by θ^I. In that derivation it was clear that the covariant transformations are a useful tool. The contributions of (6.45) and (6.50) are identical to the W_μ^I parts in the expressions of the auxiliary fields $\mathcal{A}_\mu$ and $\mathcal{V}_\mu$ in (6.52). Thus we get

$$\widehat{\delta}(\theta) \chi_i^\alpha = \theta^I \chi_i^\beta \nabla_\beta k_I{}^\alpha + \tfrac{1}{2}\kappa^2 \theta^I \left(i P_I^0 \chi_i^\alpha + P_i{}^j \chi_j^\alpha \right) ,$$

$$\widehat{\delta}(\theta) \zeta^A = \theta^I t_{IB}{}^A \zeta^B + \tfrac{1}{2}\kappa^2 \theta^I i P_I^0 \zeta^A . \tag{6.74}$$

Due to the compensating U(1) and SU(2) transformations, also the gravitino transforms under the gauge transformations. The covariant transformation is

$$\widehat{\delta}(\theta) \psi_\mu^i = -\tfrac{1}{2}\kappa^2 \theta^I \left(i P_I^0 \psi_\mu^i + P_j{}^i \psi_\mu^j \right) . \tag{6.75}$$

For the supersymmetry transformations of the fermions we need the values of the auxiliary fields, which we rewrite here in the conformal context. We obtain

---

[4]All the transformations in this section are the Poincaré transformations, indicated at some places in this book with ☐. We will use this indication only at the end of this section when we make the connection with the superconformal theory.

from (4.61)

$$A_\mu^F = -\tfrac{1}{8}i\kappa^2 g_{\alpha\bar\beta}\bar\chi^{i\bar\beta}\gamma_\mu\chi_i^\alpha + \tfrac{1}{2}i\kappa^2\bar\zeta_{\bar A}\gamma_\mu\zeta^B d^{\bar A}{}_B\,,$$

$$V_\mu^F = \tfrac{1}{8}\kappa^2 g_{\alpha\bar\beta}\bar\chi^{i\bar\beta}\gamma_\mu\chi_j^\alpha\tau_i{}^j\,. \tag{6.76}$$

These appear in the covariant derivatives. The linear part of $\mathcal{D}_\mu X^I$ in (3.19) was simplified in (5.116), and we obtain the full result:

$$\mathcal{D}_\mu X^I = \nabla_\alpha X^I\widehat\partial_\mu z^\alpha - iA_\mu^F X^I - \tfrac{1}{2}\bar\psi_\mu^i\Omega_i^I$$

$$= \nabla_\alpha X^I\left(\widehat\partial_\mu z^\alpha - \tfrac{1}{2}\bar\psi_\mu^i\chi_i^\alpha\right) + \tfrac{1}{2}\kappa^2 X^I\left(-\tfrac{1}{4}g_{\alpha\bar\beta}\bar\chi^{i\bar\beta}\gamma_\mu\chi_i^\alpha + \bar\zeta_{\bar A}\gamma_\mu\zeta^B d^{\bar A}{}_B\right)\,. \tag{6.77}$$

The covariant derivative of the hyperscalars in (3.102) are

$$\widehat{\mathcal{D}}_\mu A^{iA} = f^{iA}{}_u\mathcal{D}_\mu q^u\,,$$

$$\mathcal{D}_\mu q^u = \widehat\partial_\mu q^X + i\bar\psi_\mu^i\zeta^A - i\bar\psi_{\mu j}\zeta_{\bar B}\,\varepsilon^{ji}\rho^{\bar B A}\,. \tag{6.78}$$

With (6.76) we can also rewrite the contribution of S-supersymmetry from (4.68) as

$$\kappa^{-2}\eta^i(\epsilon) = -\tfrac{1}{2}X^I P_I^{ij}\epsilon_j + \tfrac{1}{8}\gamma^a\epsilon^j g_{\alpha\bar\beta}\bar\chi^{i\bar\beta}\gamma_a\chi_j^\alpha + \tfrac{1}{16}\gamma^{ab}\varepsilon^{ij}\epsilon_j C^{\bar A\bar B}\bar\zeta_{\bar A}\gamma_{ab}\zeta_{\bar B}$$

$$+ \gamma^a\epsilon^i\left[\tfrac{1}{16}g_{\alpha\bar\beta}\bar\chi_j^\alpha\gamma_a\chi^{j\bar\beta} + \tfrac{1}{4}\bar\zeta_{\bar A}\gamma_a\zeta^B d^{\bar A}{}_B\right]\,. \tag{6.79}$$

The value of the auxiliary field T_{ab} was already given in (6.64). It appears in the transformation of the gravitino and is therefore called the graviphoton field strength. Its bosonic part can also be written as

$$T_{ab}^-\big|_{\text{bos}} = -4\kappa^2 X^I\,\text{Im}\,N_{IJ}F_{ab}^{-J} = -2i\kappa^2\left(X^I G_{I\,ab}^- - F_I F_{ab}^{-I}\right)\,, \tag{6.80}$$

where $G_{I\,ab}$ was defined in (5.5). The graviphoton is thus a scalar-field dependent combination of the field strengths of the $(n_V + 1)$ gauge fields. The last expression in (6.80) is symplectic invariant and can thus also be used in the absence of a prepotential.

Note that this graviphoton field strength is the 'projection' of $\text{Im}\,N_{IJ}F_{ab}^{-J}$ on X^I. The gaugino field strengths are the projection of these quantities on $\nabla_{\bar\alpha}\bar X^I$

$$G_{ab}^{-\alpha} = g^{\alpha\bar\beta}\nabla_{\bar\beta}\bar X^I\,\text{Im}\,N_{IJ}F_{ab}^{-J}\,. \tag{6.81}$$

We are ready to give the supersymmetry transformations of the fermions. The most relevant part for applications is the bosonic part in the transformations, and we will here restrict to these. For the full transformations, again covariant transformations would be helpful. But the difference between the two is quadratic in fermions.

For the gravitino, we obtain from (4.64)

$$
\delta(\epsilon)\psi_\mu^i = \left(\partial_\mu + \tfrac{1}{4}\omega_\mu{}^{ab}\gamma_{ab} - \tfrac{1}{2}i\mathcal{A}_\mu\right)\epsilon^i - \mathcal{V}_\mu{}^i{}_j\,\epsilon^j - \tfrac{1}{16}\gamma^{ab}T_{ab}^-\varepsilon^{ij}\gamma_\mu\epsilon_j
$$
$$
+ \tfrac{1}{2}\kappa^2\gamma_\mu S^{ij}\epsilon_j + \dots,   \tag{6.82}
$$

where ... refers to terms of higher order in the fermions; $\mathcal{A}_\mu$, $\mathcal{V}_\mu{}^i{}_j$ and T_{ab}^- are found in (6.52) and (6.80). In (6.79) only the first term is bosonic, and is proportional to the triplet S^{ij} defined as mass matrix of the gravitini in (6.70).

For the gauginos, we insert the transformation (4.64) in (6.38) and obtain

$$
\delta(\epsilon)\chi_i^\alpha = g^{\alpha\bar{\beta}}\nabla_{\bar{\beta}}\bar{X}^I N_{IJ}\left[\slashed{D}X^J\epsilon_i + \tfrac{1}{4}\gamma^{ab}\mathcal{F}_{ab}{}^J\varepsilon_{ij}\epsilon^j + Y_{ij}{}^J\epsilon^j\right.
$$
$$
\left. + X^K\bar{X}^L f_{KL}{}^J\varepsilon_{ij}\epsilon^j + 2X^J\eta_i\right] + \dots
$$
$$
= \hat{\slashed{\partial}}z^\alpha\epsilon_i - \tfrac{1}{2}G_{ab}^{-\alpha}\gamma^{ab}\varepsilon_{ij}\epsilon^j + g^{\alpha\bar{\beta}}\epsilon^j\overline{W}_{\bar{\beta}ji} + \dots   \tag{6.83}
$$

The η term does not contribute here due to one of the expressions in (5.70). In fact, the part of Ω_i^I that transforms under η_i is the orthogonal one to χ_i^α, which has been gauge-fixed to zero. The first term follows directly from the bosonic part of (6.77). For the second term, one uses that (6.81) can also be written as

$$
G_{ab}^{-\alpha} = -\tfrac{1}{2}g^{\alpha\bar{\beta}}\nabla_{\bar{\beta}}\bar{X}^I N_{IJ}\mathcal{F}_{ab}^{-J} + \text{fermionic terms},   \tag{6.84}
$$

which follows from the definition (3.16), the expression for T^- in (6.80) and the difference between the two formulas in (5.66). For the following terms we use (6.68) and (4.52), and these combine in the expression $\overline{W}_{\bar{\beta}ji}$, the complex conjugate of the expression in (6.67)

$$
\overline{W}_{\bar{\alpha}ij} = \left(-i\varepsilon_{ij}P_I^0 - P_{Iij}\right)\nabla_{\bar{\alpha}}\bar{X}^I = -\varepsilon_{ij}g_{\beta\bar{\alpha}}k_I^\beta\bar{X}^I - P_{Iij}\nabla_{\bar{\alpha}}\bar{X}^I.   \tag{6.85}
$$

For the hyperini, we project from (4.64), use (6.78) and the S-supersymmetry again does not contribute due to (4.141) (again because the physical fermions have been chosen like this). Also (4.139) implies that only one term contributes from the Killing vectors (the index X reduces to u) and thus we get

$$
\delta(\epsilon)\zeta^A = \tfrac{1}{2}i f^{iA}{}_u\hat{\slashed{\partial}}q^u\epsilon_i + i\bar{X}^I k_I{}^X f^{iA}{}_X\varepsilon_{ij}\epsilon^j + \bar{N}_i{}^A\epsilon^i + \dots,   \tag{6.86}
$$

with again the complex conjugate of the expressions in (6.67) (using (4.142))

$$\bar{N}_i{}^{\bar{A}} = (N^i{}_A)^* = -i\varepsilon_{ij} d^{\bar{A}}{}_B f^{jB}{}_u k_I{}^u \bar{X}^I. \tag{6.87}$$

Let us introduce the name 'fermion shifts' [8, 4] for the scalar, non-derivative parts of the supersymmetry transformations of the fermions. Thus these are the last terms in (6.82), (6.83) and (6.86). The fact that these matrices appear in the mass terms of the gravitino (6.69) and in the Goldstino (6.67) is a general fact, or Ward equation of all supergravities.[5] They also lead to another interpretation of the potential. A general formula in supergravity says that the square of the fermion shifts, using the kinetic matrix of the fermions, obeys

$$\delta^i_j V = -3\kappa^2 S^{ik} S_{jk} + W_\alpha{}^{ik} g^{\alpha\bar{\beta}} \overline{W}_{\bar{\beta} jk} + 4N^i{}_A (d^{-1})^A{}_{\bar{B}} \bar{N}_j{}^{\bar{B}},$$

$$2V = -3\kappa^2 S^{ik} S_{ik} + W_\alpha{}^{ik} g^{\alpha\bar{\beta}} \overline{W}_{\bar{\beta} ik} + 4N^i{}_A (d^{-1})^A{}_{\bar{B}} \bar{N}_i{}^{\bar{B}}. \tag{6.88}$$

To prove this explicitly for the non-diagonal part, one needs the equivariance relation (4.159). For the diagonal part, inserting the definitions and e.g. (A.22) one re-obtains the form of the potential as in (6.29).

6.1.5.1 Relation Between Symmetries

A few final remarks are in order for the relation between the supersymmetry and gauge transformations. Let us first summarize how the symmetries of the Poincaré theory are related to those of the conformal theory (decomposition law):

$$\delta^\square[\epsilon, \theta] = \delta_Q[\epsilon] + \delta_G[\theta] + \delta_S[\eta(\epsilon)] + \delta_T[\lambda_T(\epsilon, \theta)] + \delta_{SU(2)}[\lambda(\epsilon, \theta)]$$

$$+ \delta_K[\lambda_K^a(0, \epsilon, \eta)]. \tag{6.89}$$

At the right-hand side of this equation are the superconformal transformations. In the third term appears (6.79). For the T and SU(2) transformations, the parameters are given by (6.43) and (6.49). When using covariant transformations, they can be reduced to (6.45) and (6.50). The final term uses (4.5), in which we can now put $\lambda_D = 0$. This last term is not often important since none of the independent fields transforms under K-symmetry.

The commutators between symmetries in the Poincaré theories are then obtained from the conformal commutators (3.18), (2.9), (2.10) combined by (6.89). We

---

[5]This can be proven by looking just at the variation of the action linear in matter fermions and with one spacetime derivative.

obtain (omitting terms of higher order in fermions)

$$[\delta^{\square}(\epsilon_1), \delta^{\square}(\epsilon_2)] = \delta_{\text{cgct}}\left(\xi_3^a(\epsilon_1, \epsilon_2)\right) + \delta_M\left(\lambda_3^{ab}(\epsilon_1, \epsilon_2)\right) + \delta_G^{\square}\left(\theta_3^I(\epsilon_1, \epsilon_2)\right),$$

$$\xi_3^a = \tfrac{1}{2}\bar{\epsilon}_2^i\gamma^a\epsilon_{1i} + \text{h.c.},$$

$$\lambda_3^{ab}(\epsilon_1, \epsilon_2) = \tfrac{1}{4}\bar{\epsilon}_1^i\left(T^{+ab}\varepsilon_{ij} + \kappa^2 X^I\gamma^{ab}P_{I\,ij}\right)\epsilon_2^j + \text{h.c.},$$

$$\theta_3^I(\epsilon_1, \epsilon_2) = \varepsilon^{ij}\bar{\epsilon}_{2i}\epsilon_{1j}X^I + \text{h.c.}, \qquad\qquad (6.90)$$

and

$$[\delta_G^{\square}(\theta), \delta_Q^{\square}(\epsilon)] = \delta_Q\left[\epsilon_3(\epsilon, \theta)\right],$$

$$\epsilon_3^i(\epsilon, \theta) = \tfrac{1}{2}\kappa^2\theta^I\left(iP_I^0\epsilon^i + P_{I\,j}{}^i\right)\epsilon^j. \qquad\qquad (6.91)$$

For example, the second term in $\lambda_3^{ab}(\epsilon_1, \epsilon_2)$ originates from the $[Q, S]$ conformal commutator where the bosonic part of the parameter (6.79) is used.

One can also see that the appearance of the last term in (6.90) is consistent with the transformation of the gauge vectors proportional to the gravitino in (6.73) following general rules as in (2.3). Similarly, the gauge transformation of the gravitino (6.75) or its supersymmetry transformation proportional to W_μ^I included in (6.82) implies that these gauge transformations do not commute the supersymmetry and also lead to (6.91). This non-vanishing is the contribution of gauge transformations to R-symmetry and is an interpretation of the moment maps.

The result (6.90) is important for interpreting solutions of the field equations. For example, for solutions with non-vanishing value of X^I and a gauged isometry group or non-vanishing value of T^{ab} (charged black hole solutions), this implies that there is a central charge in the preserved algebra.

6.2 Final Results for $D = 5$ Poincaré Supergravity

As in the previous chapter for $D = 4$, we will collect here the main results of $D = 5$ for general couplings with vector multiplets and hypermultiplets, though we will be much shorter since the results in our conventions have been collected in [9], for a large part based on previous work, e.g. [10–12]. Actually, in giving these results, we will extend the vector multiplets, which we discussed in Chaps. 3 and 4, to vector-tensor multiplets, where the vector multiplets are off-shell, but the tensor multiplets are on-shell.

We consider the theory with

- Supergravity, including the frame field e_μ^a and gravitino ψ_μ^i. The graviphoton is included in the vector multiplets.

- Vector-tensor multiplets enumerated by $\tilde{I} = 0, \ldots, n_V + n_T$ where n_V is the number of vector multiplets and n_T is the number of tensor multiplets. The index is further split as $\tilde{I} = (I, M)$, where $I = 0, \ldots, n_V$ and $M = n_V + 1, \ldots, n_V + n_T$. The vector or tensor fields are grouped in

$$H_{\mu\nu}^{\tilde{I}} \equiv \left(F_{\mu\nu}^I, \tilde{B}_{\mu\nu}^M\right), \qquad F_{\mu\nu}^I \equiv 2\partial_{[\mu} W_{\nu]}^I + f_{JK}{}^I W_\mu^J W_\nu^K, \tag{6.92}$$

where $\tilde{B}_{\mu\nu}^M$ are the fundamental tensor fields and W_μ^I are the fundamental vector fields, gauging an algebra with structure constants $f_{JK}{}^I$. The fermions of these multiplets are denoted as λ^{xi} and the real scalars as ϕ^x where $x = 1, \ldots, n_V + n_T$. In Sect. 3.3.2, we constructed actions for the vector multiplets based on a symmetric constant tensor C_{IJK}. For vector-tensor multiplets this is extended to a tensor $C_{\tilde{I}\tilde{J}\tilde{K}}$. In order to get a standard normalization, we rescale the $C_{\tilde{I}\tilde{J}\tilde{K}}$ symbol and the vector multiplet scalars $\sigma^{\tilde{I}}$ (generalizations of the σ^I introduced in Sect. 3.2.1) as follows:

$$\sigma^{\tilde{I}} \equiv \sqrt{\frac{3}{2\kappa^2}} h^{\tilde{I}}, \qquad C_{\tilde{I}\tilde{J}\tilde{K}} \equiv -2\sqrt{\frac{2\kappa^2}{3}} C_{\tilde{I}\tilde{J}\tilde{K}}. \tag{6.93}$$

Furthermore, for the tensor multiplets we need an antisymmetric and invertible metric Ω_{MN} and the structure constants $f_{IJ}{}^K$ are extended to transformation matrices $t_{I\tilde{J}}{}^{\tilde{K}}$ related by

$$C_{M\tilde{J}\tilde{K}} = \sqrt{\frac{3}{8\kappa^2}} t_{(\tilde{J}\tilde{K})}{}^P \Omega_{PM}, \qquad t_{I[M}{}^P \Omega_{N]P} = 0, \qquad t_{I(\tilde{J}}{}^{\tilde{M}} C_{\tilde{K}\tilde{L})\tilde{M}} = 0,$$

$$(t_M)_{\tilde{J}}{}^{\tilde{K}} = 0, \qquad (t_I)_{\tilde{J}}{}^{\tilde{K}} = \begin{pmatrix} f_{IJ}{}^K & (t_I)_J{}^N \\ 0 & (t_I)_M{}^N \end{pmatrix}. \tag{6.94}$$

Note that this implies that at least one index of a non-zero $C_{\tilde{I}\tilde{J}\tilde{K}}$ should correspond to a vector multiplet. The above-mentioned physical scalars ϕ^x are a parametrization of the manifold defined as solution of the constraint (4.170), which is now written as

$$C_{\tilde{I}\tilde{J}\tilde{K}} h^{\tilde{I}}(\phi) h^{\tilde{J}}(\phi) h^{\tilde{K}}(\phi) = 1. \tag{6.95}$$

- n_H hypermultiplets with scalars q^u and spinors ζ^A, where $u = 1, \ldots, 4n_H$ and $A = 1, \ldots, 2n_H$. Their interactions are determined by the frame fields, $f^{iA}{}_u$, invertible as $4n_H \times 4n_H$ matrices, identical to what we saw for $D = 4$.

Table 6.2 Multiplets and fields of the super-Poincaré theories for $D = 5$

field	pure SG	vector/tensor mult.	hypermult.	indices
e_μ^a	1			$\mu, a = 0, \ldots, 4$
ψ_μ^i	2			$i = 1, 2$
W_μ^I	1	n_V		$I = 0, \ldots, n_V$
$\tilde{B}_{\mu\nu}^M$		n_T		$M = n_V + 1, \ldots, n_V + n_T = n,$
				$\tilde{I} = 0, \ldots, n_V + n_T = n$
λ_i^x, ζ^A		$2n$	$2n_H$	$A = 1, \ldots, 2n_H$
ϕ^x, q^u		n	$4n_H$	$x = 1, \ldots, n; u = 1, \ldots, 4n_H$

This leads to the independent physical fields in Table 6.2.

As mentioned, the full results are in [9].[6] We repeat the main results. The bosonic action is

$$
\begin{aligned}
e^{-1}\mathcal{L}_{\mathrm{bos}} = {} & \frac{1}{2\kappa^2}R - \frac{1}{4}a_{\tilde{I}\tilde{J}}H_{\mu\nu}^{\tilde{I}}H^{\tilde{J}\mu\nu} - \frac{1}{2}g_{xy}\hat{D}_\mu\phi^x\hat{D}^\mu\phi^y - \frac{1}{2}g_{uv}\hat{D}_\mu q^u\hat{D}^\mu q^v - V \\
& + \frac{1}{16g}e^{-1}\varepsilon^{\mu\nu\rho\sigma\tau}\Omega_{MN}\tilde{B}_{\mu\nu}^M\left(\partial_\rho\tilde{B}_{\sigma\tau}^N + 2gt_{IJ}{}^N W_\rho^I F_{\sigma\tau}^J + t_{IP}{}^N W_\rho^I\tilde{B}_{\sigma\tau}^P\right) \\
& + \frac{\kappa}{12}\sqrt{\frac{2}{3}}e^{-1}\varepsilon^{\mu\nu\lambda\rho\sigma}C_{IJK}W_\mu^I\left[F_{\nu\lambda}^J F_{\rho\sigma}^K\right. \\
& \left. + f_{FG}{}^J W_\nu^F W_\lambda^G\left(-\frac{1}{2}F_{\rho\sigma}^K + \frac{1}{10}f_{HL}{}^K W_\rho^H W_\sigma^L\right)\right] \\
& - \frac{1}{8}e^{-1}\varepsilon^{\mu\nu\lambda\rho\sigma}\Omega_{MN}t_{IK}{}^M t_{FG}{}^N W_\mu^I W_\nu^F W_\lambda^G\left(-\frac{1}{2}F_{\rho\sigma}^K + \frac{1}{10}f_{HL}{}^K W_\rho^H W_\sigma^L\right). \\
& \hspace{11cm} (6.96)
\end{aligned}
$$

The metrics for the vectors and the vector–scalars are defined by

$$
a_{\tilde{I}\tilde{J}} = -2C_{\tilde{I}\tilde{J}\tilde{K}}h^{\tilde{K}} + 3h_{\tilde{I}}h_{\tilde{J}}, \qquad h_{\tilde{I}} \equiv C_{\tilde{I}\tilde{J}\tilde{K}}h^{\tilde{J}}h^{\tilde{K}} = a_{\tilde{I}\tilde{J}}h^{\tilde{J}},
$$

$$
g_{xy} = h_x^{\tilde{I}}h_y^{\tilde{J}}a_{\tilde{I}\tilde{J}}, \qquad h_x^{\tilde{I}} \equiv -\sqrt{\frac{3}{2\kappa^2}}\partial_x h^{\tilde{I}}(\phi). \tag{6.97}
$$

Many useful relations are given in Appendix C of [9].

The domain of the variables should be limited to $h^I(\phi) \neq 0$ and the metrics a_{IJ} and g_{xy} should be positive definite. Similar to (5.70) for $D = 4$, one can block-diagonalize the metric a_{IJ} in a singlet corresponding to the compensating multiplet, and the part corresponding to the physical scalars:

$$
\begin{pmatrix} h^I \\ h_x^I \end{pmatrix} a_{IJ} \begin{pmatrix} h^J & h_y^J \end{pmatrix} = \begin{pmatrix} 1 & 0 \\ 0 & g_{xy} \end{pmatrix}. \tag{6.98}
$$

---

[6] A few changes of notation are mentioned in (C.3).

The metric for the hyperscalars is the same as discussed for $D = 4$ in Sect. 6.1.1 with curvature as in (5.167).

All other quantities in (6.96) are related to gauged symmetries. The gauge symmetry transformations (with parameters θ^I) of the bosons are

$$\delta_G W_\mu^I = \partial_\mu \theta^I - \theta^J f_{JK}{}^I W_\mu^K, \qquad \delta_G \tilde{B}_{\mu\nu}^M = -\theta^J t_{J\tilde{K}}{}^M H_{\mu\nu}^{\tilde{K}},$$

$$\delta_G h^{\tilde{I}}(\phi) = -\theta^J t_{J\tilde{K}}{}^{\tilde{I}} h^{\tilde{K}},$$

$$\delta_G \phi^x = \theta^I k_I{}^x, \qquad k_I{}^x \equiv \frac{1}{\kappa}\sqrt{\frac{3}{2}} t_{I\tilde{J}}{}^{\tilde{K}} h^{\tilde{J}} h_{\tilde{K}}{}^x,$$

$$\delta_G q^u = \theta^I k_I{}^u, \tag{6.99}$$

where $k_I{}^u$ should be isometries of the quaternionic-Kähler metric h_{uv}, whose commutators are determined by the structure constants $f_{IJ}{}^K$:

$$2k_{[I}{}^v \partial_v k_{J]}{}^u = f_{IJ}{}^K k_K{}^u. \tag{6.100}$$

These transformations determine immediately the covariant derivatives in this bosonic truncation:

$$\hat{D}_\mu \phi^x = \partial_\mu \phi^x - W_\mu^I k_I{}^x, \qquad \hat{D}_\mu q^u = \partial_\mu q^u - W_\mu^I k_I{}^u. \tag{6.101}$$

The gauge transformations are also in one-to-one relation with triplet moment maps. For example, the moment map is obtained from the Killing vectors in (4.160), derived in the treatment of $D = 4$ and also valid here. The potential V will appear below in (6.104) after discussing the supersymmetry transformations.

The $\mathcal{N} = 2$ supersymmetry rules of the fermionic fields, up to bilinears in the fermions, are given by[7]

$$\delta(\epsilon)\psi_\mu^i = D_\mu(\omega)\epsilon^i + \frac{\mathrm{i}}{4\sqrt{6}}\left[\kappa h_{\tilde{I}} H^{\tilde{I}\nu\rho}\left(\gamma_{\mu\nu\rho} - 4g_{\mu\nu}\gamma_\rho\right)\epsilon^i + 2\kappa^{-1} P_j{}^i \gamma_\mu \epsilon^j\right],$$

$$D_\mu(\omega)\epsilon^i = \left(\partial_\mu + \tfrac{1}{4}\omega_\mu{}^{ab}\gamma_{ab}\right)\epsilon^i + \partial_\mu q^u \omega_{uj}{}^i \epsilon^j + \tfrac{1}{2}\kappa^2 W_\mu^I P_{Ij}{}^i \epsilon^j,$$

$$\delta(\epsilon)\lambda^{xi} = -\tfrac{1}{2}\mathrm{i}\widehat{\slashed{D}}\phi^x \epsilon^i + \tfrac{1}{4}\gamma \cdot H^{\tilde{I}} h_{\tilde{I}}^x \epsilon^i + \kappa^{-2} P^x{}_j{}^i \epsilon^j + \tfrac{1}{2}\kappa^{-2} W^x \epsilon^i,$$

$$\delta(\epsilon)\zeta^A = \tfrac{1}{2}\mathrm{i}\gamma^\mu \hat{D}_\mu q^u f^{iA}{}_u \epsilon_i - \kappa^{-1} \mathcal{N}^{iA} \epsilon_i. \tag{6.102}$$

---

[7]Remember that indices are lowered with the symplectic metrics ε_{ij} and C_{AB} using the NW-SE convention as in (A.15), while the index x is lowered with g_{xy}. The translation from triplet to doublet notation in (A.19) holds for all vectors such as $\mathbf{P}$ or $\mathbf{P}_x$.

Here appear quantities with some similarity between the vector and hypermultiplet sector[8]

$$W^x \equiv -\frac{\sqrt{6}}{4}\kappa h^I k_I{}^x = -\frac{3}{4}t_{J\tilde{I}}{}^{\tilde{P}}h^J h^{\tilde{I}}h^x_{\tilde{P}}, \qquad \mathcal{N}^{iA} \equiv -\frac{\sqrt{6}}{4}\kappa h^I k_I{}^u f^{iA}{}_u,$$

$$\mathbf{P} \equiv \kappa^2 h^I \mathbf{P}_I, \qquad \mathbf{P}_x \equiv \kappa^2 h^I_x. \tag{6.103}$$

See e.g. that $\mathbf{P}$ is the real analogue of S_{ij} in (6.70). As in (6.88) the supersymmetry transformations determine the potential, which is

$$V = \kappa^{-4}\left(\mathbf{P}\cdot\mathbf{P} - \tfrac{1}{2}\mathbf{P}^x\cdot\mathbf{P}_x - 2W_x W^x - 2\mathcal{N}^{iA}\varepsilon_{ij}C_{AB}\mathcal{N}^{jB}\right). \tag{6.104}$$

Let us finally note that the remarks about the gauge transformation of the gravitino (6.75) and the non-vanishing commutator between gauge and supersymmetry transformations (6.91) that we had for $D = 4$ are also applicable for $D = 5$ by just omitting P_I^0, since in $D = 5$ there is no U(1) factor in the R-symmetry group.

Also here the commutator of two supersymmetries contains a scalar-dependent gauge transformation

$$[\delta(\epsilon_1), \delta(\epsilon_2)] = \ldots + \delta_G\left(\theta^I = -\frac{\sqrt{6}}{4}i\kappa^{-1}h^I\bar{\epsilon}_2^i\epsilon_{1i}\right). \tag{6.105}$$

6.3   Final Remarks

We reviewed the constructions of the supergravity actions for $\mathcal{N} = 2$ with vector and hypermultiplets. We restricted ourselves to actions at most quadratic in spacetime derivatives in the bosonic side, linear for the fermionic parts. We used the superconformal techniques to obtain these results.[9] With the same techniques also theories with higher derivatives can be obtained. See e.g. the review [13]. See also [14–16] for various aspects of higher-derivative actions in this context and [17] for an application to get Killing spinor identities. New possibilities with higher-derivative terms and supersymmetry breaking (related to nonlinear realizations of supersymmetry) have also been explored in [18].

---

[8]One could even improve the similarity by e.g. defining $W_{xi}{}^j = \frac{1}{2}W_x\delta_i{}^j + P_{xi}{}^j$ to those in (6.67) and (6.70).

[9]There are many other approaches. We gave an overview in the introduction.

We made several further choices. For the Weyl multiplet, we used the 'standard Weyl multiplet'. An alternative would be the dilaton Weyl multiplet [19–22]. As far as we know, this does not lead to other matter couplings when considering theories quadratic in spacetime derivatives.

In Sect. 4.2 we discussed already the choices of the second compensating multiplet for $D = 4$. A similar choice exists also for $D = 5$ and $D = 6$. We have chosen in the further treatment the hypermultiplet as second compensating multiplet because this exhibits the structure of quaternionic-Kähler manifolds as projective manifolds in the same way as special Kähler manifolds are obtained. A disadvantage is that this multiplet is on-shell from the start. When one uses the tensor multiplet or the nonlinear multiplet [23, 24], one keeps the theory off-shell. See e.g. the construction for $D = 6$ in [25]. In our treatment we have chosen to eliminate auxiliary fields at an early stage, i.e. before the gauge fixing to the Poincaré theory. It may be useful to eliminate these fields only at a later stage such that off-shell Poincaré supergravity is obtained. The techniques explained in the book can then still be used.

The ideas of localization, considering rigid supersymmetric theories on a curved background, are often based on off-shell supergravity. For $\mathcal{N} = 2$ this has been considered in [26]. In view of these ideas, Euclidean versions of supergravity are studied. Special geometry in Euclidean versions is constructed in [27–30].[10] The off-shell $D = 5$ Minkowski supergravity treated in this book can also be reduced over time to an Euclidean $D = 4$ theory [33].

The main methods that we have explained in this book are still applicable for all these extensions. We hope that the extensive explanations that we gave here will be useful for future developments of the theory and for exploring the physics of solutions to the field equations.[11]

References

1. L. Andrianopoli, M. Bertolini, A. Ceresole, R. D'Auria, S. Ferrara, P. Frè, T. Magri, $N = 2$ supergravity and $N = 2$ super Yang–Mills theory on general scalar manifolds: symplectic covariance, gaugings and the momentum map. J. Geom. Phys. **23**, 111–189 (1997). http://dx. doi.org/10.1016/S0393-0440(97)00002-8. arXiv:hep-th/9605032 [hep-th]
2. A. Ceresole, R. D'Auria, S. Ferrara, The symplectic structure of $N = 2$ supergravity and its central extension. Nucl. Phys. Proc. Suppl. **46**, 67–74 (1996). http://dx.doi.org/10.1016/0920-5632(96)00008-4. arXiv:hep-th/9509160 [hep-th]
3. L. Andrianopoli, M. Bertolini, A. Ceresole, R. D'Auria, S. Ferrara, P. Frè, General matter-coupled $N = 2$ supergravity. Nucl. Phys. **B476**, 397–417 (1996). http://dx.doi.org/10.1016/ 0550-3213(96)00344-6. arXiv:hep-th/9603004 [hep-th]

---

[10]Special geometry in other signatures was considered in [31, 32].

[11]Apart from Sect. 5.5, we have refrained in this review to discuss solutions, although a lot of interesting results have been obtained, with or without supersymmetry breaking.

4. R. D'Auria, S. Ferrara, On fermion masses, gradient flows and potential in supersymmetric theories. J. High Energy Phys. **05**, 034 (2001). http://dx.doi.org/10.1088/1126-6708/2001/05/034. arXiv:hep-th/0103153 [hep-th]

5. B. de Wit, A. Van Proeyen, Potentials and symmetries of general gauged $N = 2$ supergravity-Yang–Mills models. Nucl. Phys. **B245**, 89–117 (1984). http://dx.doi.org/10.1016/0550-3213(84)90425-5

6. P. Claus, K. Van Hoof, A. Van Proeyen, A symplectic covariant formulation of special Kähler geometry in superconformal calculus. Class. Quant. Grav. **16**, 2625–2649 (1999). http://dx.doi.org/10.1088/0264-9381/16/8/305. arXiv:hep-th/9904066 [hep-th]

7. A. Ceresole, R. D'Auria, S. Ferrara, A. Van Proeyen, Duality transformations in supersymmetric Yang–Mills theories coupled to supergravity. Nucl. Phys. **B444**, 92–124 (1995). http://dx.doi.org/10.1016/0550-3213(95)00175-R. arXiv:hep-th/9502072 [hep-th]

8. S. Cecotti, L. Girardello, M. Porrati, Constraints on partial superhiggs. Nucl. Phys. **B268**, 295–316 (1986)

9. E. Bergshoeff, S. Cucu, T. de Wit, J. Gheerardyn, S. Vandoren, A. Van Proeyen, $N = 2$ supergravity in five dimensions revisited. Class. Quant. Grav. **21**, 3015–3041 (2004). http://dx.doi.org/10.1088/0264-9381/23/23/C01. http://dx.doi.org/10.1088/0264-9381/21/12/013. arXiv:hep-th/0403045 [hep-th] [Erratum: Class. Quant. Grav. **23**, 7149 (2006)]

10. M. Günaydin, G. Sierra, P.K. Townsend, The geometry of $N = 2$ Maxwell–Einstein supergravity and Jordan algebras. Nucl. Phys. **B242**, 244–268 (1984). http://dx.doi.org/10.1016/0550-3213(84)90142-1

11. M. Günaydin, M. Zagermann, The gauging of five-dimensional, $N = 2$ Maxwell–Einstein supergravity theories coupled to tensor multiplets. Nucl. Phys. **B572**, 131–150 (2000). http://dx.doi.org/10.1016/S0550-3213(99)00801-9. arXiv:hep-th/9912027 [hep-th]

12. A. Ceresole, G. Dall'Agata, General matter coupled $N = 2$, $D = 5$ gauged supergravity. Nucl. Phys. **B585**, 143–170 (2000). http://dx.doi.org/10.1016/S0550-3213(00)00339-4. arXiv:hep-th/0004111 [hep-th]

13. T. Mohaupt, Black hole entropy, special geometry and strings. Fortsch. Phys. **49**, 3–161 (2001). http://dx.doi.org/10.1002/1521-3978(200102)49:1/3<3::AID-PROP3>3.0.CO;2-#. arXiv:hep-th/0007195 [hep-th]

14. N. Banerjee, B. de Wit, S. Katmadas, The off-shell 4D/5D connection. J. High Energy Phys. **03**, 061 (2012). http://dx.doi.org/10.1007/JHEP03(2012)061. arXiv:1112.5371 [hep-th]

15. E. Bergshoeff, F. Coomans, E. Sezgin, A. Van Proeyen, Higher derivative extension of $6D$ chiral gauged supergravity. J. High Energy Phys. **1207**, 011 (2012). http://dx.doi.org/10.1007/JHEP07(2012)011. arXiv:1203.2975 [hep-th]

16. N. Banerjee, B. de Wit, S. Katmadas, The off-shell c-map. J. High Energy Phys. **01**, 156 (2016). http://dx.doi.org/10.1007/JHEP01(2016)156. arXiv:1512.06686 [hep-th]

17. F. Bonetti, D. Klemm, W. A. Sabra, P. Sloane, Spinorial geometry, off-shell Killing spinor identities and higher derivative 5D supergravities. J. High Energy Phys. **08**, 121 (2018). http://dx.doi.org/10.1007/JHEP08(2018)121. arXiv:1806.04108 [hep-th]

18. I. Antoniadis, J.-P. Derendinger, F. Farakos, G. Tartaglino-Mazzucchelli, New Fayet-Iliopoulos terms in $N = 2$ supergravity. J. High Energy Phys. **07**, 061 (2019). http://dx.doi.org/10.1007/JHEP07(2019)061. arXiv:1905.09125 [hep-th]

19. E. Bergshoeff, E. Sezgin, A. Van Proeyen, Superconformal tensor calculus and matter couplings in six dimensions. Nucl. Phys. **B264**, 653 (1986). http://dx.doi.org/10.1016/0550-3213(86)90503-1 [Erratum: Nucl. Phys. B **598**, 667 (2001)]

20. W. Siegel, Curved extended superspace from Yang–Mills theory à la strings. Phys. Rev. **D53**, 3324–3336 (1996). http://dx.doi.org/10.1103/PhysRevD.53.3324. arXiv:hep-th/9510150 [hep-th]

21. E. Bergshoeff, S. Cucu, M. Derix, T. de Wit, R. Halbersma, A. Van Proeyen, Weyl multiplets of $N = 2$ conformal supergravity in five dimensions. J. High Energy Phys. **06**, 051 (2001). http://dx.doi.org/10.1088/1126-6708/2001/06/051. arXiv:hep-th/0104113 [hep-th]

22. D. Butter, S. Hegde, I. Lodato, B. Sahoo, $N = 2$ dilaton Weyl multiplet in 4D supergravity. J. High Energy Phys. **03**, 154 (2018). http://dx.doi.org/10.1007/JHEP03(2018)154. arXiv:1712.05365 [hep-th]

23. B. de Wit, J. W. van Holten, A. Van Proeyen, Structure of $N = 2$ *supergravity*. Nucl. Phys. **B184**, 77–108 (1981) http://dx.doi.org/10.1016/0550-3213(83)90548-5. http://dx.doi.org/10.1016/0550-3213(81)90211-X [Erratum: Nucl. Phys. B **222**, 516 (1983)]

24. B. de Wit, P.G. Lauwers, R. Philippe, A. Van Proeyen, Noncompact $N = 2$ supergravity. Phys. Lett. **135B**, 295 (1984). http://dx.doi.org/10.1016/0370-2693(84)90395-2

25. F. Coomans, A. Van Proeyen, Off-shell $N = (1, 0)$, $D = 6$ supergravity from superconformal methods. J. High Energy Phys. **1102**, 049 (2011). http://dx.doi.org/10.1007/JHEP02(2011)049. arXiv:1101.2403 [hep-th]

26. C. Klare, A. Zaffaroni, Extended supersymmetry on curved spaces. J. High Energy Phys. **10**, 218 (2013). http://dx.doi.org/10.1007/JHEP10(2013)218. arXiv:1308.1102 [hep-th]

27. V. Cortés, C. Mayer, T. Mohaupt, F. Saueressig, Special geometry of Euclidean supersymmetry. I: vector multiplets. J. High Energy Phys. **03**, 028 (2004). http://dx.doi.org/10.1088/1126-6708/2004/03/028. arXiv:hep-th/0312001 [hep-th]

28. V. Cortés, C. Mayer, T. Mohaupt, F. Saueressig, Special geometry of Euclidean supersymmetry. II: hypermultiplets and the c-map. J. High Energy Phys. **06**, 025 (2005). http://dx.doi.org/10.1088/1126-6708/2005/06/025. arXiv:hep-th/0503094 [hep-th]

29. V. Cortés, T. Mohaupt, Special geometry of Euclidean supersymmetry III: the local r-map, instantons and black holes. J. High Energy Phys. **07**, 066 (2009). http://dx.doi.org/10.1088/1126-6708/2009/07/066. arXiv:0905.2844 [hep-th]

30. V. Cortés, P. Dempster, T. Mohaupt, O. Vaughan, Special geometry of Euclidean supersymmetry IV: the local c-map. J. High Energy Phys. **10**, 066 (2015). http://dx.doi.org/10.1007/JHEP10(2015)066. arXiv:1507.04620 [hep-th]

31. M.A. Lledó, Ó. Maciá, A. Van Proeyen, V.S. Varadarajan, Special geometry for arbitrary signatures, in *Handbook on pseudo-Riemannian geometry and supersymmetry*, ed. by V. Cortés. IRMA Lectures in Mathematics and Theoretical Physics, chap. 5, vol. 16 (European Mathematical Society, Zürich, 2010). hep-th/0612210

32. W.A. Sabra, Special geometry and space-time signature. Phys. Lett. **B773**, 191–195 (2017). http://dx.doi.org/10.1016/j.physletb.2017.08.021. arXiv:1706.05162 [hep-th]

33. B. de Wit, V. Reys, Euclidean supergravity. J. High Energy Phys. **12**, 011 (2017). http://dx.doi.org/10.1007/JHEP12(2017)011. arXiv:1706.04973 [hep-th]

Appendix A
Notation

We use the same conventions as in [1]. This implies that the sign factors s_1 to s_9 in Appendix A of that book, see also (C.1), are equal to $+1$. In particular, we use the metric signature $(-+\ldots+)$. If you prefer the opposite, insert a minus sign for every upper index which you see, or for an explicit metric η_{ab} or $g_{\mu\nu}$. The gamma matrices γ_a should then be multiplied by an i to implement this change of signature consistently.

We collect the indices that we use in Table A.1.

(Anti)symmetrization is done with weight one:

$$A_{[ab]} = \tfrac{1}{2}\left(A_{ab} - A_{ba}\right) \qquad \text{and} \qquad A_{(ab)} = \tfrac{1}{2}\left(A_{ab} + A_{ba}\right). \qquad (A.1)$$

The antisymmetric tensors are often contracted with γ matrices as in $\gamma \cdot T = \gamma^{ab} T_{ab}$.

A.1 Bosonic Part

For the curvatures and connections, we repeat here the main formula from the conventions of [1, Appendix A.1] in order to be able to compare with other papers in the literature.

$$R_{\mu\nu}{}^{ab}(e) = 2\partial_{[\mu}\omega_{\nu]}{}^{ab}(e) + 2\omega_{[\mu}{}^{ac}(e)\omega_{\nu]c}{}^{b}(e),$$

$$R^{\mu}{}_{\nu\rho\sigma} = R_{\rho\sigma}{}^{ab} e_a^{\mu} e_{\nu b} = 2\partial_{[\rho}\Gamma^{\mu}{}_{\sigma]\nu} + 2\Gamma^{\mu}{}_{\tau[\rho}\Gamma^{\tau}{}_{\sigma]\nu},$$

$$R_{\mu\nu} = R_{\rho\mu}{}^{ba} e_b^{\rho} e_{\nu a} = R^{\rho}{}_{\nu\rho\mu}, \qquad R = g^{\mu\nu} R_{\mu\nu},$$

$$G_{\mu\nu} = e^{-1}\frac{\delta}{\delta g^{\mu\nu}}\int \mathrm{d}^4 x\, eR = R_{\mu\nu} - \tfrac{1}{2}g_{\mu\nu}R. \qquad (A.2)$$

© Springer Nature Switzerland AG 2020
E. Lauria, A. Van Proeyen, $\mathcal{N} = 2$ *Supergravity in D = 4, 5, 6 Dimensions*,
Lecture Notes in Physics 966, https://doi.org/10.1007/978-3-030-33757-5

Table A.1 Use of indices

μ	$0, \ldots, 3$	Local spacetime
a	$0, \ldots, 3$	Tangent spacetime
i	$1, 2$	SU(2)
In Chaps. 1 and 2		
α	$1, \ldots, 4$	Spinor indices
A		All the gauge transformations
I		All gauge transformations excluding translations
From Chap. 3 onwards		
I	$0, \ldots, n = n_V$	Vector multiplets
X	$1, \ldots, 4n_H$	Scalars in hypermultiplets
$\mathcal{A}$	$1, \ldots, 2n_H$	Spinors (or USp($2n_H$) vector) in hypermultiplets
$\mathbf{V}$	$1, 2, 3$	Triplet of SU(2), see Sect. A.2.2
From Chap. 4 onwards, indices I, X, $\mathcal{A}$ are split and		
α	$1, \ldots, n = n_V$	Independent coordinates in special Kähler
X	$1, \ldots, 4(n_H + 1)$	Scalars in hypermultiplets $= \{0, r, u\}$
$\mathcal{A}$	$1, \ldots, 2(n_H + 1)$	Spinors (or USp($2(n_H + 1)$) vector) in hypermultiplets $= \{i, A\}$
r	$1, 2, 3$	Compensating directions, part of X
u	$1, \ldots, 4n_H$	Independent coordinates in quaternionic-Kähler, part of X
A	$1, \ldots, 2n_H$	Spinors (or USp($2(n_H)$) vector) in hypermultiplets, part of $\mathcal{A}$

The formulations in terms of spin connection ω and in terms of Levi-Civita connection Γ are equivalent by demanding

$$0 = \nabla_\mu e_\nu{}^a = \partial_\mu e_\nu{}^a + \omega_\mu{}^{ab}(e)e_{\nu b} - \Gamma^\rho_{\mu\nu}e_\rho{}^a , \qquad g_{\mu\nu} = e_\mu{}^a \eta_{ab} e_\nu{}^b , \qquad \text{(A.3)}$$

which leads to

$$\omega_\mu{}^{ab}(e) = 2e^{\nu[a}\partial_{[\mu}e_{\nu]}{}^{b]} - e^{\nu[a}e^{b]\sigma}e_{\mu c}\partial_\nu e_\sigma{}^c ,$$

$$\Gamma^\rho_{\mu\nu} = \tfrac{1}{2}g^{\rho\lambda}\left(2\partial_{(\mu}g_{\nu)\lambda} - \partial_\lambda g_{\mu\nu}\right) , \qquad \Gamma^\nu_{\mu\nu} = \tfrac{1}{2}\partial_\nu \ln g . \qquad \text{(A.4)}$$

Note that the Ricci tensor and scalar curvature are of opposite sign as in several older papers in this field. The sign is now chosen such that Einstein spaces with positive scalar curvature are compact.

The anticommuting Levi-Civita tensor is real, and taken to be

$$\varepsilon_{0123} = 1 , \qquad \varepsilon^{0123} = -1 , \qquad \text{(A.5)}$$

the $-$ sign is related to the one timelike direction.[1] For convenience, we give below the formulae for an arbitrary dimension D and number of timelike directions t. The contraction identity for these tensors is $(p + n = D)$

$$\varepsilon_{a_1...a_n b_1...b_p} \varepsilon^{a_1...a_n c_1...c_p} = (-)^t \, p! \, n! \, \delta^{[c_1}_{[b_1} ... \delta^{c_p]}_{b_p]} . \tag{A.6}$$

For the local case, we can still define constant tensors

$$\varepsilon_{\mu_1...\mu_D} = e^{-1} e^{a_1}_{\mu_1} ... e^{a_D}_{\mu_D} \varepsilon_{a_1...a_D} , \qquad \varepsilon^{\mu_1...\mu_D} = e \, e^{\mu_1}_{a_1} ... e^{\mu_D}_{a_D} \varepsilon^{a_1...a_D} . \tag{A.7}$$

They are thus *not* obtained from each other by raising or lowering indices with the metric. In some papers in the literature a difference is made between $\epsilon_{\mu_1...\mu_D}$ and $\varepsilon_{\mu_1...\mu_D}$, such that one of the two has the factor e^{-1} in the definition (A.7) and the other one has not. We only use the symbol defined by (A.7).

Exercise A.1 Show that the tensors in (A.7) are indeed constants, i.e. that arbitrary variations of the frame field cancel in the full expression. You can use the so-called Schouten identities, which means that antisymmetrizing in more indices than the range of the indices, gives zero. The constancy thus implies that one can have (A.5) without specifying whether the 0123 are local or flat indices. $\square$

The definition of dual tensor is in 4 and 6 dimensions, respectively,

$$D = 4 \, : \, \tilde{F}^{ab} \equiv -\tfrac{1}{2} i \varepsilon^{abcd} F_{cd} , \qquad D = 6 \, : \, \tilde{F}^{abc} \equiv -\tfrac{1}{3!} \varepsilon^{abcdef} F_{def} . \tag{A.8}$$

The minus sign in the definition of the dual is convenient for historical reasons. Indeed, when, as written in footnote 1, this ε is i times the ε in these earlier papers then this agrees with the operation that was taken there (e.g. in [2]). In 4 dimensions the dual is an imaginary operation, and the complex conjugate of a self-dual tensor is its anti-self-dual partner, while in 6 dimensions the (anti-)self-dual part of a real tensor is real.

The self-dual and anti-self-dual tensors are introduced in even dimensions as

$$F^{\pm}_{a_1...a_n} \equiv \tfrac{1}{2} \left(F_{a_1...a_n} \pm \tilde{F}_{a_1...a_n} \right) . \tag{A.9}$$

It is useful to observe relations between (anti)self-dual tensors. In 4 dimensions:

$$G^{+ab} H^-{}_{ab} = 0 , \qquad G^{\pm c(a} H^{\pm b)}{}_c = -\tfrac{1}{4} \eta^{ab} G^{\pm cd} H^{\pm}{}_{cd} ,$$

$$G^+_{c[a} H^{-c}_{b]} = 0 . \tag{A.10}$$

---

[1] Note that in many papers, e.g. [2], one takes in 4 dimensions an imaginary Levi-Civita tensor to avoid factors of i in definitions of duals (A.8).

In 6 dimensions:

$$G^{\pm abc} H^{\pm}{}_{abc} = 0 \,, \qquad G^{+cd(a} H^{-b)}{}_{cd} = \tfrac{1}{6} \eta^{ab} G^{+cde} H^{-}{}_{cde} \,,$$

$$G^{\pm}_{cd[a} H^{\pm}{}_{b]}{}^{cd} = 0 \,, \qquad G^{\pm}_{a[bc} H^{\pm}_{cd]}{}^{a} = 0 \,. \qquad\qquad (A.11)$$

A.2   SU(2) Conventions

We use the Levi-Civita ε^{ij} for which the important property is that

$$\varepsilon_{ij} \varepsilon^{jk} = -\delta_i{}^k \,, \qquad\qquad (A.12)$$

where in principle ε^{ij} is the complex conjugate of ε_{ij}, but we can use ($\varepsilon = i\sigma_2$)

$$\varepsilon_{12} = \varepsilon^{12} = 1 \,. \qquad\qquad (A.13)$$

A.2.1   Raising and Lowering Indices

There is an important difference on the use of the i, j indices between the formulae for $D = 4$ and $D = 5, 6$. For $D = 4$ these indices are raised and lowered by complex (or charge) conjugation (here A^i is used for any doublet)

$$D = 4 \, : \, (A^i)^C = A_i \,, \qquad (A_i)^C = A^i \,. \qquad\qquad (A.14)$$

For $D = 5, 6$, these indices are raised or lowered using ε^{ij}. We use NorthWest–SouthEast (NW–SE) convention, which means that this is the direction in which contracted indices should appear to raise or lower indices: see

$$D = 5, 6 \, : \, A^i = \varepsilon^{ij} A_j \,, \qquad A_i = A^j \varepsilon_{ji} \,. \qquad\qquad (A.15)$$

Note that this also implies that for these dimensions, interchanging the position of contracted indices leads to a minus sign:

$$A^i B_i = -A_i B^i \,. \qquad\qquad (A.16)$$

Another useful relation is that for any antisymmetric tensor in ij

$$A^{[ij]} = -\tfrac{1}{2} \varepsilon^{ij} A^k{}_k \,. \qquad\qquad (A.17)$$

Exercise A.2 Check that we can consider ε_{ij} as the tensor $\delta_i{}^j$ with the j index lowered. For this it is important to write $\delta_i{}^j$ and not $\delta^j{}_i$. Also ε^{ij} is the corresponding tensor with raised indices. $\qquad\square$

Implicit summation (in NW–SE direction) is also used for bilinears of fermions, e.g.

$$\bar{\lambda}\chi \equiv \bar{\lambda}^i \chi_i. \qquad (A.18)$$

A.2.2 Triplets

SU(2)-triplets can be converted into 2×2 matrices using the anti-hermitian τ, defined as $\tau_i{}^j = i\sigma_i{}^j$, being $\sigma_i{}^j$ standard Pauli matrices. We define the triplets as the traceless anti-hermitian matrices

$$Y_i{}^j \equiv \tau_i{}^j \cdot \mathbf{Y}, \qquad \mathbf{Y} = -\tfrac{1}{2}\tau_i{}^j Y_j{}^i, \qquad \lambda_i{}^j Y_j{}^i = -2\lambda \cdot \mathbf{Y}. \qquad (A.19)$$

For $D = 5, 6$ indices can be raised and lowered and thus

$$D = 5, 6 \;:\; \tau^{ij} = \varepsilon^{ik}\tau_k{}^j = \tau^{ji} = (i\sigma_3, -\mathbb{1}, -i\sigma_1) = (\tau_{ij})^*. \qquad (A.20)$$

The correspondence between real symmetric matrices and triplets is

$$Y^{ij} = \tau^{ij} \cdot \mathbf{Y} = (Y_{ij})^* = \varepsilon^{ik}\varepsilon^{j\ell}(Y^{k\ell})^*, \qquad \mathbf{Y} = \tfrac{1}{2}\tau^{ij}Y_{ij} = \tfrac{1}{2}\tau_{ij}Y^{ij}. \qquad (A.21)$$

The auxiliary field Y^{ij} of vector multiplets in $D = 4$ has the reality property (3.14), so that this is consistent with the introduction of a real vector $\mathbf{Y}$.

Useful formulae for the symmetric tensors are

$$\tau_{ij} \cdot \tau^{kl} = \delta_i^k \delta_j^l + \delta_j^k \delta_i^l,$$
$$A_{ij} B^{jk} = \delta_i^k \mathbf{A} \cdot \mathbf{B} + (\mathbf{A} \times \mathbf{B}) \cdot \tau_i{}^k. \qquad (A.22)$$

This has been written explicitly for the SU(2) generators in (1.52)–(1.55).

A.2.3 Transformations, Parameters and Gauge Fields

As mentioned in (1.52)–(1.55), the generators $U_i{}^j$ satisfy the same properties in $D = 4, 5, 6$, but in $D = 4$ the $U^i{}_j$ are defined as the complex conjugates, which implies (1.54), while for $D = 5, 6$, the property that $U_{ij} = U_{ji}$ after raising and lowering of indices implies

$$D = 4 \ : \ U_i{}^j = -U^j{}_i \,, \qquad D = 5, 6 \ : \ U_i{}^j = U^j{}_i \,. \tag{A.23}$$

The correspondence with the parameters is different with the two conventions (see (2.8))

$$D = 4 \ : \ \delta_{\mathrm{SU}(2)} = \lambda_i{}^j U_j{}^i = -2\lambda \cdot \mathbf{U} \,, \qquad D = 5, 6 \ : \ \delta_{\mathrm{SU}(2)} = \lambda^{ij} U_{ij} = 2\lambda \cdot \mathbf{U} \,. \tag{A.24}$$

Using (1.53), this implies

$$\left[\delta_{\mathrm{SU}(2)}(\lambda_1), \, \delta_{\mathrm{SU}(2)}(\lambda_2) \right] = \delta_{\mathrm{SU}(2)}(\lambda_3) \,, \qquad \lambda_3 = \mp 2\lambda_1 \times \lambda_2 \,, \tag{A.25}$$

where the upper sign is for $D = 4$ and the lower for $D = 5, 6$. Despite the difference in properties in (A.23), which can also be used for the parameters, there is one form that can be used in general $\delta = \ldots - \lambda^j{}_i U_j{}^i$.

The SU(2) transformation of a doublet, like A^i, can be written in different ways

$$D \ = 4 \ : \delta A^i = A^j \lambda_j{}^i = -\lambda^i{}_j A^j \,, \qquad \delta A_i = -\lambda_i{}^j A_j = A_j \lambda^j{}_i \,,$$
$$D = 5, 6 \ : \delta A^i = \lambda^{ij} A_j = -\lambda^i{}_j A^j \,, \qquad \delta A_i = -\lambda_{ij} A^j = A_j \lambda^j{}_i \,. \tag{A.26}$$

A.3 Gamma Matrices and Spinors

A general treatment of gamma matrices and spinors is given in [3, Sect. 3]. In that review general spacetime signatures are treated. Of course, that material is not original, and is rather a convenient reformulation of earlier works [4–7]. Another approach to the theory of spinors has been presented in [8]. For Lorentzian signatures, the conventions that we use are given in [1, Chap. 3].

Coefficients $t_r = \pm 1$ appear in Majorana flip relations and relations between spinors and conjugated spinors

$$\bar{\lambda} \gamma_{a_1 \ldots a_r} \chi = t_r \bar{\chi} \gamma_{a_1 \ldots a_r} \lambda \,,$$
$$\chi_{a_1 \ldots a_r} = \gamma_{a_1 \ldots a_r} \lambda \ \rightarrow \ \bar{\chi}_{a_1 \ldots a_r} = t_0 t_r \bar{\lambda} \gamma_{a_1 \ldots a_r} \,, \tag{A.27}$$

and they are given by

	t_0	t_1	t_2	t_3	t_4	t_5
$D = 4$	$+$	$-$	$-$	$+$	$+$	
$D = 5$	$+$	$+$	$-$	$-$	$+$	$+$
$D = 6$	$-$	$+$	$+$	$-$	$-$	$+$

$$\tag{A.28}$$

However, take into account that often the i, j indices are hidden, see (A.18), such that e.g. in 5 dimensions

$$D = 5 : \bar{\lambda}\chi \equiv \bar{\lambda}^i \chi_i = \bar{\chi}_i \lambda^i = -\bar{\chi}^i \lambda_i = -\bar{\chi}\lambda . \tag{A.29}$$

A.3.1 $D = 4$

One useful tool, which is explained in [1, Chap. 3], is that complex conjugation can be performed by using charge conjugation, since for a Lagrangian the two are equal. The most important rules are as follows[2]:

- For scalars in spinor space, charge conjugation is equal to complex conjugation.
- For matrices in SU(2) space (and scalars in spinor space), the choice of the position of the indices is chosen such that (complex and) charge conjugation changes the height of the index: e.g. $(M^{ij})^C = M_{ij}$, or $(M^i{}_j)^C = M_i{}^j$.
- Majorana spinors in a Weyl basis satisfy $(\chi^i)^C = \chi_i$. The position of the i index indicates the chirality, see Tables 2.3 and 3.1, and thus charge conjugation changes the chirality.
- For a bispinor built from two spinors and a matrix M in spinor (and possible SU(2)) space:

$$(\bar{\lambda} M \psi)^C = \overline{\lambda^C} M^C \psi^C , \qquad \text{e.g. } (\bar{\lambda}^i M_i{}^j \psi_j)^C = \bar{\lambda}_i M^i{}_j \psi^j , \tag{A.30}$$

and $(MN)^C = M^C N^C$.
- The action in spinor space can be derived from $\gamma_a^C = \gamma_a$, and $\gamma_*^C = -\gamma_*$.
- For the fermions of hypermultiplets, the index $\mathcal{A}$ is by charge conjugation raised or lowered and at the same time replaced by its barred one, $\bar{\mathcal{A}}$, e.g. the charge conjugate of $\zeta^{\mathcal{A}} = P_L \zeta^{\mathcal{A}}$ is $\zeta_{\bar{\mathcal{A}}} = P_R \zeta_{\bar{\mathcal{A}}}$.

For the frame fields of hypermultiplets in 4 dimensions:

$$\left(f^X{}_{i\mathcal{A}}\right)^C = \left(f^X{}_{i\mathcal{A}}\right)^* = f^{Xi\bar{\mathcal{A}}} = \varepsilon^{ij} \rho^{\bar{\mathcal{A}}\mathcal{B}} f^X{}_{j\mathcal{B}} , \qquad f_{i\bar{\mathcal{A}}X} = \left(f^{i\mathcal{A}}{}_X\right)^* . \tag{A.31}$$

For the USp(2r) connection one has

$$\left(\omega_{XA}{}^B\right)^C = \left(\omega_{XA}{}^B\right)^* = \bar{\omega}^{\bar{\mathcal{A}}}{}_{\bar{\mathcal{B}}} = -\rho^{\bar{\mathcal{A}}\mathcal{D}} \omega_{XD}{}^{\mathcal{E}} \rho_{\mathcal{E}\bar{\mathcal{B}}} . \tag{A.32}$$

---

[2]These properties follow from the definition that the charge conjugate of a spinor χ is $\chi^C = iC\gamma^0 \chi$ and $\bar{\chi} = \chi^T C$.

The charge conjugate of the antisymmetric metric $C_{\mathcal{A}\mathcal{B}}$ is $C^{\bar{\mathcal{A}}\bar{\mathcal{B}}}$:

$$(C_{\mathcal{A}\mathcal{B}})^* = C^{\bar{\mathcal{A}}\bar{\mathcal{B}}} = \rho^{\bar{\mathcal{A}}C}\rho^{\bar{\mathcal{B}}\mathcal{D}}C_{C\mathcal{D}}. \tag{A.33}$$

These matrices are also used to raise and lower indices on bosonic quantities with the NW–SE convention.

A.3.2 $D = 5$

In odd dimensions the antisymmetric product of all the matrices is proportional to the unit matrix. We choose the sign as follows:

$$\gamma^{abcde} = i\varepsilon^{abcde}. \tag{A.34}$$

Exercise A.3 Check that this implies

$$\gamma^{abcd} = i\varepsilon^{abcde}\gamma_e,$$
$$2\gamma^{abc} = i\varepsilon^{abcde}\gamma_{ed},$$
$$3!\gamma^{ab} = i\varepsilon^{abcde}\gamma_{edc},$$
$$4!\gamma^{a} = i\varepsilon^{abcde}\gamma_{edcb},$$
$$5! = i\varepsilon^{abcde}\gamma_{edcba},$$
$$i\varepsilon^{abcde}\gamma_{ef} = 4\gamma^{[abc}\delta_f^{d]},$$
$$i\varepsilon^{abcde}\gamma_{efg} = 12\gamma^{[ab}\delta_{gf}^{cd]}. \tag{A.35}$$

□

The charge conjugation C and $C\gamma_a$ are antisymmetric in $D = 5$. We include a multiplication by ε^{ij} in the definition of charge conjugation[3]:

$$(\lambda^i)^C \equiv -i\gamma^0 C^{-1}(\lambda^j)^*\varepsilon^{ji}. \tag{A.36}$$

The elementary spinors are 'symplectic Majorana', which means that they are invariant under this C operation.

---

[3]The choice of sign is motivated by the requirement that the translation parameter in the supersymmetry commutation relation reduces from $D = 5$ to $D = 4$ as in (A.59) below.

The practical rules are as follows:

- For bosons charge conjugation is equal to complex conjugation. However, a bispinor $\bar{\lambda}^i \chi_i$ built from symplectic Majorana spinors is pure imaginary.
- For a bosonic matrix in SU(2) space, M, the charge conjugate is defined as $M^C \equiv \sigma_2 M^* \sigma_2$, or explicitly $(M_i{}^j)^C = -\varepsilon_{ik}(M_k{}^\ell)^* \varepsilon^{\ell j}$.
- For a bispinor that is also a scalar in SU(2) space: $(\bar{\lambda} M \psi)^C = -\bar{\lambda}^C M^C \psi^C$, where M can be a matrix in spinor space and/or SU(2) space.
- Symplectic Majorana spinors are invariant under C.
- For matrices in spinor space: $\gamma_a^C = -\gamma_a$.

As an example, see that the expression

$$\bar{\lambda}^i \gamma^\mu \partial_\mu \lambda_i \tag{A.37}$$

is real for symplectic Majorana spinors.

For hypermultiplets the reality condition depends on a matrix $\rho_{\mathcal{A}\bar{\mathcal{B}}}$. The charge conjugation under which the symplectic Majorana spinors are invariant is then similar to (A.36):

$$\left(\zeta^{\mathcal{A}}\right)^C \equiv -\mathrm{i}\gamma^0 C^{-1}\left(\zeta^{\mathcal{B}}\right)^* \rho^{\bar{\mathcal{B}}\mathcal{A}} = \zeta^{\mathcal{A}}. \tag{A.38}$$

A.3.3 $D = 6$

Here there is again chirality, as for every even dimension, but moreover, there are real self-dual tensors. Similarly, we define

$$\gamma_* = \gamma_0 \dots \gamma_5 = -\gamma^0 \dots \gamma^5, \tag{A.39}$$

without a factor i. The essential formula is

$$\gamma_{abc}\gamma_* = -\tilde{\gamma}_{abc}, \tag{A.40}$$

Exercise A.4 Show that $\gamma_{abc} P_L = \gamma^-_{abc}$ as in 4 dimensions where $\gamma_{ab} P_L = \gamma^-_{ab}$. $\square$

Spinors can satisfy the (symplectic) Majorana condition and be chiral at the same time, and we generally use such 'symplectic Majorana–Weyl' spinors. Thus complex conjugation does not change chirality, and we can raise and lower i indices as in (A.15).

We choose the charge conjugation matrix to be symmetric.[4] We use the same definition for charge conjugation as in (A.36), and thus this also applies for symplectic Majorana spinors.

The rules for charge conjugation are

- For bosons charge conjugation is equal to complex conjugation. Also a bispinor $\bar{\lambda}^i \chi_i$ built from symplectic Majorana spinors is real.
- For a bosonic matrix in SU(2) space, M, the charge conjugate is defined as $M^C \equiv \sigma_2 M^* \sigma_2$, or explicitly $(M_i{}^j)^C = -\varepsilon_{ik}(M_k{}^\ell)^* \varepsilon^{\ell j}$.
- For a bispinor that is also a scalar in SU(2) space: $(\bar{\lambda} M \psi)^C = -\overline{\lambda^C} M^C \psi^C$, where M can be a matrix in spinor space and/or SU(2) space.
- Symplectic Majorana spinors are invariant under C.
- For matrices in spinor space: $\gamma_a^C = \gamma_a$ and $\gamma_*^C = \gamma_*$.

A.3.4   Products of γ Matrices and Fierzing

There are some useful identities for calculations in arbitrary dimensions [9]. For a product of two antisymmetrized gamma matrices, one can use

$$\gamma_{a_1...a_i} \gamma^{b_1...b_j} = \sum_{k=|i-j|}^{i+j} \frac{i! j!}{s! t! u!} \delta_{[a_i}^{[b_1} \cdots \delta_{a_{t+1}}^{b_s} \gamma_{a_1...a_t]}{}^{b_{s+1}...b_j]}, \tag{A.41}$$

$$s = \tfrac{1}{2}(i+j-k), \qquad t = \tfrac{1}{2}(i-j+k), \qquad u = \tfrac{1}{2}(-i+j+k).$$

The numeric factor can be understood as follows. Between the i indices of the first gamma factor, we select $s = i - t$ of them for the contraction. That choice is a factor $\binom{i}{s}$. The same number of indices s is chosen between the j indices of the second factor. That is a factor $\binom{j}{s}$. Finally, we can contract these s indices in $s!$ ways. In [9] a few extra rules are given and a diagrammatic technique is explained that is based on the work of Kennedy [10].

For contractions of repeated gamma matrices, one has the formula

$$\gamma_{b_1...b_k} \gamma_{a_1...a_\ell} \gamma^{b_1...b_k} = c_{k,\ell} \gamma_{a_1...a_\ell}$$

$$c_{k,\ell} = (-)^{k(k-1)/2} k! (-)^{k\ell} \sum_{i=0}^{\min(k,\ell)} \binom{\ell}{i} \binom{D-\ell}{k-i} (-)^i, \tag{A.42}$$

---

[4]That is a choice in 6 dimensions, as we can also use the antisymmetric charge conjugation matrix $C' = C\gamma_*$.

for which tables were given in [9] in dimensions 4, 10, 11 and 12. Useful examples in 4 dimensions are

$$\gamma^a \gamma^{bc} \gamma_a = \gamma^{ab} \gamma_c \gamma_{ab} = 0, \qquad \gamma^{ab} \gamma_{cd} \gamma_{ab} = 4\gamma_{cd}. \tag{A.43}$$

Further, there is the Fierz relation. We know that the gamma matrices are matrices in dimension $\Delta = 2^{\text{Int } D/2}$, and that a basis of $\Delta \times \Delta$ matrices is given by the set

$$\left\{ \mathbb{1}, \gamma_a, \gamma_{a_1 a_2}, \ldots, \gamma^{a_1 \ldots a_{[D]}} \right\} \qquad \text{where} \qquad \begin{cases} [D] = D & \text{for even } D, \\ [D] = (D-1)/2 & \text{for odd } D, \end{cases} \tag{A.44}$$

of which only the first has non-zero trace. This is the basis of the general Fierz formula for an arbitrary matrix M in spinor space:

$$2^{\text{Int}(D/2)} M_\alpha{}^\beta = \sum_{k=0}^{[D]} (-)^{k(k-1)/2} \frac{1}{k!} \left(\gamma_{a_1 \ldots a_k} \right)_\alpha{}^\beta \, \text{Tr} \left(\gamma^{a_1 \ldots a_k} M \right). \tag{A.45}$$

Further Fierz identities can be found in [10].

Exercise A.5 Check that Fierz identities for chiral spinors in 4 dimensions lead to

$$\bar{\psi}_L \phi_L \, \bar{\chi}_L \lambda_L = -\frac{1}{2} \bar{\psi}_L \lambda_L \, \bar{\chi}_L \phi_L + \frac{1}{8} \bar{\psi}_L \gamma^{ab} \lambda_L \, \bar{\chi}_L \gamma_{ab} \phi_L,$$

$$\bar{\psi}_L \phi_L \, \bar{\chi}_R \lambda_R = -\frac{1}{2} \bar{\psi}_L \gamma^a \lambda_R \, \bar{\chi}_R \gamma_a \phi_L. \tag{A.46}$$

An extra minus sign w.r.t. (A.45) appears here because fermions λ and ϕ are interchanged. In 5 dimensions the Fierz equation is

$$\bar{\psi}^i \phi^j \, \bar{\chi}^k \lambda^\ell = -\frac{1}{4} \bar{\psi}^i \lambda^\ell \bar{\chi}^k \phi^j - \frac{1}{4} \bar{\psi}^i \gamma^a \lambda^\ell \bar{\chi}^k \gamma_a \phi^j + \frac{1}{2} \bar{\psi}^i \gamma^{ab} \lambda^\ell \bar{\chi}^k \gamma_{ab} \phi^j. \tag{A.47}$$

In 6 dimensions ($\bar{\psi}_R = \bar{\psi} P_L$) one has

$$\bar{\psi}_R \phi_L \, \bar{\chi}_L \lambda_R = -\frac{1}{4} \bar{\psi}_R \gamma^a \lambda_R \, \bar{\chi}_L \gamma_a \phi_L + \frac{1}{48} \bar{\psi}_R \gamma^{abc} \lambda_R \, \bar{\chi}_L \gamma_{abc} \phi_L,$$

$$\bar{\psi}_R \phi_L \, \bar{\chi}_R \lambda_L = -\frac{1}{4} \bar{\psi}_R \lambda_L \, \bar{\chi}_R \phi_L + \frac{1}{8} \bar{\psi}_R \gamma^{ab} \lambda_L \, \bar{\chi}_R \gamma_{ab} \phi_L. \tag{A.48}$$

For those that want to go further, a more complicated identity for doublet spinors in 4 dimensions is

$$\varepsilon^{jk} \gamma^{ab} \lambda_i \bar{\lambda}_j \gamma_{ab} \lambda_k = 8\varepsilon^{jk} \lambda_k \bar{\lambda}_i \lambda_j. \tag{A.49}$$

It uses the first equation of (A.46), symmetries of the bilinears and manipulations between the ε symbols.

$\square$

A.4   Spinors from 5 to 6 and 4 Dimensions

For some part of this text (especially for hypermultiplets) we started from explicit formulae for symplectic Majorana spinors in $D = 5$ and these formulae are useful also in 4 and 6 dimensions. The spinors have different properties in 4 and 6 dimensions, but we can easily translate the formulae.

Let us temporarily *write $\tilde{\zeta}^A$ for the symplectic Majorana–Weyl spinors in $D = 6$*, which are right-handed. We consider them now as 4-component spinors, and translate to the spinors ζ^A of $D = 5$ by

$$\zeta^A = i\tilde{\zeta}^A, \qquad \bar{\zeta}^A = -i\bar{\tilde{\zeta}}^A. \tag{A.50}$$

For spinors that are left-handed in 6 dimensions, e.g. the supersymmetry parameter ϵ^i, we have

$$\epsilon^i = \tilde{\epsilon}^i, \qquad \bar{\epsilon}^i = \bar{\tilde{\epsilon}}^i. \tag{A.51}$$

Finally, the last γ-matrix from 6 dimensions, γ_5 gets translated to an imaginary unit: i on a left-handed spinor, and $-i$ on a right-handed spinor. Combined with the above factors, we thus have

$$\gamma_5\tilde{\epsilon}^i = i\epsilon^i, \qquad \gamma_5\tilde{\zeta}^A = -\zeta^A. \tag{A.52}$$

Reduction from 5 to 4 dimensions is a bit more involved [11]. We give here the rules and some illustrative examples. *Now the tilde indicates quantities for $D = 5$.*

The rules are based on the fact that the γ_4 matrix in 5 dimensions is identified with γ_* in 4 dimensions, and the charge conjugation matrix of 5 dimensions is $\tilde{C} = C\gamma_*$ in 4 dimensions. An important difference is that in 5 dimensions i, j indices are raised and lowered as in (A.15), while this is related to charge conjugation and change of chirality in 4 dimensions, see Sect. A.3.1. The symplectic Majorana condition of $D = 5$, (A.36), can be written as

$$(\tilde{\lambda}^i)^* = i\tilde{C}\gamma^0\tilde{\lambda}^j\varepsilon_{ji}. \tag{A.53}$$

This is now translated in terms of the 4-dimensional charge conjugation as

$$(\tilde{\lambda}^i)^C = -\gamma_*\tilde{\lambda}^j\varepsilon_{ji}. \tag{A.54}$$

Now translation rules depend on the choice in 4 dimensions whether λ^i is left- or right-handed:

$$\lambda^i = \pm\gamma_*\lambda^i, \qquad \lambda_i = \mp\gamma_*\lambda_i. \tag{A.55}$$

In order to be consistent with (A.54) we can then write

$$\tilde{\lambda}^i = \lambda^i \mp \lambda_j \varepsilon^{ji} . \tag{A.56}$$

This then also implies

$$\overline{\tilde{\lambda}^i} = (\tilde{\lambda}^i)^T C = \left(\lambda^i \mp \lambda_j \varepsilon^{ji} \right)^T C \gamma^* = \pm \bar{\lambda}^i + \bar{\lambda}_j \varepsilon^{ji} . \tag{A.57}$$

Similar rules apply to the spinors of the hypermultiplets, replacing ε_{ij} by $\rho_{\mathcal{A}\bar{\mathcal{B}}}$. We took in $D = 4$ the choice $\zeta^{\mathcal{A}} = P_L \zeta^{\mathcal{A}}$, hence we have

$$\tilde{\zeta}^{\mathcal{A}} = \zeta^{\mathcal{A}} - \zeta_{\bar{\mathcal{B}}} \rho^{\bar{\mathcal{B}}\mathcal{A}} , \qquad \overline{\tilde{\zeta}}^{\mathcal{A}} = \bar{\zeta}^{\mathcal{A}} + \bar{\zeta}_{\bar{\mathcal{B}}} \rho^{\bar{\mathcal{B}}\mathcal{A}} . \tag{A.58}$$

As a first example, let us start from the translation parameter in terms of two supersymmetries, which are left-handed in $D = 4$, as in (1.6):

$$\begin{aligned}
\tilde{\xi}^a(\tilde{\epsilon}_1, \tilde{\epsilon}_2) &= \tfrac{1}{2} \bar{\tilde{\epsilon}}_2^i \gamma^a \tilde{\epsilon}_1^j \varepsilon_{ji} \\
&= \tfrac{1}{2} \left(\bar{\epsilon}_2^i + \bar{\epsilon}_{2k} \varepsilon^{ki} \right) \gamma^a \left(\epsilon_1^j - \epsilon_{1\ell} \varepsilon^{\ell j} \right) \varepsilon_{ji} \\
&= \tfrac{1}{2} \bar{\epsilon}_2^i \gamma^a \epsilon_{1i} + \tfrac{1}{2} \bar{\epsilon}_{2i} \gamma^a \epsilon_1^i = \xi^a ,
\end{aligned} \tag{A.59}$$

since the other terms disappear by chirality. The right-hand side is the correct expression in 4 dimensions. The 5-dimensional commutator contains another component:

$$\begin{aligned}
\tilde{\xi}^4(\tilde{\epsilon}_1, \tilde{\epsilon}_2) &= \tfrac{1}{2} \bar{\tilde{\epsilon}}_2^i \gamma^4 \tilde{\epsilon}_1^j \varepsilon_{ji} \\
&= \tfrac{1}{2} \left(\bar{\epsilon}_2^i + \bar{\epsilon}_{2k} \varepsilon^{ki} \right) \gamma_* \left(\epsilon_1^j - \epsilon_{1\ell} \varepsilon^{\ell j} \right) \varepsilon_{ji} \\
&= -\tfrac{1}{2} \bar{\epsilon}_2^i \epsilon_1^j \varepsilon_{ij} + \tfrac{1}{2} \bar{\epsilon}_{2i} \epsilon_{1j} \varepsilon^{ij} .
\end{aligned} \tag{A.60}$$

As a second example we translate the non-closure relation obtained in (3.62)

$$\left[\delta(\tilde{\epsilon}_1), \delta(\tilde{\epsilon}_2) \right] \tilde{\zeta}^{\mathcal{A}} = \tilde{\xi}^\mu \partial_\mu \tilde{\zeta}^{\mathcal{A}} + \tfrac{1}{4} \left[(\bar{\tilde{\epsilon}}_2^i \tilde{\epsilon}_1^j) - \gamma^\nu (\bar{\tilde{\epsilon}}_2^i \gamma_\nu \tilde{\epsilon}_1^j) \right] \varepsilon_{ji} \mathrm{i} \tilde{\Gamma}^{\mathcal{A}} ,$$

$$\mathrm{i} \tilde{\Gamma}^{\mathcal{A}} \equiv \slashed{\nabla} \tilde{\zeta}^{\mathcal{A}} + \tfrac{1}{2} W_{\mathcal{B}\mathcal{C}\mathcal{D}}{}^{\mathcal{A}} \tilde{\zeta}^{\mathcal{B}} \overline{\tilde{\zeta}}^{\mathcal{C}} \tilde{\zeta}^{\mathcal{D}} . \tag{A.61}$$

We define in 4 dimensions the $\Gamma^{\mathcal{A}}$ as right-handed, and for convenience redefine the Γ^A in $D = 4$ from the rule with the lower sign in (A.56) with a factor i (and thus $-$i for the left-handed component in order to define Majorana spinors in $D = 4$), i.e.

$$\mathrm{i} \tilde{\Gamma}^{\mathcal{A}} = -\Gamma^{\mathcal{A}} + \Gamma_{\bar{\mathcal{B}}} \rho^{\bar{\mathcal{B}}\mathcal{A}} , \tag{A.62}$$

which leads to

$$- \Gamma^{\mathcal{A}} = -P_R \Gamma^{\mathcal{A}} = P_R \mathrm{i} \tilde{\Gamma}^{\mathcal{A}}, \qquad \Gamma_{\bar{\mathcal{A}}} = P_L \Gamma_{\bar{\mathcal{A}}} = -P_L \mathrm{i} \tilde{\Gamma}^{\mathcal{B}} \rho_{\mathcal{B}\bar{\mathcal{A}}}. \qquad (A.63)$$

We thus obtain the expression $\Gamma^{\mathcal{A}}$ from projecting to the right-handed part:

$$- \Gamma^{\mathcal{A}} = \not{\nabla} \zeta^{\mathcal{A}} + \tfrac{1}{2} W_{\mathcal{BCD}}{}^{\mathcal{A}} (-\zeta_{\bar{\mathcal{E}}} \rho^{\bar{\mathcal{E}}\mathcal{B}}) \zeta^{\mathcal{C}} \zeta^{\mathcal{D}}, \qquad (A.64)$$

since Fierz identities together with the symmetry of the W tensor annihilate the part with 3 antichiral fermions. We thus obtain

$$\Gamma^A = -\not{\nabla} \zeta^{\mathcal{A}} + \tfrac{1}{2} W_{\mathcal{BC}}{}^{\bar{\mathcal{D}}\mathcal{A}} \zeta_{\bar{\mathcal{D}}} \zeta^{\mathcal{B}} \zeta^{\mathcal{C}}, \qquad W_{\mathcal{BC}}{}^{\bar{\mathcal{D}}\mathcal{A}} \equiv \rho^{\bar{\mathcal{D}}\mathcal{E}} W_{\mathcal{BCE}}{}^{\mathcal{A}}. \qquad (A.65)$$

In the same way, the left-handed part of the commutator (A.61) is

$$[\delta(\epsilon_1), \delta(\epsilon_2)] \zeta^{\mathcal{A}} = \xi^{\mu} \partial_{\mu} \zeta^{\mathcal{A}} + \tfrac{1}{2} \varepsilon^{ij} \bar{\epsilon}_{2i} \epsilon_{1j} \Gamma_{\bar{\mathcal{B}}} \rho^{\bar{\mathcal{B}}\mathcal{A}} + \tfrac{1}{4} \gamma^{\nu} \Gamma^A \left(\bar{\epsilon}_2^i \gamma_{\nu} \epsilon_{1i} + \bar{\epsilon}_{2i} \gamma_{\nu} \epsilon_1^i \right). $$
$$(A.66)$$

Note that the reduction to 4 dimensions implies that the spacetime derivative ∂_4 vanishes. However the $\nu = *$ part in (A.61) contributes such that the left-handed parts cancel a similar contribution from $\bar{\epsilon}_2^i \tilde{\epsilon}_1^j$.

Appendix B
Superalgebras

Lie superalgebras have been classified in [12]. We cannot go through the full classification mechanism of course, but will consider the most important superalgebras, the 'simple Lie superalgebras', which have no non-trivial invariant subalgebra. However, one should know that for superalgebras there are more subtle issues, as e.g. not any semi-simple superalgebra is the direct sum of simple superalgebras. A good review is [13]. The fermionic generators of such superalgebras are in representations of the bosonic part. If that 'defining representation' of the bosonic algebra in the fermionic generators is completely reducible, the algebra is said to be 'of classical type'. The others are 'Cartan type superalgebras' $W(n)$, $S(n)$, $\tilde{S}(n)$ and $H(n)$, which we will further neglect. For further reference, we give in Table B.1 the list of the real forms of superalgebras 'of classical type' [14–16]. In this table 'defining representation' gives the fermionic generators as a representation of the bosonic subalgebra. The 'number of generators' gives the numbers of (bosonic,fermionic) generators in the superalgebra. We mention first the algebra as an algebra over $\mathbb{C}$, and then give different real forms of these algebras. With this information, you can reconstruct all properties of these algebras, up to a few exceptions. The names which we use for the real forms is for some algebras different from those in the mathematical literature [14–16] and chosen such that it is most suggestive of its bosonic content. There are isomorphisms as $SU(2|1) = OSp(2^*|2, 0)$, and $SU(1, 1|1) = S\ell(2|1) = OSp(2|2)$. In the algebra $D(2, 1, \alpha)$ the three $S\ell(2)$ factors of the bosonic group in the anticommutator of the fermionic generators appear with relative weights 1, α and $-1 - \alpha$. The real forms contain, respectively, $SO(4) = SU(2) \times SU(2)$, $SO(3, 1) = S\ell(2, \mathbb{C})$ and $SO(2, 2) = S\ell(2) \times S\ell(2)$. In the first and last case α should be real, while $\alpha = 1 + ia$ with real a for $p = 1$. In the limit $\alpha = 1$ one has the isomorphisms $D^p(2, 1, 1) = OSp(4 - p, p|2)$.

For the real forms of $SU(m|m)$, the one-dimensional subalgebra of the bosonic algebra is not part of the irreducible algebra. In some papers these are called $PSU(m|m)$, keeping the name $SU(m|m)$ for the algebra including the abelian factor.

© Springer Nature Switzerland AG 2020
E. Lauria, A. Van Proeyen, $\mathcal{N} = 2$ *Supergravity in D = 4, 5, 6 Dimensions*,
Lecture Notes in Physics 966, https://doi.org/10.1007/978-3-030-33757-5

Table B.1 Lie superalgebras of classical type

Name	Range	Bosonic algebra	Defining repres.	Number of generators
$SU(m\mid n)$	$m \geq 2$ $m \neq n$ $m = n$	$SU(m) \oplus SU(n)$ $\oplus U(1)$ no $U(1)$	$(m, \bar{n})\oplus$ $(\bar{m}, n)$	$m^2 + n^2 - 1,$ $2mn$ $2(m^2 - 1), 2m^2$
$S\ell(m\mid n)$ $SU(m - p, p\mid n - q, q)$ $SU^*(2m\mid 2n)$ $S\ell'(n\mid n)$		$S\ell(m) \oplus S\ell(n) \quad\oplus SO(1, 1)$ $SU(m - p, p) \oplus SU(n - q, q) \;\oplus U(1)$ $SU^*(2m) \oplus SU^*(2n) \quad\oplus SO(1, 1)$ $S\ell(n, \mathbb{C})$		$\left.\begin{array}{l}\text{if}\\ m \neq n\end{array}\right\}$
$OSp(m\mid n)$	$m \geq 1$ $n = 2, 4, ..$	$SO(m) \oplus Sp(n)$	(m, n)	$\frac{1}{2}(m^2 - m+$ $n^2 + n), mn$
$OSp(m - p, p\mid n)$ $OSp(m^*\mid n - q, q)$		$SO(m - p, p) \oplus Sp(n)$ $SO^*(m) \oplus USp(n - q, q)$		n even m, n, q even
$D(2, 1, \alpha)$	$0 < \alpha \leq 1$	$SO(4) \oplus S\ell(2)$	$(2, 2, 2)$	$9, 8$
$D^p(2, 1, \alpha)$		$SO(4 - p, p) \oplus S\ell(2)$		$p = 0, 1, 2$
$F(4)$		$SO(7) \oplus S\ell(2)$	$(8, 2)$	$24, 16$
$F^p(4)$ $F^p(4)$		$SO(7 - p, p) \oplus S\ell(2)$ $SO(7 - p, p) \oplus SU(2)$		$p = 0, 3$ $p = 1, 2$
$G(3)$		$G_2 \oplus S\ell(2)$	$(7, 2)$	$17, 14$
$G_p(3)$		$G_{2,p} \oplus S\ell(2)$		$p = -14, 2$
$P(m - 1)$	$m \geq 3$	$S\ell(m)$	$(m \otimes m)$	$m^2 - 1, m^2$
$Q(m - 1)$	$m \geq 3$	$SU(m)$	Adjoint	$m^2 - 1, m^2 - 1$
$Q(m - 1)$ $Q((m - 1)^*)$ $UQ(p, m - 1 - p)$		$S\ell(m)$ $SU^*(m)$ $SU(p, m - p)$		

Note that the superalgebras $SU(m\mid m)$, which are indicated in this table with bosonic subalgebra $SU(m)\oplus SU(m)$ are often indicated as $PSU(m\mid m)$, while then $SU(m\mid m)$ refers to the (non-simple) algebra including the $U(1)$ factor

Furthermore, in that case there are subalgebras obtained from projections of those mentioned here with only one factor $SU(n)$, $S\ell(n)$, $SU^*(n)$ or $SU(n - p, p)$ as bosonic algebra.

Appendix C
Comparison of Notations

Unfortunately the normalization of F and various other functions vary in the $\mathcal{N} = 2$ literature. A lot of standard notations can be summarized in factors $s_1, \ldots, s_9$. The first three of these were already identified in the book of Misner–Thorne–Wheeler [17]. More explanations are given in [1, Appendix A]. We summarize here a few relevant formulae

$$\eta_{ab} = s_1 \text{diag}(-+++) \,,$$

$$R_{\mu\nu}{}^{\rho}{}_{\sigma} = s_2 \left(\partial_\mu \Gamma^\rho_{\nu\sigma} - \partial_\nu \Gamma^\rho_{\mu\sigma} + \Gamma^\rho_{\mu\tau} \Gamma^\tau_{\nu\sigma} - \Gamma^\rho_{\nu\tau} \Gamma^\tau_{\mu\sigma} \right) \,,$$

$$s_2 s_3 R_{\mu\nu} = R^\rho{}_{\nu\rho\mu} \,,$$

$$D_\mu \psi = \partial_\mu + s_2 s_4 \tfrac{1}{4} \omega_\mu{}^{ab} \gamma_{ab} \,, \qquad D_\mu V^a = \partial_\mu V^a + s_2 s_4 \omega_\mu{}^{ab} V_b \,,$$

$$\varepsilon_{0123} = s_5 \,, \qquad \varepsilon^{0123} = -s_5 \,,$$

$$\gamma_\mu \gamma_\nu + \gamma_\nu \gamma_\mu = 2 s_6 g_{\mu\nu} \,,$$

$$\gamma_* = \gamma_5 = s_7 \mathrm{i} \gamma_0 \gamma_1 \gamma_2 \gamma_3 \,, \qquad \tfrac{1}{2} \varepsilon_{abcd} \gamma^{cd} = -\mathrm{i} \frac{s_5}{s_7} \mathrm{i} \gamma_* \gamma_{ab} \,,$$

$$(\bar{\psi} \lambda)^* = s_8 (-t_0 t_1) \bar{\psi} \lambda \,,$$

$$[\delta(\epsilon_1), \delta(\epsilon_2)] \phi = s_9 \tfrac{1}{2} \bar{\epsilon}_2 \gamma^\mu \epsilon_1 \partial_\mu \phi \,. \tag{C.1}$$

Note that the factor $(-t_0 t_1) = 1$ for $D = 4$ and $D = 6$, see (A.28).

In Table C.1, we compare notations between this book and some of the original papers in the development of $\mathcal{N} = 2$ matter-coupled supergravity in $D = 4$.

Here : means the notation used in this book, and for most part also used in[1] [18–20].

---

[1]The exception is the Ricci sign (sign of R) that has been chosen negative in these papers.

© Springer Nature Switzerland AG 2020

E. Lauria, A. Van Proeyen, $\mathcal{N} = 2$ *Supergravity in D = 4, 5, 6 Dimensions*,
Lecture Notes in Physics 966, https://doi.org/10.1007/978-3-030-33757-5

Table C.1 Comparison of notations

Here	I	II
$s_i = 1$		
$s_{1,2,3,4,5} =$	$(+1, -1, -1, 1, i)$	$(-1, -1, -2, 1, -1)$
$s_{6,7,8,9} =$	$(+1, -1, +1, +4)$	$(+1, +1, +1, 2i)$
γ_μ	γ_μ	$i\gamma_\mu$
$g_{\alpha\bar\beta}$	$-g_{A\bar B}$	g_{ij^*}
$\mathcal{K}$	$-K$	$\mathcal{K}$
F	$-\frac{i}{4}F$	
X^I	X^I	L^Λ
Z^I	Z^I	X^Λ
F_I	$-\frac{i}{4}F_I$	M_Λ
N_{IJ}	$-N_{IJ}$	$-N_{\Lambda\Sigma}$
N_{IJ}	iN_{IJ}	$N_{\Lambda\Sigma}$
$C_{\alpha\beta\gamma}$	$e^{-K}Q_{ABC}$	iC_{ijk}
$F_{\mu\nu}^I$	$F_{\mu\nu}^I$	$-2\mathcal{F}_{\mu\nu}^\Lambda$
$f_{IJ}{}^K$	$-g f_{IJ}{}^K$	$f_{\Lambda\Sigma}^\Delta$
$\mathcal{P}_I$		$-\mathcal{P}_\Lambda$
ϵ	2ϵ	$\sqrt{2}\epsilon$
ψ_μ	ψ_μ	$\sqrt{2}\psi_\mu$
Ω^I	Ω^I	
η_i	η_i	
ϕ_μ	$\frac{1}{2}\phi_\mu$	
$f_\mu{}^a$	$\frac{1}{2}f_\mu{}^a$	
λ_K	$\frac{1}{2}\Lambda_K$	
$V_\mu{}^i{}_j$	$\frac{1}{2}\mathcal{V}_\mu{}^i{}_j$	
χ^i	χ^i	
$T_{ab}^-\epsilon^{ij}$	T_{ab}^{-ij}	

I : The second column refers to many of the original papers on $D = 4$ $\mathcal{N} = 2$ superconformal tensor calculus and special geometry. These are e.g. [21–36].

II : The third column contains the notation of the the matter-coupled Lagrangians of [37, 38].

A few comments are in order. In the original papers often the Pauli signature was used, which means that there is an index $a = 4$ or $\mu = 4$, we translated for vectors to $V_4 = iV_0$ in order to put them under the general scheme. The factor $s_2 s_3$ determines the normalization of the Ricci scalar $R = g^{\mu\nu} R_{\mu\nu}$.

Note also that in the older papers an antisymmetrization $[ab]$ was $(ab - ba)$ without the factor as in (A.1). Also one used

$$\sigma_{ab} = \tfrac{1}{2}\gamma_{ab}. \tag{C.2}$$

Finally, we want to point out that there are also a few changes in notation between this book and the notation the papers [39, 40]. To use the latter papers in the present notation, one has to make the following substitutions:

$$k_I{}^X \leftarrow -k_I{}^X, \qquad A_\mu^I \leftarrow W_\mu^I, \qquad \mathcal{D}_\mu \leftarrow \hat{D}_\mu,$$

$$t_{IA}{}^B \leftarrow -t_{IA}{}^B, \qquad \mathbf{P}_I \leftarrow \tfrac{1}{2}\mathbf{P}_I, \qquad \sigma_i{}^j \leftarrow i\tau_i{}^j. \tag{C.3}$$

References

1. D.Z. Freedman, A. Van Proeyen, *Supergravity* (Cambridge University Press, Cambridge, 2012). http://www.cambridge.org/mw/academic/subjects/physics/theoretical-physics-and-mathematical-physics/supergravity?format=AR
2. P. Van Nieuwenhuizen, Supergravity. Phys. Rept. **68**, 189–398 (1981). http://dx.doi.org/10.1016/0370-1573(81)90157-5
3. A. Van Proeyen, Tools for supersymmetry, Ann. Univ. Craiova, Phys. **AUC 9** (part I), 1–48 (1999). hep-th/9910030
4. J. Scherk, Extended supersymmetry and extended supergravity theories, in *Recent developments in gravitation*, ed. M. Lévy, S. Deser (Plenum Press, New York, 1979), p. 479
5. T. Kugo, P.K. Townsend, Supersymmetry and the division algebras. Nucl. Phys. **B221**, 357–380 (1983). http://dx.doi.org/10.1016/0550-3213(83)90584-9
6. P. Van Nieuwenhuizen, An introduction to simple supergravity and the Kaluza–Klein program, in *Les Houches 1983, Proceedings, Relativity, Groups and Topology II*, ed. by B.S. DeWitt, R. Stora (North-Holland, Amsterdam, 1984), p. 823
7. T. Regge, The group manifold approach to unified gravity, in *Les Houches 1983, Proceedings, Relativity, Groups and Topology II*, ed. by B.S. DeWitt, R. Stora (North-Holland, Amsterdam, 1984), p. 933
8. R. D'Auria, S. Ferrara, M.A. Lledó, V.S. Varadarajan, Spinor algebras. J. Geom. Phys. **40**, 101–128 (2001). http://dx.doi.org/10.1016/S0393-0440(01)00023-7. arXiv:hep-th/0010124 [hep-th]
9. J. W. van Holten, A. Van Proeyen, $N = 1$ supersymmetry algebras in $d = 2, 3, 4$ mod. 8. J. Phys. **A15**, 3763 (1982). http://dx.doi.org/10.1088/0305-4470/15/12/028
10. A.D. Kennedy, Clifford algebras in two ω dimensions. J. Math. Phys. **22**, 1330–1337 (1981). http://dx.doi.org/10.1063/1.525069
11. J. Rosseel, A. Van Proeyen, Hypermultiplets and hypercomplex geometry from 6 to 3 dimensions. Class. Quant. Grav. **21**, 5503–5518 (2004). http://dx.doi.org/10.1088/0264-9381/21/23/013. arXiv:hep-th/0405158 [hep-th]
12. V.G. Kac, A sketch of Lie superalgebra theory. Commun. Math. Phys. **53**, 31–64 (1977). http://dx.doi.org/10.1007/BF01609166
13. L. Frappat, P. Sorba, A. Sciarrino, Dictionary on Lie superalgebras (1996). arXiv:hep-th/9607161 [hep-th]
14. V.G. Kac, Lie superalgebras. Adv. Math. **26**, 8–96 (1977). http://dx.doi.org/10.1016/0001-8708(77)90017-2

15. M. Parker, Classification of real simple Lie superalgebras of classical type. J. Math. Phys. **21**, 689–697 (1980). http://dx.doi.org/10.1063/1.524487

16. W. Nahm, Supersymmetries and their representations. Nucl. Phys. **B135**, 149 (1978) http://dx. doi.org/10.1016/0550-3213(78)90218-3 [**7** (1977)]

17. C.W. Misner, K.S. Thorne, J.A. Wheeler, *Gravitation*. (W.H. Freeman, Princeton, 1970)

18. B. de Wit, A. Van Proeyen, Isometries of special manifolds, in *Proceedings of the Meeting on Quaternionic Structures in Mathematics and Physics*, Trieste, September 1994. Available on http://www.emis.de/proceedings/QSMP94/

19. A. Van Proeyen, Vector multiplets in $N = 2$ supersymmetry and its associated moduli spaces, in *1995 Summer school in High Energy Physics and Cosmology*, The ICTP Series in Theoretical Physics, ed. by E. Gava et al., vol.12 (World Scientific, Singapore, 1997), p.256. hep-th/9512139

20. B. de Wit, A. Van Proeyen, Special geometry and symplectic transformations. Nucl. Phys. Proc. Suppl. **45BC**, 196–206 (1996). http://dx.doi.org/10.1016/0920-5632(95)00637-0 arXiv:hep-th/9510186 [hep-th]

21. B. de Wit, J.W. van Holten, Multiplets of linearized SO(2) supergravity. Nucl. Phys. **B155**, 530–542 (1979). http://dx.doi.org/10.1016/0550-3213(79)90285-2

22. B. de Wit, J.W. van Holten, A. Van Proeyen, Transformation rules of $N = 2$ supergravity multiplets. Nucl. Phys. **B167**, 186–204 (1980). http://dx.doi.org/10.1016/0550-3213(80)90125-X

23. M. de Roo, J.W. van Holten, B. de Wit, A. Van Proeyen, Chiral superfields in $N = 2$ supergravity. Nucl. Phys. **B173**, 175–188 (1980). http://dx.doi.org/10.1016/0550-3213(80)90449-6

24. B. de Wit, J.W. van Holten, A. Van Proeyen, Central charges and conformal supergravity, Phys. Lett. **95B**, 51–55 (1980). http://dx.doi.org/10.1016/0370-2693(80)90397-4

25. B. de Wit, J.W. van Holten, A. Van Proeyen, Structure of $N = 2$ supergravity. Nucl. Phys. **B184**, 77–108 (1981). http://dx.doi.org/10.1016/0550-3213(83)90548-5. http://dx.doi.org/10. 1016/0550-3213(81)90211-X [Erratum: Nucl. Phys. **B222**, 516 (1983)]

26. A. Van Proeyen, Superconformal tensor calculus in $N = 1$ and $N = 2$ supergravity, in *Supersymmetry and Supergravity 1983, 19th Winter School and Workshop of Theoretical Physics*, Karpacz, ed. by B. Milewski (World Scientific, Singapore, 1983)

27. B. de Wit, P.G. Lauwers, R. Philippe, A. Van Proeyen, Noncompact $N = 2$ supergravity. Phys. Lett. **135B**, 295 (1984). http://dx.doi.org/10.1016/0370-2693(84)90395-2

28. B. de Wit, P.G. Lauwers, R. Philippe, S.Q. Su, A. Van Proeyen, Gauge and matter fields coupled to $N = 2$ supergravity. Phys. Lett. **B134**, 37–43 (1984). http://dx.doi.org/10.1016/ 0370-2693(84)90979-1

29. B. de Wit, A. Van Proeyen, Potentials and symmetries of general gauged $N = 2$ supergravity – Yang–Mills models. Nucl. Phys. **B245**, 89–117 (1984). http://dx.doi.org/10.1016/0550-3213(84)90425-5

30. B. de Wit, P.G. Lauwers, A. Van Proeyen, Lagrangians of $N = 2$ supergravity–matter systems. Nucl. Phys. **B255**, 569–608 (1985). http://dx.doi.org/10.1016/0550-3213(85)90154-3

31. E. Cremmer, C. Kounnas, A. Van Proeyen, J. Derendinger, S. Ferrara, B. de Wit, L. Girardello, Vector multiplets coupled to $N = 2$ supergravity: superhiggs effect, flat potentials and geometric structure. Nucl. Phys. **B250**, 385–426 (1985). http://dx.doi.org/10.1016/0550-3213(85)90488-2

32. S. Cecotti, S. Ferrara, L. Girardello, Geometry of type II superstrings and the moduli of superconformal field theories. Int. J. Mod. Phys. **A4**, 2475 (1989). http://dx.doi.org/10.1142/ S0217751X89000972

33. B. de Wit, A. Van Proeyen, Broken sigma model isometries in very special geometry. Phys. Lett. **B293**, 94–99 (1992). http://dx.doi.org/10.1016/0370-2693(92)91485-R. arXiv:hep-th/9207091 [hep-th]

34. B. de Wit, A. Van Proeyen, Special geometry, cubic polynomials and homogeneous quaternionic spaces. Commun. Math. Phys. **149**, 307–334 (1992). http://dx.doi.org/10.1007/BF02097627 arXiv:hep-th/9112027 [hep-th]

35. B. de Wit, F. Vanderseypen, A. Van Proeyen, Symmetry structure of special geometries. Nucl. Phys. **B400**, 463–524 (1993). http://dx.doi.org/10.1016/0550-3213(93)90413-J. arXiv:hep-th/9210068 [hep-th]

36. P. Claus, B. de Wit, B. Kleijn, R. Siebelink, P. Termonia, $N = 2$ supergravity Lagrangians with vector–tensor multiplets. Nucl. Phys. **B512**, 148–178 (1998). http://dx.doi.org/10.1016/S0550-3213(97)00781-5. arXiv:hep-th/9710212 [hep-th]

37. L. Andrianopoli, M. Bertolini, A. Ceresole, R. D'Auria, S. Ferrara, P. Frè, General matter-coupled $N = 2$ supergravity. Nucl. Phys. **B476**, 397–417 (1996) http://dx.doi.org/10.1016/0550-3213(96)00344-6. arXiv:hep-th/9603004 [hep-th]

38. L. Andrianopoli, M. Bertolini, A. Ceresole, R. D'Auria, S. Ferrara, P. Frè, T. Magri, $N = 2$ supergravity and $N = 2$ super Yang–Mills theory on general scalar manifolds: symplectic covariance, gaugings and the momentum map. J. Geom. Phys. **23**, 111–189 (1997). http://dx.doi.org/10.1016/S0393-0440(97)00002-8. arXiv:hep-th/9605032 [hep-th]

39. E. Bergshoeff, S. Cucu, T. de Wit, J. Gheerardyn, R. Halbersma, S. Vandoren, A. Van Proeyen, Superconformal $N = 2$, $D = 5$ matter with and without actions. JHEP **10**, 045 (2002). http://dx.doi.org/10.1088/1126-6708/2002/10/045. arXiv:hep-th/0205230 [hep-th]

40. E. Bergshoeff, S. Cucu, T. de Wit, J. Gheerardyn, S. Vandoren, A. Van Proeyen, $N = 2$ supergravity in five dimensions revisited. Class. Quant. Grav. **21**, 3015–3041 (2004). http://dx.doi.org/10.1088/0264-9381/23/23/C01. http://dx.doi.org/10.1088/0264-9381/21/12/013. arXiv:hep-th/0403045 [hep-th] [Erratum: Class. Quant. Grav. **23** (2006) 7149]

Index

© Springer Nature Switzerland AG 2020
E. Lauria, A. Van Proeyen, $\mathcal{N} = 2$ *Supergravity in D* $= 4, 5, 6$ *Dimensions*,
Lecture Notes in Physics 966, https://doi.org/10.1007/978-3-030-33757-5

Printed in the United States
By Bookmasters